Sören Östlund and Kaarlo Niskanen (Eds.)
Mechanics of Paper Products

Also of interest

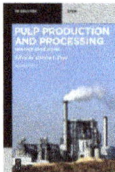

Pulp Production and Processing.
High-Tech Applications
Popa, 2020
ISBN 978-3-11-065883-5, e-ISBN 978-3-11-065884-2

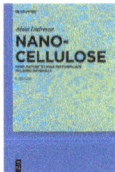

Nanocellulose.
From Nature to High Performance Tailored Materials
Dufresne, 2017
ISBN 978-3-11-047848-8, e-ISBN 978-3-11-048041-2

Nordic Pulp & Paper Research Journal.
The international research journal on sustainable utilization of forest
bioresources
Tom Lindström (Editor in Chief)
ISSN 0283-2631, e-ISSN 2000-0669

Holzforschung.
International Journal of the Biology, Chemistry, Physics, and
Technology of Wood
Lennart Salmén (Editor in Chief)
ISSN 0018-3830, e-ISSN 1437-434X

Mechanics of Paper Products

Edited by
Sören Östlund and Kaarlo Niskanen

2nd Edition

DE GRUYTER

Editors
Prof. Sören Östlund
KTH Royal Institute of Technology
School of Engineering Sciences
Department of Engineering Mechanics
SE-100 44 Stockholm
Sweden
soren@kth.se

Prof. Kaarlo Niskanen
Mid Sweden University
FSCN
SE-851 70 Sundsvall
Sweden
kaarlo.niskanen@miun.se

ISBN 978-3-11-061741-2
e-ISBN (PDF) 978-3-11-061938-6
e-ISBN (EPUB) 978-3-11-061950-8

Library of Congress Control Number: 2020945568

Bibliographic information published by the Deutsche Nationalbibliothek
The Deutsche Nationalbibliothek lists this publication in the Deutsche Nationalbibliografie;
detailed bibliographic data are available on the Internet at http://dnb.dnb.de.

© 2021 Walter de Gruyter GmbH, Berlin/Boston
Cover image: Ratchat/iStock/Getty Images Plus
Typesetting: Integra Software Services Pvt. Ltd.
Printing and binding: CPI books GmbH, Leck

www.degruyter.com

This book is dedicated to Professor Per Johan Gustafsson who sadly passed away during preparation of the final manuscript.

Preface

This book was initiated by the Alf de Ruvo Memorial Foundation.

The career of Alf de Ruvo may be described in a few lines, but his efforts for the Swedish forest products industry in general as a scientist, technical innovator, mentor, and source of inspiration were unparalleled. Alf de Ruvo had a degree in Chemical Engineering, but his own research covered paper physics, composite materials, product properties, and converting of paperboard materials. His work was characterized by his cross-disciplinary approach to science. It is not an exaggeration to say that he brought science to the art of paper mechanics and brought paper mechanics in Sweden to the absolute frontier of paper technology. Hence, there has been a move from a trial-and-error approach to a well-funded methodology based on fracture mechanics to an analysis of statistical variations as a systematic tool in the evolution of the science of paper mechanics.

Hence, following the tradition of Alf de Ruvo, this book is aimed at paper mechanics from a solid and continuum mechanics point of view and not from a paper technology perspective. It is hoped that this book fills a knowledge gap, considering the essential role of solid and continuum mechanics in understanding papermaking, converting, and the end-use of paper and board materials. All the authors have contributed on a voluntary basis, with a never-failing enthusiasm and solid belief in the value of this book.

Professor Tom Lindström
On behalf of the Alf de Ruvo Memorial Foundation

https://doi.org/10.1515/9783110619386-202

Contents

List of contributing authors

Professor Lars Berglund,
Department of Fiber and Polymer Technology,
Wallenberg Wood Science Centre,
KTH Royal Institute of Technology, SE-100 44
Stockholm, Sweden,
email: blund@kth.se

Professor Leif A. Carlsson,
Department of Ocean and Mechanical
Engineering, Engineering Building 36,
Florida Atlantic University, Boca Raton,
FL 33431-0991, USA,
email: carlsson@fau.edu

Professor Douglas W. Coffin,
Department of Chemical, Paper, and
Biomedical Engineering, 64 Engineering
Building, Miami University, Oxford, OH
45056, USA,
email: coffindw@miamiOH.edu

Professor Per-Johan Gustafsson,[1]
Department of Construction Sciences,
Lund University, Box 118, SE-221 00 Lund,
Sweden

Dr. Rickard Hägglund,
SCA, Box 716, SE-851 21 Sundsvall, Sweden,
email: rickard.hagglund@sca.com

Professor Artem Kulachenko,
Department of Engineering Mechanics,
Solid Mechanics, KTH Royal Institute of
Technology, SE-100 44 Stockholm, Sweden,
email: artem@kth.se

Dr. Petri Mäkelä,
Department of Engineering Mechanics, Solid
Mechanics, KTH Royal Institute of
Technology, SE-100 44 Stockholm, Sweden,
email: petri.makela@telia.com

Professor Kaarlo Niskanen,
FSCN, Mid Sweden University, SE-851 70
Sundsvall, Sweden,
email: kaarlo.niskanen@miun.se

Dr. Mikael Nygårds,
Department of Engineering Mechanics, Solid
Mechanics, KTH Royal Institute of
Technology, SE-100 44 Stockholm, Sweden,
email: mikael.nygards@billerudkorsnas.com

Professor Sören Östlund,
Department of Engineering Mechanics, Solid
Mechanics, KTH Royal Institute of
Technology, SE-100 44 Stockholm, Sweden,
email: soren@kth.se

Professor Emeritus Tetsu Uesaka,
Department of Chemical Engineering and
FSCN, Mid Sweden University, SE-851 70
Sundsvall, Sweden,
email: tetsu.uesaka@miun.se

---

[1] Professor Gustafsson sadly passed away during the preparation of the second edition of this book, but, prior to that, he was able to finalize Chapter 2.

https://doi.org/10.1515/9783110619386-204

Kaarlo Niskanen, Sören Östlund

1 The challenge

This book discusses the mechanical properties of products made of wood fibres. Printing papers and office papers are the ubiquitous examples of such products. Packaging made of paper or board is another, and tissue papers are a third very common use of wood fibres. A little further from paper products, one has fibreboards that are used in furniture and construction. When compared with paper, fibreboards have similar structure but different physical properties. The renewable wood raw material used in all these products makes them attractive to the modern society, and therefore many believe that wood fibres will increasingly find also other uses. We have written this textbook in part to anticipate such developments.

Since – by definition – future products do not exist today, we use some of today's paper and board products as concrete examples. The properties seen in these products today help one understand what may be required in other applications tomorrow. The cases we discuss in this book give a broad picture of the various challenges that face anyone working with materials based on wood fibres. We expect that the reader will learn how to solve analogous challenges in other situations.

The fact that we focus on the mechanical properties of paper and board products means that this textbook discusses practical applications of mechanical engineering. The reader is assumed to have good understanding of engineering mechanics, or to be able to acquire such as needed. One should also feel comfortable with mathematical concepts and be able to appreciate the power of numerical modelling methods. The reader is not expected to know much of paper or papermaking. We will describe the main features in Chapter 2, exemplifying what paper and board are as materials. Building a deeper understanding of materials based on wood fibres is in itself a goal of this book.

Before one can solve a practical performance problem or develop a better product, one has to understand what the problem really is. In an effort to underline the importance of problem definition, we have chosen a somewhat unorthodox and admittedly difficult approach. We have tried to build a book that starts from two practical cases (the "box" and the "web"), identifies what are the important process requirements and the related material or systems properties, before exploring the material properties, and ultimately discussing how the material properties or systems behaviour could be improved. The result is not perfect because, as it turns out, little work has been done on the really crucial issues.

The approach outlined means that this book overlooks lots of the material traditionally discussed in the context of the mechanical properties of paper and board.

Kaarlo Niskanen, Mid Sweden University, Sundsvall, Sweden
Sören Östlund, KTH Royal Institute of Technology, Stockholm, Sweden

https://doi.org/10.1515/9783110619386-001

There are many excellent books on papermaking, paper physics, and the mechanical properties of paper products, explaining what is already known. We have been more interested in unveiling what should be known.

With all that said, we are confident that this book will be very useful not only when developing new products and materials based on wood fibres but also – and most readily – when solving performance problems of today's paper and board products.

A short outline of the book contents is the following. After the introduction to paper materials in Chapter 2, the main body is divided into four parts that clarify the underlying logic. Part I considers the basic structural strength of paper products. Wood fibres are particularly beneficial in applications where stiffness at low weight is required – as in packaging applications. The first practical case introduced in Chapter 3 is the "box", the manufacturing and usage requirements of paper-based containers and boxes. The performance of a box depends on the structure of the box and on the changes in the material that are created in the process of converting a board into a box. Chapter 4 illustrates the latter aspect by considering the corners of paperboard boxes. The quality of corners is crucial, for example, in liquid packaging applications where defects can allow leakage or trigger box fracture. Chapter 5 then provides the machinery that has been developed specifically for the analysis of fracture that is triggered by a defect or other discontinuity in a structure.

The discussion in Part I concerned primarily the "static" or "instantaneous" strength of products. Often in reality, the dynamics of the process is crucial. This extension to structural strength is discussed in Part II. A major advantage of paper materials is the high speed at which the web can be manufactured and converted – but only provided that the web can be held stable in the fast process. This is the second practical case, the "web". Chapter 6 explains what determines the short-time dynamic stability of "running" paper or board webs. The requirements posed by the fast dynamical loading are different from the requirements on boxes. A stack of boxes should hold through the entire logistic chain that may last for months. The rheological properties of paper materials – applicable in both short-time web stability and long-time box endurance – are then explored in Chapter 7.

From Chapters 6 and 7, we conclude that strength problems of paper products are also a systemic issue. Product performance cannot be comprehensively explained by just the average material properties. Spatial and temporal variability in the material properties and process conditions is often crucial. Therefore, Chapter 8 frames out the general methodology for obtaining statistically reliable information about box failures and web breaks.

After tackling the "static strength" and "dynamic strength" of paper products, it is time in Part III to consider the dimensional instability caused by changes in the moisture content of paper. The moisture sensitivity of paper is a direct result of the biological origin of the wood fibres. It enables the preparation of paper even without any added adhesives, and equivalently makes it possible to recycle paper by simply soaking it in water. Thus, moisture sensitivity is a challenge in all uses of wood fibres. One just has to learn to minimize the problems. This is the motivation for Chapters 9 and 10.

Moisture changes are practically impossible to avoid in normal usage of paper products. Chapter 7 discusses the effect of moisture on the creep rate of boxes. In Chapter 9, we explain how changes in moisture can lead to permanent out-of-plane deformations, especially if the paper surface is exposed to liquid water. Chapter 10 then explains the mechanics of a printing nip where water and other liquids are applied on paper or board.

Finally, having illustrated some of the requirements that usage poses on the paper material, we are ready to ask how the relevant material properties could be controlled. This is considered in Part IV. In Chapter 11, we discuss the relations between the stiffness properties of paper, the papermaking fibres and the papermaking process using micromechanical concepts and numerical simulations. Chapter 12 continues from there, asking what functionality can be achieved if one goes beyond paper, using wood as an ingredient in bio-composites.

The book is by no means a comprehensive account of the performance of paper products or materials based on wood fibres. We have not intended to give such an account. What we hope to have accomplished is a book that illustrates how one can systematically tackle challenges in the development of better products. Finding out the right questions is critical if one wants to give relevant answers. We have certainly not uncovered all questions!

Per-Johan Gustafsson, Kaarlo Niskanen

2 Paper as an engineering material

2.1 Introduction

This textbook concerns the mechanical properties of products that are made of paper or paperboard. Throughout the book, therefore, it is important to know the properties of paper as an engineering material. In this chapter, we describe the general mechanical properties of paper and the papermaking process. The manufacturing process, including the wood raw material used, naturally governs these mechanical properties, but here we limit the discussion to what is most important to know when tackling problems of product performance. An overview of the underlying mechanisms and raw material effects is presented in Chapters 11 and 12. Those interested in learning more are referred to the many textbooks on papermaking, such as the series *Papermaking Science and Technology*, published by the Finnish Paper Engineers' Association.

Paper is a thin, almost two-dimensional material. Everyday papers, such as office paper and newsprint, have a thickness of about 0.1 mm. The mass per unit area of such papers, called the *basis weight*, is usually between 40 and 100 g/m² depending on the type of paper. Specially prepared paper can have a thickness as low as 0.01 mm and a basis weight of a few grams per square meter. On the other end, paper material used for book covers or fixtures to display products in stores can be more than 1 mm thick. Thick paper grades are called *board* or *paperboard*, and typically they have a basis weight between 150 and 500 g/m². The trade terminology for paper and board grades refers primarily to the applications where the materials are used, not to their structure. In this book, *paper* is a general term used for all kinds of paper and paperboard materials.

As will be described in Chapter 12, many wood-based panels used for building and furniture can resemble thick boards. Products made of such panels could be analysed in the same way as we analyse paper-based products in this book.

A paper machine creates a continuous web that is 5–10 m wide. A finished roll may contain 10 km of paper. The coordinate system used throughout this book is defined in Fig. 2.1, where x is the running direction of the web. The orientation of this direction is customarily referred to as the machine direction (MD), and the lateral direction as the cross-machine direction (CD). The thickness direction is the z-direction (ZD).

The main constituent of paper is wood or plant fibres that are specially prepared into a pulp, as outlined in Section 2.5. The water suspension of the fibres

Per-Johan Gustafsson, Lund University, Lund, Sweden
Kaarlo Niskanen, Mid-Sweden University, Sundsvall, Sweden

https://doi.org/10.1515/9783110619386-002

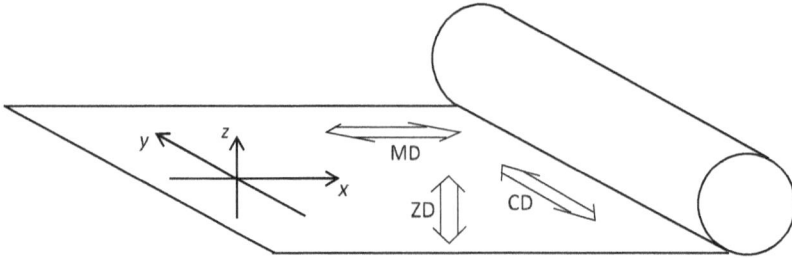

Fig. 2.1: The coordinate system for paper that is used in this book.

and various additives form what is called *furnish*. When a thin layer of furnish is dried, the fibres bond to each other and paper is formed. Most published data on the mechanical properties of paper comes from measurements done on paper sheets made in a laboratory. The properties of these laboratory sheets, or hand-sheets, differ from machine-made commercial paper even if the same raw material is used.

Constitutive equations are used to analyse the mechanical properties of paper. When a constitutive equation is combined with the equations for equilibrium, geo-metrical compatibility, and boundary conditions, one can calculate the magnitude, distribution, and time variation of stress and deformations in a specific product structure. In general, constitutive laws or theories describe how stress and tempera-ture at a point interact with the deformation or straining of the material. In the case of paper, the effect of temperature is relatively small, but the effect of moisture is very important.

Stresses and *strains* are quantities that are defined for continuous materials. Usually paper can be regarded as continuous all the way down to the centimetre scale, after which the fibres network structure starts to dominate. Thus, we can use phenomenological models derived from macroscopic experiments to represent ma-terial behaviour in products made of paper. However, special care is needed when determining phenomenological models for ZD deformations and loads and for frac-ture processes in general. A typical paper sheet consists of only 10 fibre layers (here layer refers to the typical number of fibres in the thickness direction of paper but does not mean that fibres would actually form clear layers), so that experimental boundary effects can be large in ZD. In fracture processes, deformations take place at very small scales close to a crack, and the material behaviour at that length scale must be determined. Chapters 11 and 12 examine the effects of microscopic structure and composition on the constitutive behaviour of paper. More information on the mechanical properties of paper and their measurement can be found in Mark et al. (2002) and Niskanen (2008).

2.2 Linear elasticity of paper

2.2.1 Paper as an engineering material elastic constants

The *elastic constants* give the stress to strain relation for paper when the performance is linear elastic. In the general three-dimensional (3D) case, the state of stress is defined by the six independent stress components: the three normal stresses σ_x, σ_y, and σ_z, and the three shear stresses $\tau_{xy} = \tau_{yx}$, $\tau_{xz} = \tau_{zx}$, and $\tau_{yz} = \tau_{zy}$ (see Fig. 2.2). A positive value of shear stress, τ_{ij}, corresponds to stress acting on the surface with its normal in the positive i-direction and directed in the positive j-direction. Because of moment equilibrium, shear stresses are equal in pairs, $\tau_{ij} = \tau_{ji}$. The alternative notations, σ_{MD}, σ_1, σ_{11}, σ_{xx}, and so on, are sometimes used instead of σ_x and others when the coordinates coincide with MD and others. The state of strain in the 3D case is also defined by six components, the normal strains ε_i, and the shear strains y_{ij}. The strains referred to in this book are the conventional small strain theory engineering strains. This means that the normal strains are defined as elongation Δl divided by the initial length l, and the shear strains are defined as the change of an originally right angle (Fig. 2.3). The strains can also be defined by means of the local displacements u_x, u_y, and u_z of the material: $\varepsilon_x = \delta u_x / \delta x$, and so on, and $y_{xz} = \delta u_z / \delta x + \delta u_x / \delta z$, and so on.

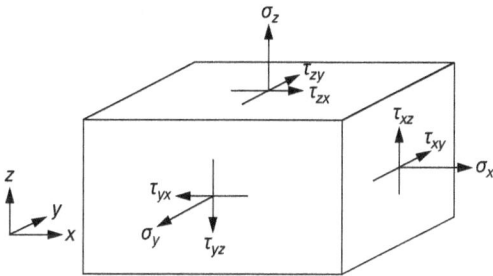

Fig. 2.2: Stress components in a 3D state of stress.

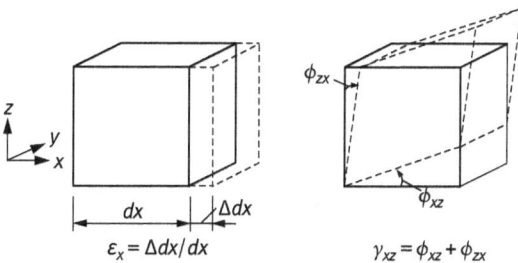

Fig. 2.3: Normal strain ε_x and the shear strain y_{xz} in a 3D state of small strain.

Unlike many other materials, the elastic modulus E of paper is significantly aniso-tropic. This arises from the manufacturing process, giving $E_x > E_y > E_z$ (see Section 2.5). The anisotropy between the in-plane and thickness directions comes from the low thickness of paper. Because a typical fibre's length, 1–3 mm, is more than 10 times larger than paper thickness, fibres must be aligned in the plane of the paper. The ZD straining of paper creates primarily transverse stresses in fibres, whereas in-plane straining creates longitudinal stresses. The longitudinal elastic modulus of fibres is larger than the transverse modulus (see Chapter 12).

A reasonable approximation is that the anisotropy of paper is orthotropic, that is, the stiffness properties are symmetric with respect to the x-, y-, and z-axes, even though there may be a slight deviation in the symmetry axes because of skewness in fibre orientation (see Section 2.5). The components of the stress and strain vec-tors $\boldsymbol{\sigma}$ and $\boldsymbol{\varepsilon}$ are coupled by the elastic compliance tensor $\boldsymbol{\varepsilon} = \boldsymbol{S\sigma}$, or

$$
\begin{bmatrix} \varepsilon_x \\ \varepsilon_y \\ \varepsilon_z \\ \gamma_{xy} \\ \gamma_{xz} \\ \gamma_{yz} \end{bmatrix} = \begin{bmatrix} 1/E_x & -v_{yx}/E_y & -v_{zx}/E_z & 0 & 0 & 0 \\ -v_{xy}/E_x & 1/E_y & -v_{zy}/E_z & 0 & 0 & 0 \\ -v_{xz}/E_x & -v_{yz}/E_y & 1/E_z & 0 & 0 & 0 \\ 0 & 0 & 0 & 1/G_{xy} & 0 & 0 \\ 0 & 0 & 0 & 0 & 1/G_{xz} & 0 \\ 0 & 0 & 0 & 0 & 0 & 1/G_{yz} \end{bmatrix} \begin{bmatrix} \sigma_x \\ \sigma_y \\ \sigma_z \\ \tau_{xy} \\ \tau_{xz} \\ \tau_{yz} \end{bmatrix}
\tag{2.1}
$$

or by the stiffness tensor $\boldsymbol{\sigma} = \boldsymbol{C\varepsilon}$,

$$
\begin{bmatrix} \sigma_x \\ \sigma_y \\ \sigma_z \\ \tau_{xy} \\ \tau_{xz} \\ \tau_{yz} \end{bmatrix} = A \begin{bmatrix} E_x(1-v_{yz}v_{zy}) & E_x(v_{yx}+v_{zx}v_{yz}) & E_x(v_{zx}+v_{yx}v_{zy}) & 0 & 0 & 0 \\ E_y(v_{xy}+v_{zy}v_{xz}) & E_y(1-v_{xz}v_{zx}) & E_y(v_{zy}+v_{xy}v_{zx}) & 0 & 0 & 0 \\ E_z(v_{xz}+v_{xy}v_{yz}) & E_z(v_{yz}+v_{xz}v_{yx}) & E_z(1-v_{yx}v_{xy}) & 0 & 0 & 0 \\ 0 & 0 & 0 & G_{xy}/A & 0 & 0 \\ 0 & 0 & 0 & 0 & G_{xz}/A & 0 \\ 0 & 0 & 0 & 0 & 0 & G_{yz}/A \end{bmatrix} \begin{bmatrix} \varepsilon_x \\ \varepsilon_y \\ \varepsilon_z \\ \gamma_{xy} \\ \gamma_{xz} \\ \gamma_{yz} \end{bmatrix},
\tag{2.2}
$$

where $A = 1/(1 - v_{xy}v_{yx} - v_{xz}v_{zx} - v_{yz}v_{zy} - v_{xy}v_{zx}v_{yz} - v_{yx}v_{xz}v_{zy})$. Linear elasticity implies that the compliance tensor in eq. (2.1) is symmetric (Malvern, 1969) and accordingly

$$
\begin{cases} v_{xy} = v_{yx}E_x/E_y \\ v_{xz} = v_{zx}E_x/E_z \\ v_{yz} = v_{zy}E_y/E_z \end{cases}
\tag{2.3}
$$

Thus, the number of independent material parameters is reduced to 9. Equations (2.1)–(2.3) define the generalized Hooke's law for paper. Strictly speaking, the elas-tic moduli E, shear moduli G, and Poisson's ratios v are applicable only in the ideal case of linear elastic behaviour (i.e. reversible and loading-rate independent).

Nevertheless, it is customary to apply the notation in eqs. (2.1)–(2.3) for measurement results, and, for example, to denote the slope of a measured stress–strain curve as elastic modulus E even when the result may depend on the experimental conditions. In the case of paper, measured stress–strain curves are linear elastic within measurement accuracy when the load level is well below the failure load. In many practical situations, paper is exposed to pure plane stresses ($\sigma_z = \tau_{xz} = \tau_{yz} = 0$). In this special case,

$$\begin{bmatrix} \sigma_x \\ \sigma_y \\ \tau_{xy} \end{bmatrix} = \frac{1}{1 - v_{xy}v_{yx}} \begin{bmatrix} E_x & E_x v_{yx} & 0 \\ E_y v_{xy} & E_y & 0 \\ 0 & 0 & G_{xy} \end{bmatrix} \begin{bmatrix} \varepsilon_x \\ \varepsilon_y \\ \gamma_{xy} \end{bmatrix}. \tag{2.4}$$

The influence of an eventual change of temperature Θ and moisture content (MC) χ can be taken into account by adding temperature- and moisture-induced strains to eqs. (2.1) and (2.2), that is,

$$\varepsilon = \mathbf{S}\sigma + \alpha\Delta\Theta + \beta\Delta\chi \quad \text{and} \quad \sigma = \mathbf{C}(\varepsilon - \alpha\Delta\Theta - \beta\Delta\chi). \tag{2.5}$$

This formulation ignores the moisture and temperature dependence of the elastic constants. In orthotropic materials, the thermal expansion and hygroexpansion coefficients in the principal directions are

$$\alpha = \begin{bmatrix} \alpha_x & \alpha_y & \alpha_z & 0 & 0 & 0 \end{bmatrix}^T \quad \text{and} \quad \beta = \begin{bmatrix} \beta_x & \beta_y & \beta_z & 0 & 0 & 0 \end{bmatrix}^T. \tag{2.6}$$

The thermal strains $\alpha\Delta\Theta$ of paper are, in general, much smaller than the hygroscopic strains. For example, a 20 °C temperature change may lead to $\alpha\Delta\Theta \approx 0.01\%$, while a change in the relative humidity (RH) of air from 10% to 60% leads to $\beta\Delta\chi \geq 0.1\%$. Thus, thermal expansion can usually be excluded. RH gives the concentration of water vapour in air relative to the saturation concentration that is possible at the given temperature.

2.2.2 Typical stiffness values for paper

Table 2.1 shows a collection of directly measured values of the elastic stiffness parameters for a few paper grades. Many values are missing because of measurement difficulties caused by the small thickness of paper. Various estimation schemes have been developed to escape direct measurement (Baum, 1987).

Table 2.1 demonstrates that the ZD stiffness of paper is generally low compared to the in-plane stiffnesses. The negative value of Poisson's ratio v_{xz} for the paperboard shows that uniaxial tensile loading in MD increased thickness in this case, which is not uncommon. The elastic moduli measured in compression are usually equal to the corresponding tensile values.

The high density of the coated paper is caused by the coating. Without the coating, the paper would have similar density as the other samples. In general, the

Table 2.1: Measured values of elastic stiffness parameters in tensile loading for some machine-made papers.

	Paperboard (Persson, 1991)	Carton board (Baum, 1984)	Linerboard (Baum, 1984)	Coated paper, middle of web (Stålne, 2006)	Coated paper, web edge (Stålne, 2006)
Density (kg/m^3)	640	780	691	1,140	1,140
MD modulus E_x (MPa)	5,420	7,440	7,460	7,690	7,660
CD modulus E_y (MPa)	1,900	3,470	3,010	3,050	2,570
ZD modulus E_z (MPa)	17	40	29		140
Poisson's ratio ν_{xy}	0.38	0.15	0.12	0.33	0.27
Poisson's ratio ν_{xz}	−2.20	0.008	0.011		
Poisson's ratio ν_{yx}	0.14			0.07	0.10
Poisson's ratio ν_{yz}	0.54	0.021	0.021		
Poisson's ratio ν_{zx}	0.05				−0.04
Poisson's ratio ν_{zy}	0.05				0.03
Shear modulus G_{xy} (MPa)	1,230	2,040	1,800	1,910	1,820
Shear modulus G_{xz} (MPa)	8.8	137	129		
Shear modulus G_{yz} (MPa)	8.0	99	104		

density of paper is between 300 and 900 kg/m³. The in-plane elastic modulus usually increases with density (cf. Section 11.2.5) and ranges from 1 to 9 GPa when the effect of anisotropy is removed by averaging over MD and CD. Typical values of the MD/CD ratio of elastic moduli in machine-made papers are 2–4, but can be as high as 5–6.

From the variation of the ZD elastic modulus against density in Fig. 2.4, one can see that the data for different pulps (cf. Section 2.5) can fall on one single line. The same is not true for in-plane elastic moduli.

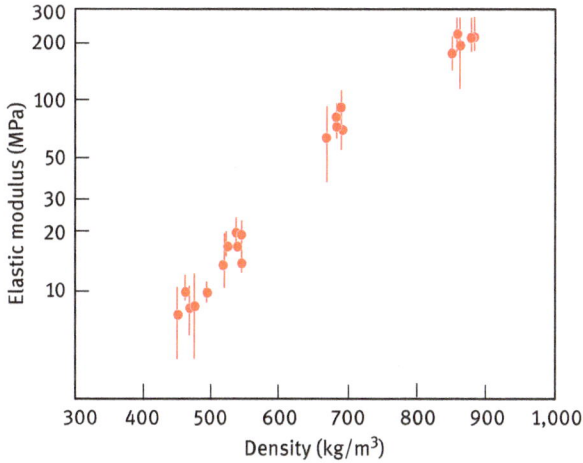

Fig. 2.4: Elastic modulus in ZD (logarithmic scale) of laboratory sheets against density for mechanical pulps (density ca. 500 kg/m³) and chemical pulps (density > 700 kg/m³), using data from Girlanda and Fellers (2007).

Water acts as a softener of paper. Thus, the elastic modulus of paper depends on the MC (Figs. 2.5 and 2.6). Ultimately, at high MC, the modulus of paper goes to zero as the bonding between fibres opens, and one returns to a state that prevailed when

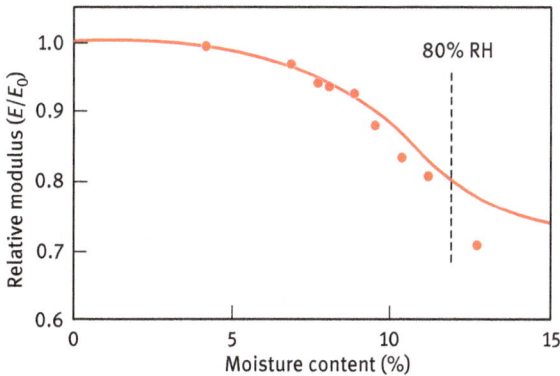

Fig. 2.5: Elastic modulus against moisture content for a set of laboratory sheets. The modulus values are given relative to the value in dry paper (dots). The curve shows a theoretical prediction. Reprinted from Salmén et al. (1984) with permission from Elsevier.

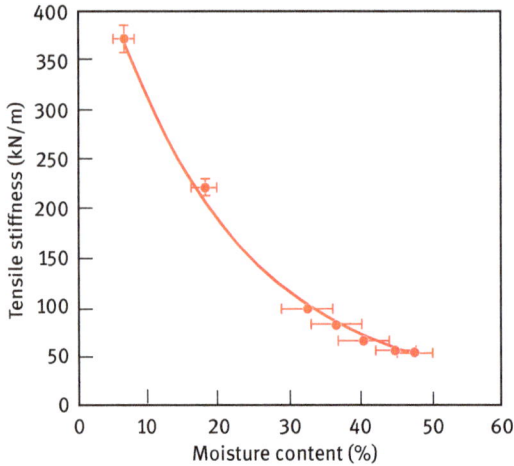

Fig. 2.6: Tensile stiffness against moisture content for a machine-made paper, measured with a cyclic small-strain excitation. Tensile stiffness is equal to elastic modulus multiplied by paper thickness, the latter being a slightly increasing function of moisture content. Drawn using data of Ketoja et al. (2007).

drying started in the papermaking process. We note in passing that it is this reversibility of the papermaking process that makes recycling of paper possible.

The softening effect makes paper increasingly viscoelastic and viscoplastic, which means that, especially at higher MC, the slope of the measured stress–strain curves depends on the strain rate. The apparent modulus (slope of the stress–strain curve) increases if strain rate is increased. At MC of 50% or higher, it is governed by interactions between fibres that are mediated by liquid water. Therefore, any stress created by constrained deformations would rapidly relax to zero. In addition to elastic modulus, the softening effect of moisture is evident in the stress–strain behaviour of paper, discussed next.

2.3 Stress–strain behaviour of paper

2.3.1 In-plane tensile loading

In principle, a stress increment may cause an instant or delayed and reversible (i.e. elastic) or irreversible (i.e. inelastic or plastic) strain increment (Fig. 2.7). The presence of a delayed response implies that the stress–strain behaviour is time dependent or rate dependent. Furthermore, the relationship between stresses and strains can be linear or non-linear. The stress–strain curve of paper exhibits all these behaviours. The time dependence seen in the creep and stress relaxation of paper is

Load $\sigma(t)$

Instant response: t_4-t_1 small

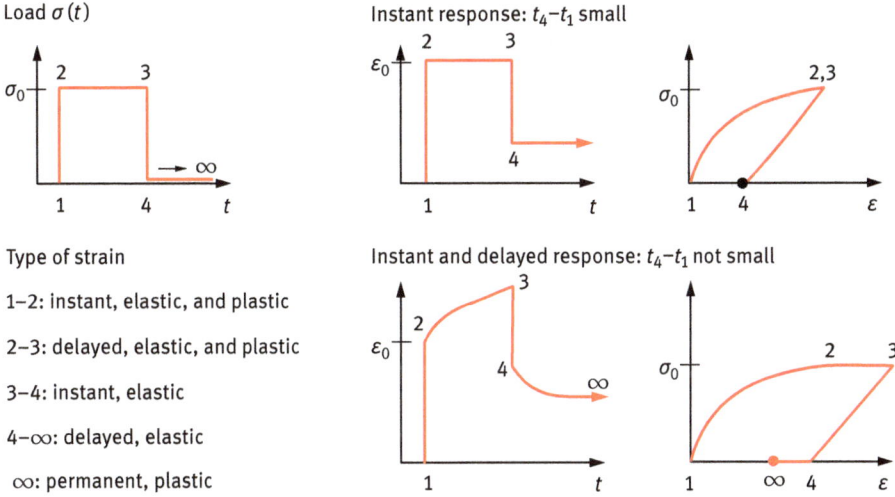

Type of strain

1–2: instant, elastic, and plastic

2–3: delayed, elastic, and plastic

3–4: instant, elastic

4–∞: delayed, elastic

∞: permanent, plastic

Instant and delayed response: t_4-t_1 not small

Fig. 2.7: Instant and delayed response to load.

discussed in Chapter 7. This section gives a general overview of the different aspects of the 3D stress–strain behaviour of paper.

A recursive tensile in-plane stress–strain measurement of paper usually gives a result of the type shown in Fig. 2.8. One can see that the elastic modulus changes very little even though part of the strain is irreversible or plastic. This is typical of almost all paper grades: the elastic modulus decreases by a maximum of 10% before the breaking point is reached. Brittle paper grades, such as baking paper or

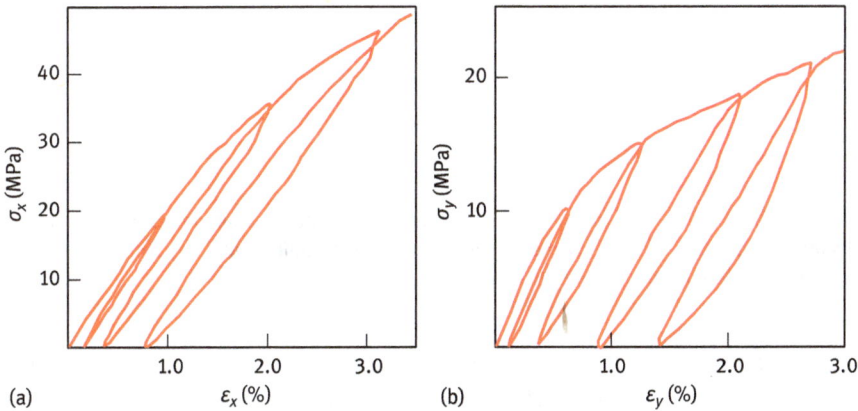

Fig. 2.8: Recursive stress–strain curves of a paperboard in MD (a) and Co^{max}D (b), from Persson (1991). Reproduced with permission from the author.

glassine, exhibit a larger loss in the elastic modulus, while ductile paper grades, such as sack paper, show a modest increase. Corresponding stress–strain curves are illustrated in Fig. 2.9.

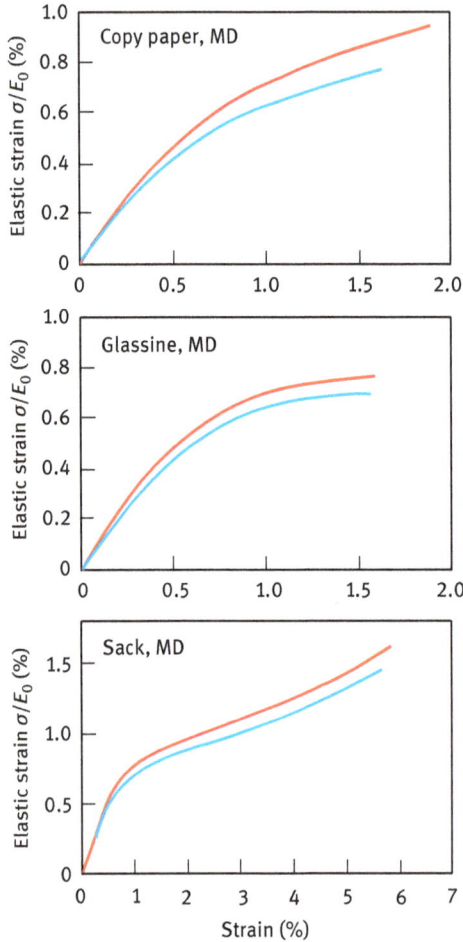

Fig. 2.9: Examples of MD stress–strain curves of some machine-made paper grades. Stress values are divided by the elastic modulus E_0 measured initially at zero strain, giving an estimate of the elastic strain. Data courtesy of Lauri Salminen.

The fact that the elastic modulus of paper changes only a little before the peak stress suggests that the microscopic fibres' network structure undergoes permanent plastic deformations that do not weaken elastic stiffness of the fibres. However, after the peak stress the elastic modulus decreases. This is apparent in the post-peak unloading–reloading cycles shown in Fig. 2.10. The post-peak behaviour in general can be recorded only when short specimens are used (Hillerborg et al., 1976; Tryding and Gustafsson, 2000); long specimens show sudden failure at the peak stress, which is discussed in Section 5.3.4.

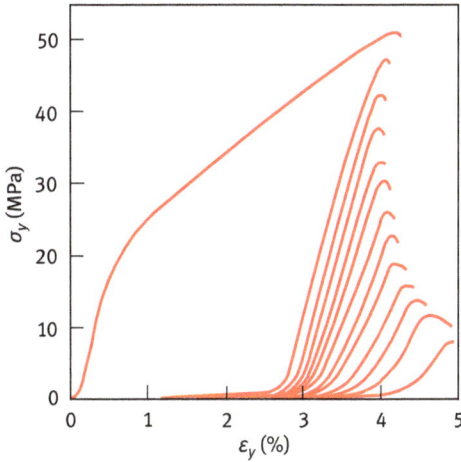

Fig. 2.10: Post-peak reloading stress–strain behaviour of paperboard in CD using 5 mm long and 15 mm wide specimens. Data of Tryding (1996). Reproduced with permission from the author.

If one measures how the post-peak cohesive stress decays as the fracture process zone widens, one gets the results illustrated in the right-hand side of Fig. 2.11. The fracture zone widens to more than 1 mm before all stiffness is lost. The large ductility presumably comes from the length of fibres. Stiffness does not reach zero before the fibres crossing the fracture line have been pulled out of either half of the specimen. In Fig. 2.11, the newsprint is made of mechanical pulp, which has shorter fibres than the chemical pulp used in the strong kraft paper. In brittle paper grades, such as glassine and baking paper, inter-fibre bonding is unusually strong, and therefore, fibres break during the fracture process, and the fracture zone widening is small.

2.3.2 Viscoelastic effects

The effect of strain rate on the in-plane tensile stress–strain curve is illustrated in Fig. 2.12. In this particular example, the breaking strain does not decrease when the strain rate is increased. However, often the increase in strain rate leads to a reduction in the breaking strain of paper, which is expected if the material becomes more brittle at high strain rates.

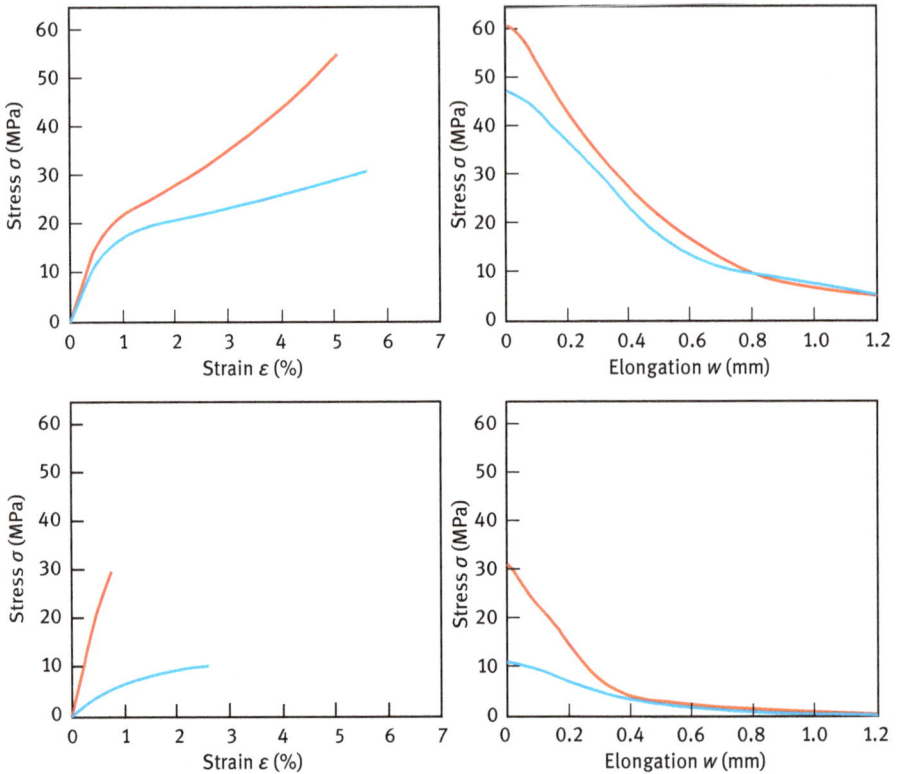

Fig. 2.11: Ordinary tensile stress–strain curves (left) and post-peak cohesive stress versus fracture zone widening (right) in MD (red) and CD (blue) in a 70 g/m^2 kraft paper (top) and 45 g/m^2 newsprint (bottom). The stress–strain measurements used 150 mm long and 15 mm wide specimens, and the post-peak measurement used 5 mm long and 15 mm wide specimens. Data of Tryding (1996). Reproduced with permission from the author.

The softening effect of moisture is shown in Fig. 2.13. The elastic modulus and breaking stress are lower and the breaking strain is higher at the higher relative humidity, corresponding to the higher MC in the paper.

In the in-plane tensile stress–strain curves displayed in Figs. 2.9 and 2.13, the breaking strain of paper ranges from 1% to 5%, which is quite typical. The values increase with increasing MC, and they may decrease with increasing strain rate. The breaking strain of paper falls below 1% only in very special cases. One can also see that the apparently linear part of the curves ends somewhere in the neighbourhood of 0.5%. The breaking stress is usually strongly correlated with the elastic modulus so that the ratio of the two is close to 1%, and the in-plane tensile breaking stress values range from 10 to 100 MPa.

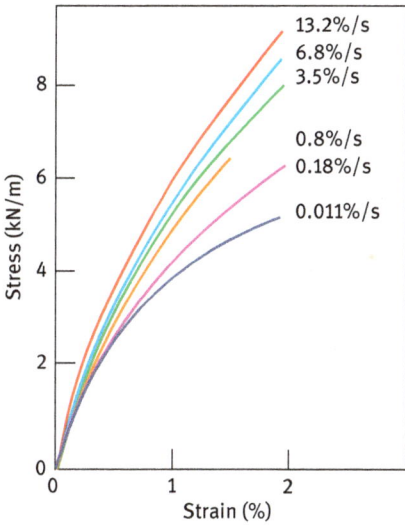

Fig. 2.12: Stress–strain curves at different strain rates for a wrapping paper, after Andersson and Sjöberg (1953). Stress values are multiplied by the thickness of paper. Reproduced with permission from Svenska Papper soch Cellulosa ingeniörsföreningen (SPCI).

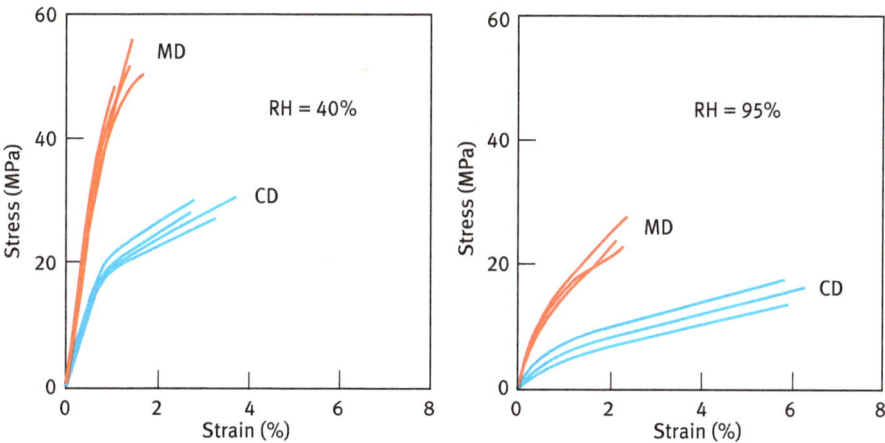

Fig. 2.13: Stress–strain curves in MD and CD of a paperboard at relative humidity of 40% and 95%. The corresponding moisture contents are 6.6% and 20%. Drawn using data of Yeh et al. (1991).

2.3.3 Other loading modes

Because paper is a thin planar material, the measurement of in-plane compression is complicated. Buckling of the specimen must be prevented with some fixture that creates in-plane forces. Even if buckling is prevented, paper fails under compressive stress much sooner than under tensile stress (Fig. 2.14). In the ZD testing, the situation is the opposite, and the compressive behaviour is easy to measure. The ZD

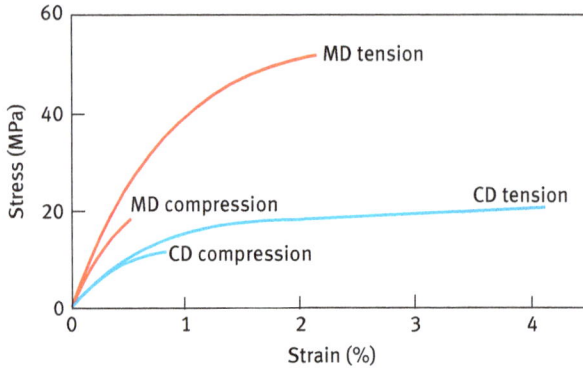

Fig. 2.14: Comparison of compressive and tensile behaviour of a paperboard in MD and CD, after Fellers (1980). Reproduced with permission from the author.

compressive strain is determined by the pore volume fraction and surface roughness, which are both pressed away by the applied stress. As the pore volume closes, the apparent stiffness of the material increases rapidly towards infinity, giving an exponential compressive stress–strain curve in ZD.

Tensile and shear testing in ZD requires an adhesive layer on the surfaces to transfer stress to the specimen. Because paper is thin, the adhesive contributes to the measured displacement and influences the deformation of paper. Tensile strain values in the ZD of paper are large, and stress values are small compared to the in-plane directions (Fig. 2.15). The maximum stress is observed relatively early in the stress–strain curve, before a long post-peak tail. Typical values for the maximum tensile stress in ZD are 0.1–1 MPa, two orders of magnitude smaller than in MD. Three-dimensional strains measured for tensile loading in MD and CD and shear loading in the MD–CD plane are illustrated in Fig. 2.16.

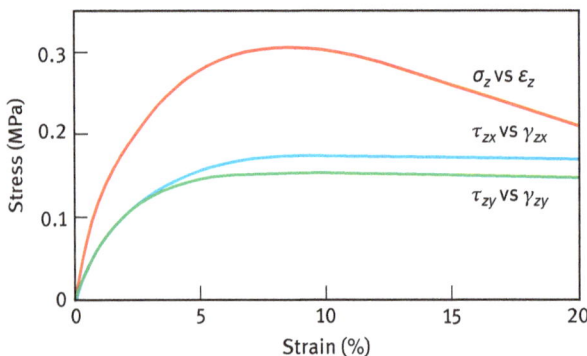

Fig. 2.15: Out-of-plane stress versus strain for tensile and shear loading of a paperboard. Data of Persson (1991). Reproduced with permission from the author.

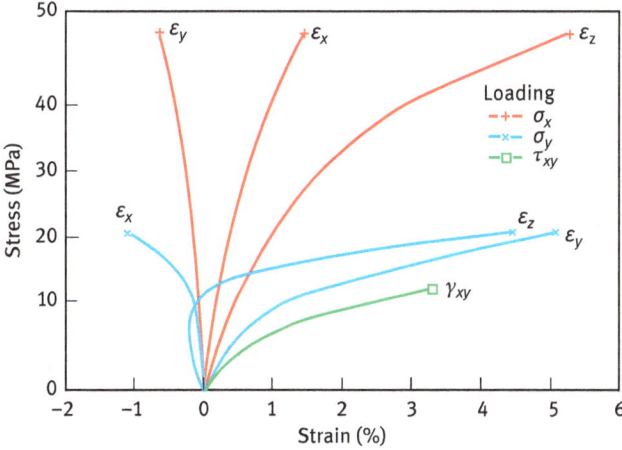

Fig. 2.16: Strains in the three principal directions of a paperboard, created by tensile stress in MD or CD and the shear strain created by a MD–CD shear stress, after Persson (1991). The shear behaviour was calculated from tensile loading at 45° angle from MD, assuming orthotropy. Reproduced with permission from the author.

2.4 Multi-axial strength

In packages and other structural uses, multi-axial stresses in the material often arise from contents and stacking. In order to analyse the material performance in such situations, it is necessary to model the multi-axial constitutive behaviour and ultimate strength of the material. As shown previously, the constitutive laws of paper can be highly non-linear, especially in ZD. For a representative description of the multi-axial non-linear loading–unloading behaviour, the elastic and plastic strain components must be separated (Xia et al., 2002).

In spite of the complicated constitutive behaviour of the paper materials, the ultimate strength values under multi-axial stresses can be estimated in a relatively simple manner from the Tsai–Wu criterion (Tsai and Wu, 1971). One calculates an effective scalar stress index f^{TW} according to

$$
f^{\text{TW}} =
\begin{bmatrix} \sigma_x \\ \sigma_y \\ \sigma_z \\ \tau_{xy} \\ \tau_{xz} \\ \tau_{yz} \end{bmatrix}^T
\begin{bmatrix} a_x \\ a_y \\ a_z \\ 0 \\ 0 \\ 0 \end{bmatrix}
+
\begin{bmatrix} \sigma_x \\ \sigma_y \\ \sigma_z \\ \tau_{xy} \\ \tau_{xz} \\ \tau_{yz} \end{bmatrix}^T
\begin{bmatrix} b_{xx} & b_{xy} & b_{xz} & 0 & 0 & 0 \\ b_{yx} & b_{yy} & b_{yz} & 0 & 0 & 0 \\ b_{zx} & b_{zy} & b_{zz} & 0 & 0 & 0 \\ 0 & 0 & 0 & b_{xyxy} & 0 & 0 \\ 0 & 0 & 0 & 0 & b_{xzxz} & 0 \\ 0 & 0 & 0 & 0 & 0 & b_{yzyz} \end{bmatrix}
\begin{bmatrix} \sigma_x \\ \sigma_y \\ \sigma_z \\ \tau_{xy} \\ \tau_{xz} \\ \tau_{yz} \end{bmatrix},
\qquad (2.7)
$$

and assumes that failure occurs when $f^{TW} = 1$. Here a_i and b_{ij} are material parameters, and b is symmetric, $b_{ij} = b_{ji}$. The zeros in eq. (2.7) arise because the sign of shear stress cannot affect the effective stress for an orthotropic material, and therefore, shear stresses are decoupled from other stresses. The 12 independent material parameters in eq. (2.7) must then be determined from experiments under different loading modes.

In the case where only xy-plane stresses are applied, the Tsai–Wu criterion simplifies into

$$f^{TW} = a_x\sigma_x + a_y\sigma_y + b_{xx}\sigma_x^2 + 2b_{xy}\sigma_x\sigma_y + b_{yy}\sigma_y^2 + b_{xyxy}\tau_{xy}^2 = 1. \tag{2.8}$$

In the biaxial σ_x–σ_y stress plane, this equation corresponds to an ellipse. Comparison with measured data for $\tau_{xy} = 0$ is shown in Fig. 2.17. If the shear stress τ_{xy} is increased, the size of the ellipse predicted by eq. (2.8) decreases. For the linerboard material in Fig. 2.17, the failure stress in pure shear mode can be estimated from network mechanics (Gustafsson et al., 2001) to occur at $\tau_{xy} \approx 2.4$ MPa.

If only out-of-plane stresses are applied, the Tsai–Wu criterion (eq. (2.7)) predicts failure when

$$f^{TW} = a_z\sigma_z + b_{zz}\sigma_z^2 + 2b_{xzxz}\tau_{xz}^2 + b_{yzyz}\tau_{yz}^2 = 1. \tag{2.9}$$

This is compared with measured values in Fig. 2.18. Clearly, the Tsai–Wu criterion does not work for compressive normal stresses σ_z larger than 1 MPa (compressive stresses are negative in Fig. 2.18). For larger compressive normal stresses, the Coulomb's friction law seems to apply, or

$$\sqrt{\tau_{xz}^2 + \tau_{yz}^2} = c - \mu\sigma_z, \tag{2.10}$$

where μ is the coefficient of friction, and c is the maximum cohesive stress.

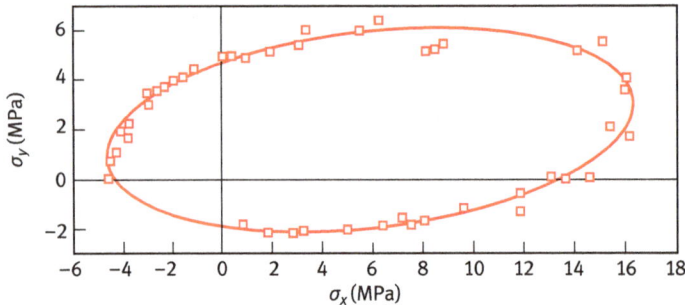

Fig. 2.17: Biaxial strength data (squares) for a linerboard compared with the Tsai–Wu criterion (eq. (2.8), line), after Fellers et al. (1983). The material parameters are $a_x = -0.158$ MPa^{-1}, $a_y = -0.314$ MPa^{-1}, $b_{xx} = 0.0172$ MPa^{-2}, $b_{yy} = 0.1120$ MPa^{-2}, and $b_{xy} = -0.011$ MPa^{-2}. Reproduced with permission from the Pulp and Paper Fundamental Research Society (www.ppfrs.org).

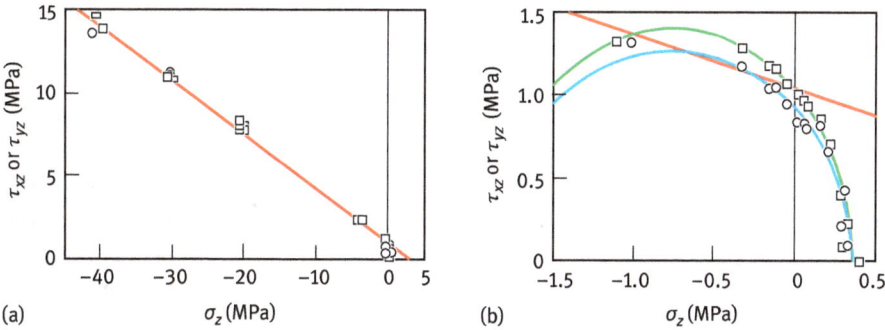

Fig. 2.18: Failure stress in out-of-plane loading of a paperboard for $\tau_{xz} = 0$ (circles) and $\tau_{yz} = 0$ (squares), compared with the Tsai–Wu criterion (eq. (2.9)) and Coulomb criterion (eq. (2.10)) at large and small normal stresses, (a) and (b), respectively (Stenberg, 2002). The parameter values are $a_z = 2.25$ MPa^{-1}, $b_{zz} = 1.48$ MPa^{-2}, $b_{xzxz} = 0.94$ MPa^{-2}, $b_{yzyz} = 1.16$ MPa^{-2}; and $\mu = 0.32$, $c = 1.05$ MPa. Reproduced with permission from the author.

2.5 Mechanical properties in relation to the papermaking process

2.5.1 Preparation of papermaking fibres

When paper is manufactured from wood, one usually starts by cutting the wood into chips (Fig. 2.19). The chips are then disintegrated in fibres either mechanically or chemically to prepare what is called pulp; depending on the process, the end product is mechanical pulp fibres or chemical pulp fibres. In the native wood,

Fig. 2.19: Pine wood chips prepared for papermaking. The grainy structure comes from the annual variation in wood growth. Chip length in the grain direction is 25 mm. Figure courtesy of Lisbeth Hellström, FSCN.

typical fibre dimensions are 1–3 mm in length and 20–40 µm in cross-sectional width. Different pulp fibres, other plant fibres, and man-made fibres are compared in Section 12.2.1.

In a typical mechanical pulp manufacturing process, the wood chips are sheared between rotating steel plates. The pattern of the plate surface is designed to optimize the pulp quality. Steam is applied to soften the lignin that holds fibres together in the wood material, and the cellulose of the fibre cell wall, to reduce fibre damage in the disintegration process. Pressure and chemicals may also be used for this purpose. Depending on the specific manufacturing process, different types of mechanical pulp are obtained, such as thermomechanical pulp, chemi-thermomechanical pulp, and their variants. One can also start with solid wood and grind fibres of the wood surface, which lead to pulps such as groundwood and pressure groundwood. A thorough coverage of mechanical pulping can be found in Lönnberg (2009).

In the chemical pulping process, wood chips are cooked with chemicals to dissolve the lignin that holds fibres together in wood. In addition, water-soluble hemicelluloses are extracted in the process. Depending on the chemicals used, different types of chemical pulps are obtained, such as the kraft pulp (also called sulphate pulp) and sulphite pulp (Gullichsen and Fogelholm, 1999). The cooking process leaves the structure of fibres rather intact (Fig. 2.20) except for the removal of lignin and hemicelluloses (see Section 12.2.1 for more information). A typical *yield* of the chemical pulping process is slightly more than 50%, meaning that about half of the original dry mass of wood is retained in the fibres (compared with mechanical pulping where the yield is usually more than 90%). In chemical pulping, the chemicals dissolved from the wood material have been traditionally used for energy. However, recently increasing development efforts have been directed to various bio-refinery concepts that convert the extracted chemicals to renewable fuels or polymeric raw materials.

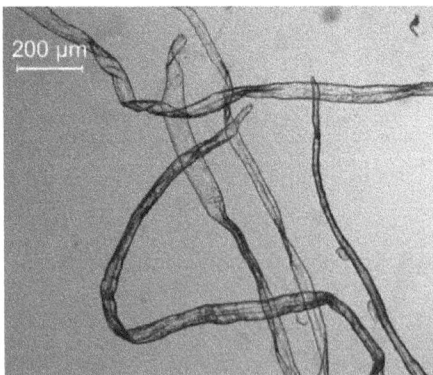

Fig. 2.20: Mildly refined chemical pulp fibres made of pine wood; courtesy of Boel Nilsson, SCA.

After the fibres are separated, both chemical and mechanical pulp can be bleached with chemicals to increase the whiteness of the final product. Then the pulp is treated further in a mechanical process called *refining* or *beating*, which increases the flexibility and conformability of the fibres and opens up the fibrillar structure of the fibre surface. This is necessary to achieve good bonding between the fibres so that the resulting paper has sufficient strength. Especially in the mechanical pulping process, a large fraction of the fibres is damaged and broken into fragments (Fig. 2.21). Sections of the fibre wall structure break off as flat lamella and in narrow ribbon-like fibrils; these small fibre fragments are called fines.

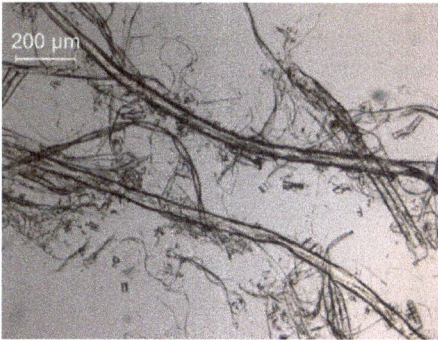

Fig. 2.21: Spruce wood mechanical TMP pulp fibres, fibre fragments, and fines; courtesy of Boel Nilsson, SCA.

The fragmentation of fibres in mechanical pulping is intentional. Smaller particles give a more uniform paper structure, reduce transparency, and improve the sharpness of print on the paper. Chemical pulping causes some fibre damage also, but not as much as in mechanical pulping. Thus, chemical pulp is generally stronger than mechanical pulp. Also, because lignin and some of the hemicelluloses are removed, chemical pulp does not turn yellow as easily as mechanical pulp when exposed to light or heat.

After the pulp manufacture, one adds other components, such as mineral fillers and chemicals, to obtain a water-based furnish ready for papermaking. Fillers increase the opacity and whiteness of paper. Chemicals are added to help retain the fines particles of pulp along with the fibres when water is removed on the paper machine ("retention aids"), to improve bonding between fibres ("bonding aids"), and to control ink penetration into the paper ("sizing"). Tuning of the pulp properties and furnish composition is the main method used to tailor paper properties.

2.5.2 Effect of the paper machine

Figure 2.22 demonstrates the structure of a paper machine, here cut in two parts to fit the figure on the page. The total length of a paper machine is typically a few hundred meters. The machine begins at top left with a forming section where the furnish at about 1% solids content is spread from a "headbox" (red in Fig. 2.23) on a moving wire. The low solids content is necessary so that fibres can be spread uniformly on the wire. Water is then removed by suction units (yellow) through a top and bottom wire and by wet pressing, where cylinders (dark green) press the wet paper web between two wires or felts. The forming and pressing sections together are called the "wet end" of the paper machine (see Paulapuro, 2008, for more information). When leaving the wet end, the solids content of the web is around 50%. Then the paper web is moved to the dryer section and pressed against hot cylinders so that water evaporates (red). A large number of dryer cylinders are needed because of the high speed of the process, which can reach 2,000 m/min.

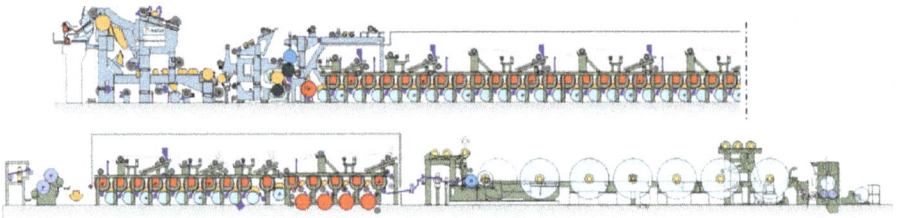

Fig. 2.22: Containerboard machine equipped with a gap former and two-layer headbox; courtesy of Metso Paper Inc.

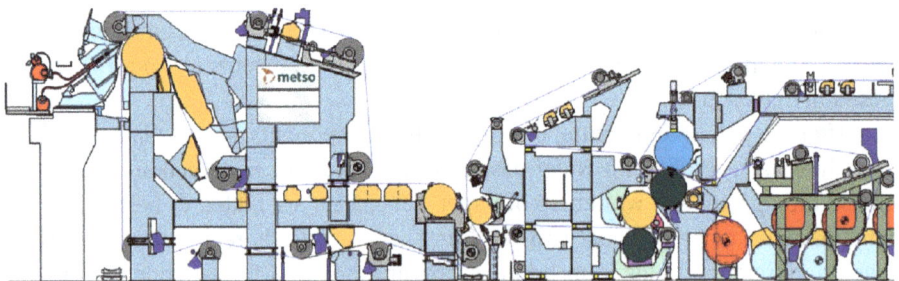

Fig. 2.23: Details of the forming section of the machine in Fig. 2.22.

In the "dry" end of the paper machine (bottom half of Fig. 2.22), water suspensions of sizing and pigments can be spread on the web surfaces to improve paper appearance and performance in printing (the blue rolls on the left). Further drying is then needed before winding to paper rolls. There is no coating or calendaring in the paper-making line of Fig. 2.22 because it is designed for containerboards that do not need high surface quality. In printing papers and packaging boards, one or several mineral coating layers can be used to maximize the product quality.

The initial forming section of the paper machine determines the network structure of fibres in the paper (Fig. 2.24). The fibre distribution is disordered but not completely random because the fibres have a tendency to form bundles or *flocs*. Furnish is diluted to less than 1%, and turbulence is induced in the headbox and on the wire to reduce the flocculation of fibres. Nevertheless, the mass distribution of paper always shows some variability, starting from a few centimetres down to fractions of a millimetre. At small length scales similar to the paper thickness (0.1 mm and higher), one has to consider the microscopic 3D structure, which is also disordered.

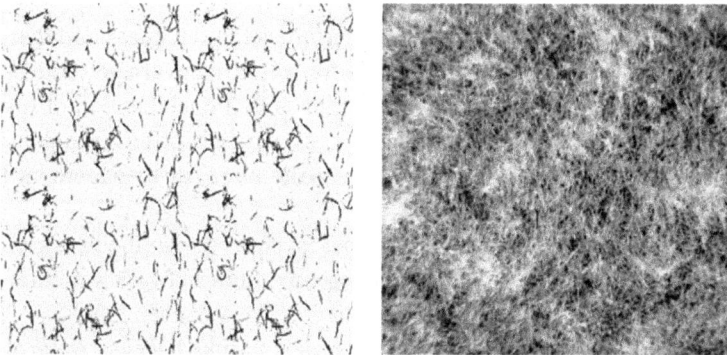

Fig. 2.24: Surface image of a paper sheet containing a small fraction of fibres dyed black (left), and a layer split from a sheet, showing fibres and fibre bundles (right, courtesy of Pekka Pakarinen).

The non-uniform in-plane mass distribution of paper is called *formation*. It can best be seen with bare eyes in thin paper grades such as newsprint or some office papers. Aside from being a visual imperfection, formation can cause out-of-plane deformations to paper if the MC changes (Section 9.3), and in rare cases, it can reduce the strength properties of paper, such as creep resistance. The non-uniform paper formation also shows up in the orientation distribution of the fibres (Fig. 2.24) and the mass density, thickness, and porosity of paper from the microscopic to macroscopic. The non-uniformity of fibre orientation can accentuate eventual problems with out-of-plane deformations of paper because fibre orientation has a strong effect on the hygroscopic strains.

Fibre orientation in paper arises from the forming process. A small speed difference is generally needed between the wire and the furnish jet that comes from the headbox. At the same time, the speed difference creates a shear field through the thickness of the furnish layer, which in turn rotates fibres more parallel to the MD. The anisotropy of fibre orientation is one of two factors that cause anisotropy in the mechanical properties and hygroexpansion of paper (Niskanen, 1993), the other being the drying effect discussed later. On a board paper machine, slight CD deviations can occur in the flow direction of the furnish so that the local symmetry axis of the fibre orientation distribution may be a few degrees off the MD. The non-zero fibre orientation angle can cause diagonal curl (Section 9.2) in products where paper or board is used in sheet form, and this can lead to, for example, paper jam in a copy machine.

The direction of initial water removal creates a ZD profile of fines and filler concentration in paper. The smaller particles are flushed with water, generally towards the wires. Because of the flushing, layers close to the wire surfaces can be depleted of fines and fillers. A ZD variation arises even in the fibre orientation distribution because the shear field in the suspension layer changes as the water removal progresses.

The wet pressing stage of a paper machine determines the thickness and density of paper. Intensive wet pressing is favourable for water removal and reduces the energy needed in the dryer section, but it also leads to a densification of the paper. Low paper thickness gives low bending stiffness, which is often a problem, for example, in the handling of paper sheets or in the strength of paperboard boxes.

The forming and wet pressing stages of the paper machine determine the structure of paper from centimetres down to the microscopic fibre network structure. Information that is more detailed can be found in chapter 1 of Niskanen (2008).

On the dryer section of the machine, the paper web shrinks because water is removed from the fibres. Earlier in the process, water is removed only from the pore space between fibres. At this stage of drying, a tension must be applied on the web to prevent fluttering and to improve contact with the drying cylinders. The drying tension on the paper machine prevents paper shrinkage in the MD, which occurs almost exclusively in the CD. Drying tension versus shrinkage is the second factor that influences the anisotropy in the mechanical properties and hygroexpansion of paper.

The CD drying shrinkage has a parabolic profile across the web (Fig. 2.25). The edges of the web shrink without any constraint, while in the centre of the web shrinkage is partially prevented by adhesion or friction against the drying cylinders. This results in parabolic profiles of material properties and web tension. For example, the CD elastic modulus has the lowest values and the CD hygroexpansion has the largest values at the web edges.

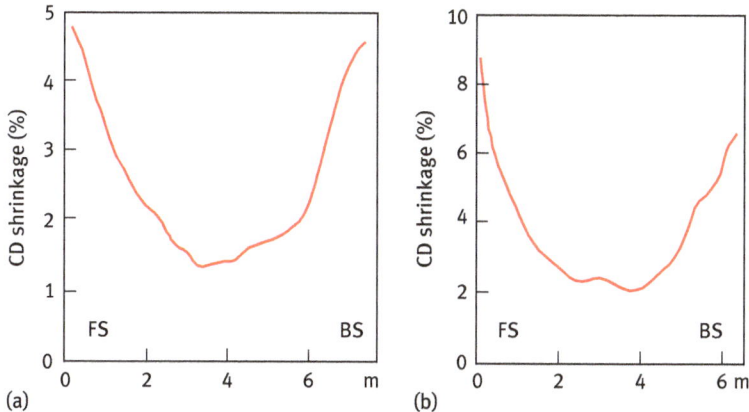

Fig. 2.25: Cross-machine direction shrinkage that occurred from the forming section to dry paper on a commercial newsprint paper machine (a) and office paper machine (b). Reproduced from Niskanen (1993) with permission from Paperi ja Puu Oy.

Literature references

Andersson, O. and Sjöberg, L. (1953). Tensile studies of paper at different rates of elongation. Sv. Papperstidn. 56(16), 615–624.

Baum, G.A. (1984). The elastic properties of paper, a review, In: Design criteria for paper performance, Kolseth, P., Fellers, C., Salmén, L. and Rigdahl, M., eds, 1984, Seminar on Progress in Paper Physics, Stockholm, Svenska Träforskningsinstitutet (STFI)-Meddelande A969, 1–27.

Fellers, C. (1980). The significance of structure on the compression behavior of paper. Doctoral dissertation. Royal Institute of Technology, Stockholm, Sweden.

Fellers, C., Westerlind, B. and de Ruvo, A. (1983). An investigation of the biaxial failure envelope of paper. Presented at the British Paper and Board Industry Federation, "The Role of Fundamental Research in Papermaking Symposium," 2, September 1981, Cambridge, UK (London: Mechanical Engineering Publications), pp. 527–559.

Girlanda, O. and Fellers, C. (2007). Evaluation of the tensile stress-strain properties in the thickness direction of paper materials. Nordic Pulp Pap. Res. J. 22(1),49–56.

Gullichsen, J. and Fogelholm, C.-J. (1999). Chemical pulping, Helsinki. Finland: Fapet Oy.

Gustafsson, P.J., Nyman, U. and Heyden, S. (2001). A network mechanics failure criterion. Report TVSM-7128, Div. of Structural Mechanics, Lund University.

Hillerborg, A., Modéer, M. and Petersson, P.-E. (1976). Analysis of crack formation and crack growth in concrete by means of fracture mechanics and finite elements. Cem. Concr. Res. 6, 773–782.

Ketoja, J.A., Tanaka, A., Asikainen, J. and Lehti, S.T. (2007). Creep of wet paper. Presented at 61st Appita Annual Conference and Exhibition, Gold Coast, Australia, May 6–9.International Paper Physics Conference (Carlton, Australia: Appita).

Lönnberg, B. ed (2009). Mechanical pulping. 2nd Edition, Helsinki, Finland: Paperi ja Puu Oy.

Malvern, L.E. (1969). Introduction to the mechanics of a continuous medium. Englewood Cliffs, NJ, USA: Prentice-Hall.

Mark, R.E., Habeger, C.H., Borch, J. and Lyne, M.B. (2002). Handbook of physical testing of paper. 1, New York: Marcel Decker.

Niskanen, K. ed (2008). Paper physics. 2nd Edition, Helsinki, Finland: Paperi ja Puu Oy.

Niskanen, K.J. (1993). Anisotropy of laser paper. Paperi. ja. Puu. 75, 321–328.

Paulapuro, H. ed (2008). Papermaking, part i stock preparation and wet end. 2nd Edition, Helsinki, Finland: Paperi ja Puu.

Persson, K. (1991). Material model for paper; experimental and theoretical aspects. Diploma Thesis. Division of Solid Mechanics, Lund, Sweden: Lund Institute of Technology.

Salmén, L., Carlsson, L., de Ruvo, A. et al. (1984). Treatise on the elastic and hygroexpansional properties of paper by a composite laminate approach. Fibre Sci. Technol. 20(4), 283–296.

Stålne, K. (2006). Testing of elastic properties of printing paper. Report. Div. Structural Mechanics, Lund University.

Stenberg, N. (2002). On the out-of-plane mechanical behavior of paper materials. Doctoral Thesis. Department of Solid Mechanics, Royal Institute of Technology, Stockholm, Sweden.

Tryding, J. (1996). In-plane fracture of paper. Doctoral Thesis, Division of Structural Mechanics, Lund University.

Tryding, J. and Gustafsson, P.J. (2000). Characterization of tensile fracture properties of paper. Tappi J. 83(2), 84–89.

Tsai, S. and Wu, E. (1971). A general theory of strength for anisotropic materials. J. of Compos. Mater. 5, 58–80.

Xia, Q., Boyce, M. and Parks, D. (2002). A constitutive model for the anisotropic elastic- plastic deformation of paper and paperboard. Int. J. Solids Struct. 39, 4053–4071.

Yeh, K.C., Considine, J.M. and Suhling, J.C. (1991). The influence of moisture content on the nonlinear constitutive behavior of cellulosic materials. In: Proceedings of the International Paper Physics Conference (Atlanta, GA, USA: Tappi Press), Book 2, p. 695.

Part I: **Structural strength**

Rickard Hägglund, Leif A. Carlsson

3 Packaging performance

3.1 Introduction

Packaging is an important application of paper materials. The main purpose of packaging is to facilitate shipment of goods from the producer to the consumer. Packaging has many other important functions, such as protecting the packaged goods from hazards such as contamination in the distribution environment, facilitating transportation and storing of products, and carrying printed information and graphics. The design and development of packaging are often linked to the product to be shipped. The packaging design process starts by establishing mechanical loads, marketing aspects, shelf life, quality assurance, distribution environment, and legal, regulatory, graphic design, end-use, and environmental factors. The design criteria, packaging performance, completion time targets, resources, and cost constraints need to be established and agreed upon.

This chapter focuses on the performance requirements that relate to the strength of the package. An important example of such requirements is the strength of the container in compression (stacking strength) during high humidity exposure. The packaging must withstand the load during the specified time of storage. Packaging for cooler/freezer products must be able to support loads at low temperatures in a moist environment without failing.

Packaging may be categorized in one of three groups depending on its role in the distribution chain:

- Primary packaging, or consumer packaging, is the material that first envelops the product. This is usually the smallest unit of distribution or use, and it is the package in direct contact with the contents. Examples are carton board packaging of small items, a glass jar, or a plastic bottle. Carton board boxes are often referred to as folding boxes or simply cartons.
- Secondary packaging, or transport packaging, is outside the primary packaging and is often used to group primary packages together. One of the most common forms of secondary packaging is the corrugated board box.
- Tertiary packaging refers to materials used in bulk handling, warehouse storage, and transport and normally includes materials such as pallets and stretch film.

This chapter discusses mainly corrugated board packaging and its performance, but parallels are made with carton board packaging. The secondary packaging

Rickard Hägglund, SCA R&D Centre AB, Sundsvall, Sweden
Leif A. Carlsson, Florida Atlantic University, Boca Raton, FL, USA

https://doi.org/10.1515/9783110619386-003

carries most of the load during transport, although in some systems the primary packaging supports significant load and requires strength.

The first commercial carton board box was produced in England in 1817. Folding cartons were introduced in the 1860s. They were shipped flat to save space, ready to be erected by customers when required. Mechanical die cutting and creasing was developed soon after. In 1915, the first milk carton was patented, and in 1929, machinery was developed for commercial production of cartons. In 1935, the milk carton was implemented in a dairy plant. The first corrugating machine, that is, the machine that makes corrugated board, appeared in the United States in the early 1900s. Until the year 1919, the majority of products were shipped in wooden crates and often transported by train. The corrugated box was relatively new, and the railroads had no experience in handling and transporting them. Railroad companies did not assume the liability for damage while shipping items in corrugated boxes until Rule 41 was established in 1919. In the original version of Rule 41, the most important requirement for the box was that it contains and protects the shipped goods. As a result, the box materials and structures were designed primarily for burst and puncture resistance, which are measures of the tensile strength and ability to resist penetration. Later, World War II contributed to establishing corrugated board packaging when it was called upon to deliver goods to all corners of the earth. After World War II, the market for corrugated board expanded rapidly, and the range of sizes and capabilities of packaging grew to fit the large number of new products developed.

Today, corrugated paperboard is the most popular material for transport packages for a wide variety of products, varying from fresh fruit and vegetables, consumer electronics, and industrial machinery to semi-bulk transports of various commodities in large bins. The global corrugated board market in 2019 was $263 billion, and is forecast to grow significantly in the coming years. More than 60% of corrugated board is used to package non-food products. The largest single end-use sector is the electrical goods market. It is equally suitable for any mode of transport, for example, shipping by sea, railroad, truck, or air. This versatility is largely due to the possibility of combining different materials and thereby accommodating particular requirements in the distribution system in a tailor-made way. Corrugated board is considered as environmentally friendly. Currently more than 80% of corrugated board is recycled.

3.2 Paper-based packaging materials

Packaging paper materials are made from cellulose fibres produced either from virgin wood or recycled fibres, or both. The paper surface is sometimes coated in order to enhance printability. Paperboard is often combined with polymers or metal foils to form laminates for packaging of products such as juices and dairy products. This section describes the structure and manufacture of corrugated board and the process of converting corrugated board into boxes. Carton board materials are also discussed. The manufacturing process of carton board differs from that of corrugated board, but the converting process of the carton board is similar.

3.2.1 Corrugated board

Corrugated board is a sandwich construction with a web core and face sheets made from paper. *Container board* is the common name for the paper materials used to manufacture corrugated board, and includes linerboard, used for the facings, and fluting, which is the paper used in the core. The face sheets and core are typically glued together with a starch-based adhesive. The core must also provide shear transfer between the face sheets to minimize sliding deformation during bending. Furthermore, the corrugated board should be stiff and strong in the out-of-plane direction in order to keep the face sheets apart and parallel at the correct distance during in-plane and transverse normal loading. A weak core may fail due to its inability to support the face sheets against local buckling or wrinkling.

Corrugated board is made of a number of layers depending on the packaging requirements. The single-face corrugated board consists of two layers: one linerboard and the corrugated fluting layer (Fig. 3.1). The single-wall corrugated board is a true sandwich consisting of three layers: two linerboards and the corrugated fluting core. The double- and triple-wall corrugated boards consist of two and four additional layers, respectively. About 80% of corrugated board is produced as single-wall board.

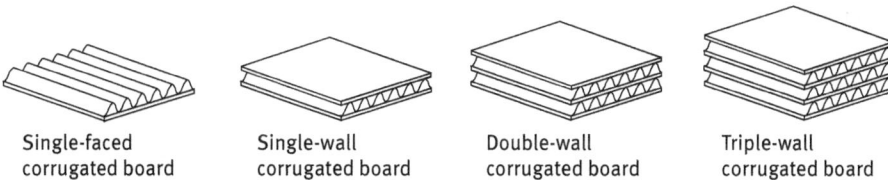

| Single-faced corrugated board | Single-wall corrugated board | Double-wall corrugated board | Triple-wall corrugated board |

Fig. 3.1: Single-wall, double-wall, and triple-wall corrugated boards.

Figure 3.2 provides the geometrical parameters of a corrugated sandwich panel. In this case, the board is symmetric, but asymmetric board configurations are also used. The core has a sine-wave shape. The main dimensions of a corrugated board are its thickness d_{corr}, as well as the wavelength λ_{core}, and height h_{core} of the flutings. The cross-machine direction (CD) of the face sheets and core is oriented parallel to the flutes.

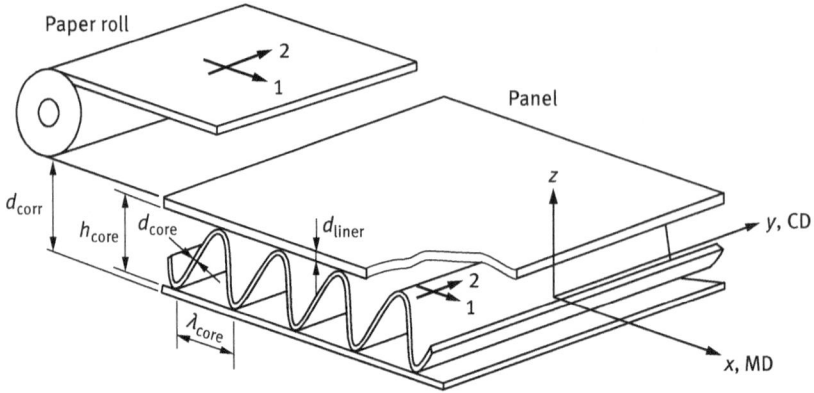

Fig. 3.2: Geometric parameters of a corrugated board. The wavelength and height of the fluting waves in the core are λ_{core} and h_{core}; the thicknesses of the fluting and linerboard are $d_{fluting}$ and d_{liner}, and the total thickness of the corrugated board is $d_{corr} = 2d_{liner} + h_{core}$.

Figure 3.3 lists wavelengths and flute heights for the most common commercial single-wall corrugated boards. The letter designations have been assigned in the order the different geometries were introduced to the market, not in the order of size. The wave length and flute height slightly vary between production sites. Boards with smaller flutes have more flutes per unit length. Generally, large flute height improves box compression strength because it gives high bending stiffness of the board. Small flutes are used mainly for high-quality print and when the requirement on box strength is lower. E- and F-flutes are often used as a replacement for carton board. C-flute is typical in conventional transport packages where compression strength is required. B-flute is used, for example, for canned goods and other products not requiring high compression strength. A-flute is used as cushion pads or in heavy double- or triple-wall boards.

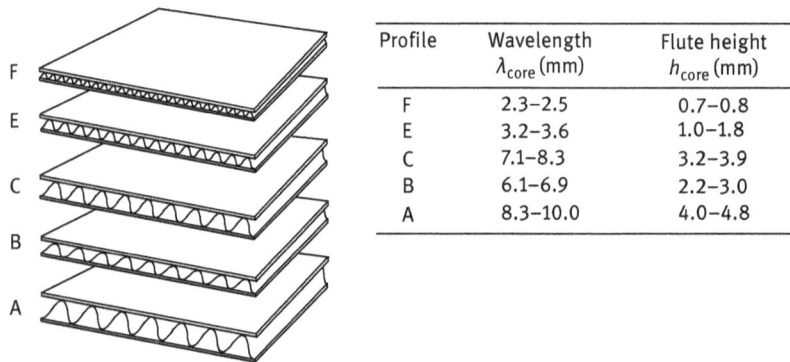

Profile	Wavelength λ_{core} (mm)	Flute height h_{core} (mm)
F	2.3–2.5	0.7–0.8
E	3.2–3.6	1.0–1.8
C	7.1–8.3	3.2–3.9
B	6.1–6.9	2.2–3.0
A	8.3–10.0	4.0–4.8

Fig. 3.3: Wavelength and height intervals of common flute profiles.

Many package styles and structural designs are possible and often specified using the international standard of the Federation of Corrugated Board Manufacturers (FEFCO, 2011). One of the most common boxes is the regular slotted container (RSC) denoted FEFCO 0201 (Fig. 3.4). It is manufactured from a single sheet of corrugated board that is scored and slotted to permit folding. Flaps attached to the side and end panels become the top and bottom panels of the box. A flap is defined by the area from the edge of the sheet to the flap folding lines. Opposite pairs of flaps are long enough so that they can close the bottom and top of the box.

Fig. 3.4: Regular slotted container (RSC).

Corrugated board is manufactured in a continuous operation where the fluting is first plasticized by hot steam, then fed between a pair of sine-wave-shaped gears to form the wavy web that is finally glued to the face sheets (Kuusipalo, 2008). The direction of manufacture in Fig. 3.5 is from left to right: (1) the fluting unwinds from the reel stand and is heated and moistened with steam. (2) The fluting then passes between a pair of hot gears to get its sine-wave shape. (3) Glue is applied onto the flute tips that are glued to the preheated liner (single facer) to form a single-faced web. (4) This is then heated again, and glue is applied on the opposite flute tips. (5) The second liner (double backer) is heated and bonded to the single-faced web to form a corrugated board sandwich panel. (6) The board is dried under pressure between hot plates. For double-wall and triple-wall boards, the process is repeated, normally in line with the original single back operation. (7) Finally, the corrugated board is cut to shape conforming to the package design.

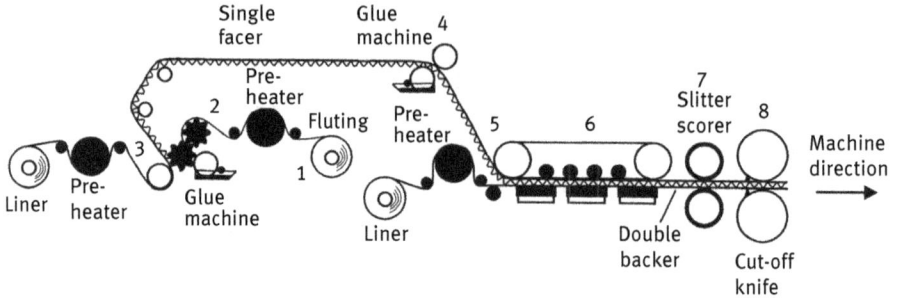

Fig. 3.5: The manufacturing process of corrugated board.

3.2.2 Box manufacturing process

The converting of corrugated board refers to all the process stages that transform the corrugated board into the final corrugated container (Kuusipalo, 2008). Typically, this involves printing, slotting, creasing, die cutting, folding, and gluing. Printing can be post-printing, directly to the corrugated packaging, or pre-printing, in which case the liner is printed before it is assembled into the corrugated board. Most of the RSCs are produced with one in-line operation that converts a flat sheet into its final form, ready to be shipped to the customer.

Die cutting is used when the package demands very precise cutting or has a complicated shape. The die-cut tool may also be equipped with creasing rules (dies) that make scores to the board to define folding lines in the box. The principles of flat-bed cutting and creasing are shown in Fig. 3.6. In rotary die cutting (Fig. 3.7), the cutting and creasing rules are mounted on the surface of a wooden cylinder, enabling continuous operation.

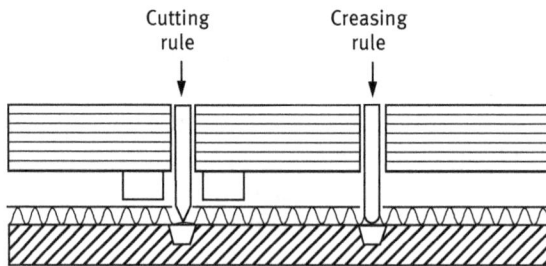

Fig. 3.6: Flat-bed die cutting and creasing.

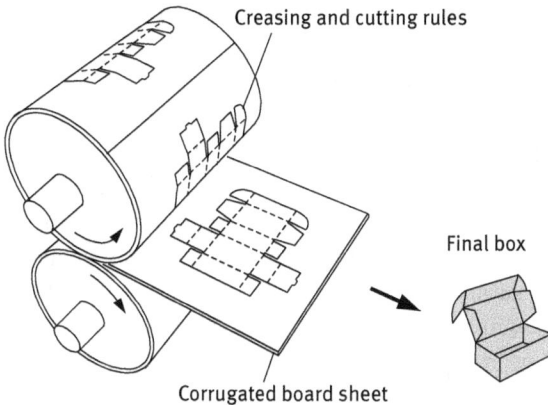

Fig. 3.7: Rotary cutting and creasing tool.

In the manufacture of the corrugated board, the material is subject to various lateral loads in, for example, feed roller and printing nips. Depending on board properties such as thickness and strength, the die-cutting process may damage the corrugated board (Cavlin and Edholm, 1988). As the male die (Fig. 3.6) penetrates the board, tensile, compressive, and shear stresses are created in the board. Shear deformation in particular may fail the bond between layers in the board. This delamination alters the strength properties of the creases (discussed in more detail in Chapter 4). If the crease is too deep, the tension may break the upper liner. On the other hand, if the crease is too shallow, then the bottom linerboard may fail in tension during the box folding phase (Hägglund and Isaksson, 2008; Isaksson and Hägglund, 2005; Thakkar et al., 2008).

3.2.3 Carton board

Carton board is the common name for paper used in packaging cartons. The material consists of three or more furnish layers manufactured simultaneously on a multi-layer paperboard machine (see, e.g. Fig. 2.22). Cardboard may be coated with polymers to achieve a material that can be used in ovens, microwaves, and other demanding conditions, or it may be laminated with metal films to enhance appearance and protect the content. According to the composition and thickness of the layers, the main types of carton boards are (Confederation of European Paper Industries [CEPI], 2011) as follows:

– Solid bleached board is typically made from pure bleached chemical pulp with two or three layers of coating on the top and one layer on the bottom. It is a medium-density board with good printing properties for graphical and packaging end uses, and it is white throughout the thickness. It is used in such markets as cosmetics, graphics, pharmaceuticals, tobacco, and luxury packaging. It can also be combined with other materials for liquid packaging applications.

- Solid unbleached board (SUB) is typically made from pure unbleached chemical pulp with two or three layers of coating on the top surface. In some cases, a white bottom surface layer is applied. SUB is primarily used in packaging of beverages such as bottles and cans because it is very strong and can be made resistant to water. It is also used in a wide variety of general packaging areas where strength is important.
- Folding boxboard is typically made from a number of layers of mechanical pulp sandwiched between two layers of chemical pulp with up to three layers of coating on the top printing surface and one layer of coating on the bottom. This is a low-density material with high bending stiffness with a yellowish colour on the inside. It is used in markets such as pharmaceuticals, frozen and chilled foods, and confectionary.
- White-lined chipboard (WLC) consists predominantly of recycled pulp. WLC typically has three layers of coating on the top printing surface and one layer on the bottom. Because of its recycled content, the back side is grey. WLC is used in a range of applications such as packages for frozen and chilled foods, breakfast cereals, shoes, tissues, and toys.

The main types of cartons are tube and tray. In a tube, the machine direction of the board runs horizontally around the package in order to maximize resistance against bulging of the side panels. The European Carton Makers Association (ECMA) Code of Folding Carton Design Styles is a reference standard for folding carton package designs (ECMA, 2011). The process of converting carton board to packaging is analogous to the converting of corrugated board discussed previously, except that the carton board material is continuous. The carton board material behaviour is discussed in detail in Chapter 4.

3.3 Loads imposed on boxes

Boxes are subject to a number of different loading conditions in the filling, stacking, transportation, and storage operations. The loads are especially important for corrugated boxes that are used as secondary packages to support loads and protect primary packages, which in turn typically are not supporting loads. The loads discussed in this section, however, may apply also for primary packages.

The package content imposes internal loads that affect the service life of the box. When a box is filled with a fluid or a granular material, pressure exerts on the box walls (Fig. 3.8a). The pressure caused by a fluid increases linearly with depth. The pressure caused by a granular material grows non-linearly, reaching asymptotically a maximum value. The pressure causes the box panels to bulge. The amount of bulging is governed by the bending stiffness and dimensions of the box panels. Bulging has a negative impact on the load carrying capability of the box. Excessive bulging may also cause undesirable contacts with surrounding objects and reduce the perceived quality of the box. In trays, the weight of the content acts on the bottom (Fig. 3.8b), causing the bottom panel to sag. In tight stacks, this may cause problems if the bottom of a tray gets in contact with the contents of the tray below.

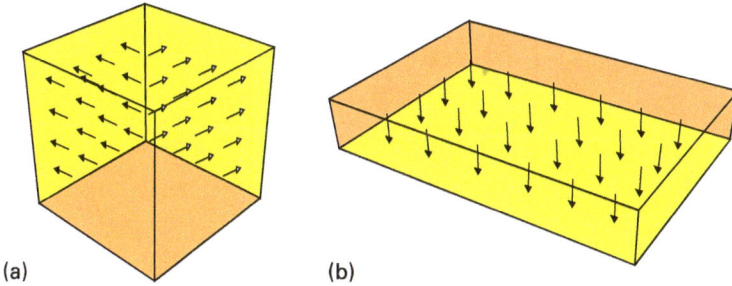

(a)   (b)

Fig. 3.8: Internal pressure on box walls (a) and tray bottom (b).

Boxes are often stacked on a pallet (Fig. 3.9). The boxes at the bottom are thus subject to a top-to-bottom compressive loading by the overlying boxes. If the content is rigid and of dimensions similar to the box, the content may carry part of the load, increasing the load carrying capability. The distribution of loads on the boxes depends on the pattern in which the boxes are stacked on the transport pallet. Kellicutt (1963) found that the column stacking pattern in Fig. 3.9a gives maximum strength utilization because high corner forces are transferred effectively to the neighbouring boxes. Interlock stacking (Fig. 3.9b) provides better stability, but the load carrying capability is significantly less because rigid corners rest on flexible panel centres.

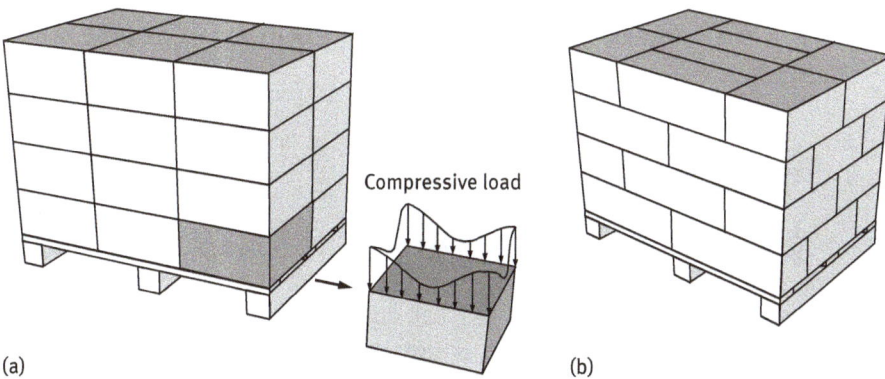

Compressive load

(a)   (b)

Fig. 3.9: Column stacking (a) and interlock stacking (b), with the top-to-bottom compressive load on a box shown in (a).

Deviations from perfect stacking reduce the load carrying capability. Misalignment of the stack by a few centimetres (Fig. 3.10a) or a tilt of a few degrees (Fig. 3.10b) reduces the load carrying capacity significantly. The same is true with overhangs, that is, when the pallet is smaller than the stack area (Fig. 3.10c). Underhangs (Fig. 3.10d) do not influence strength, but they occupy more space.

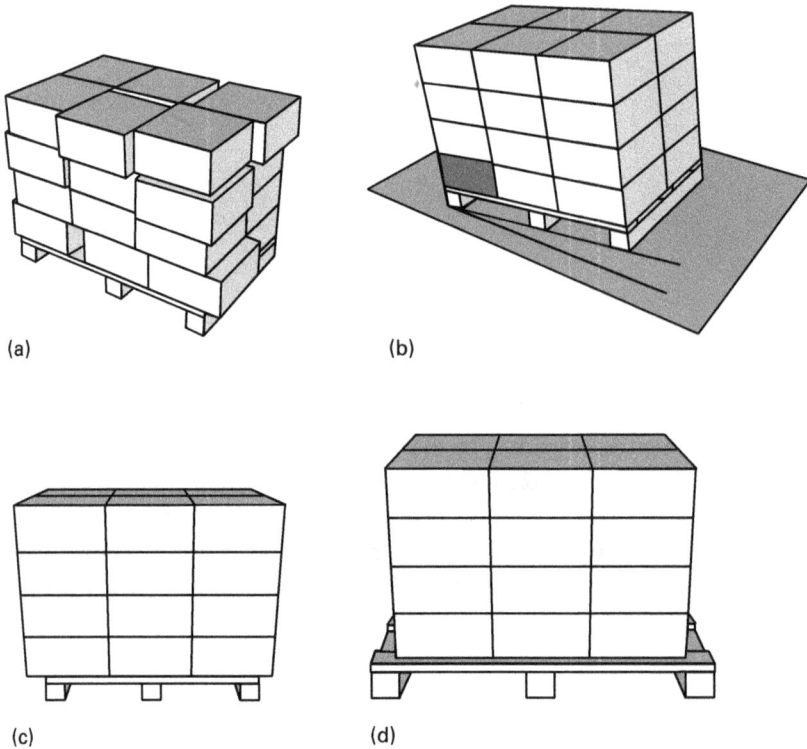

(a)

(b)

(c)

(d)

Fig. 3.10: Offset in column stacking (a), inclined stack (b), pallet "overhang" (c), and pallet "underhang" (d).

3.4 Strength of boxes

Knowledge about the compression loading of corrugated board panels and structures has increased considerably in the last decades (Maltenfort, 1956; McKee and Gander, 1957; Westerlind and Carlsson, 1992). This section describes the strength of corrugated board and carton board boxes under compressive loading.

3.4.1 Short-term compressive loading

The short-term compression strength is the single most important property specification of a corrugated board box. It is a direct measure of a box performance in a stack. The standardized box compression test (BCT) measures the failure load. BCT is a pure top-to-bottom compression test of a box placed between flat parallel steel plates and displaced at a constant deformation speed (Fig. 3.11). The compressive load and cross-head displacement are recorded continuously until collapse occurs. The maximum load attained is reported as the BCT value. Testing is conducted at a constant 23 °C temperature and 50% relative humidity (RH).

Fig. 3.11: Box compression test (BCT).

Several parameters are used to characterize the mechanical performance of the corrugated board material. The most important properties are the edge crush test (ECT) and the bending stiffness. The ECT value is a measure of the compressive strength of a small corrugated board specimen in the direction of the flutes, while bending stiffness is a measure of resistance to bending. Similarly, for carton board, the most important mechanical properties are the bending stiffness and its compressive strength. The compressive strength of carton board is measured at a short distance between the grips, in order to avoid buckling of the specimen. This test is referred to as the short-span compression test (SCT), ISO 9895.

The manner in which a corrugated box fails in compression depends on the board constituents and the box dimensions. Buckling and compression failure are the two principal failure modes of a corrugated board box. Usually, the vertical panels of boxes fail through a post-buckling collapse that occurs after buckling of the panels (Fig. 3.12a). This failure mode is particularly common for tall boxes and boxes made of relatively thin board of low bending stiffness. Typically, two of the four panels bow outwards and the other two inwards. Boxes that are short and very stiff in bending do not buckle and fail by forming horizontal wrinkles (Fig. 3.12b). The failure of carton board boxes is dominated by the corners as discussed in Chapter 4.

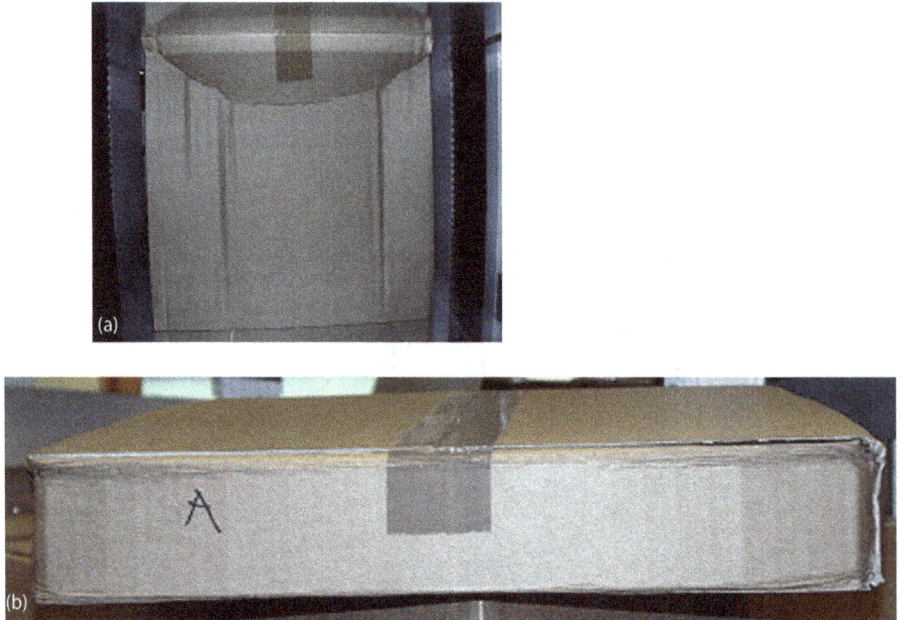

Fig. 3.12: Compressive failure of a C-flute corrugated board box of height 400 mm (a) and 70 mm (b).

Figure 3.13 shows the load–deformation responses of the boxes in Fig. 3.12. The initial stiffness is very low due to the creases. The shallow box is stiffer and stronger than the tall box. Once the faces of the tall box have buckled, the corners carry most of the applied load. Buckling of the tall box occurs approximately at a load two-thirds of the maximum load. If the fluting is strong and stiff enough to support the linerboards, then the failure typically initiates on the concave side linerboard (facing), close to the corners (Fig. 3.14). The shallow box that does not buckle has a fairly uniform load distribution along its perimeter, which explains its higher strength.

It is necessary to distinguish between local and global buckling. Local buckling (Fig. 3.14) occurs when the linerboard buckles between the flute crests. Global buckling occurs when the entire panel buckles. At collapse, the local buckles coalesce into a wrinkle that starts near a corner and runs diagonally towards the centre of the panel (Fig. 3.14). Figure 3.15 illustrates the evolution of local strain distribution on the concave side of a panel. Figure 3.16 shows the corresponding load–displacement curve of the panel under in-plane compression. The strain field is initially fairly uniform, but failure becomes localized when a zone of diagonal strain concentration emerges, and finally a wrinkle is formed after which the board fails.

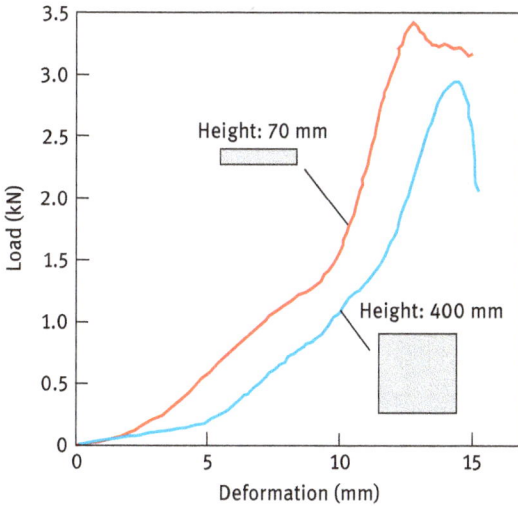

Fig. 3.13: Load–deformation response in compression of the C-flute corrugated board boxes shown in Fig. 3.12.

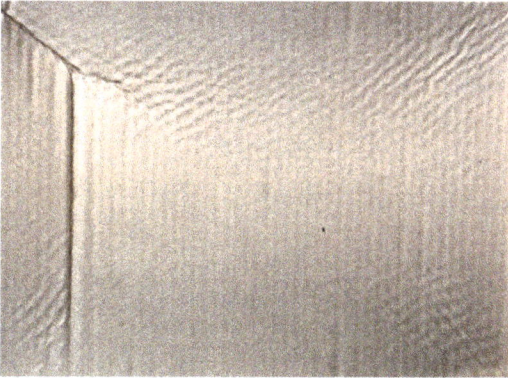

Fig. 3.14: Local buckling and wrinkling pattern on the concave side of a collapsing panel of a corrugated board box.

The compression test results shown in Figs. 3.15 and 3.16 refer to panel tests because box compression tests are relatively costly and involve a number of uncontrolled variables that result from the creasing and folding operations. Therefore, it has become customary to conduct compression testing of panels to estimate the strength of corrugated board boxes. The test rig in Fig. 3.17 simulates the boundary conditions and loading of the panels in a corrugated board box under top-to-bottom loading. Out-of-plane

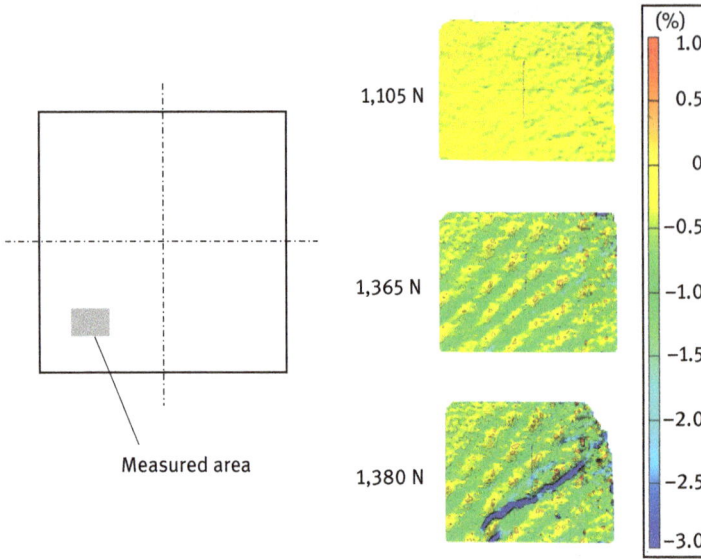

Fig. 3.15: Strain distributions in a 60 × 40 mm² area on the concave side of a 400 × 400 mm² C-flute panel under compressive loading, measured using digital image correlation. The three load levels are indicated on the load–displacement curve shown in Figure 3.16.

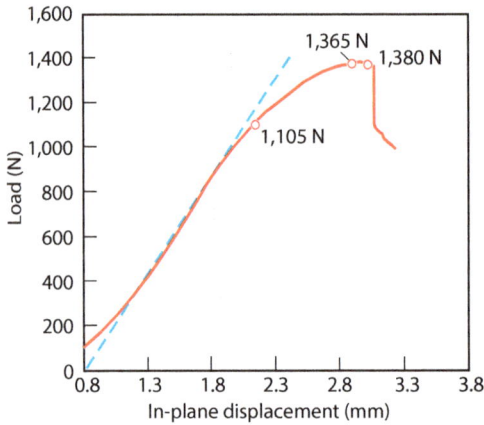

Fig. 3.16: Load versus in-plane displacement in an edge-wise compression-loaded C-flute panel. The three points at loads of 1,105 N, 1,365 N, and 1,380 N are the same for which strain distribution is shown in Fig. 3.15.

deformations are prevented along the perimeter, but rotations are allowed. In a panel compression test, the in-plane deformations are generally much smaller than in a box because load is introduced more uniformly on the edges in the panel test.

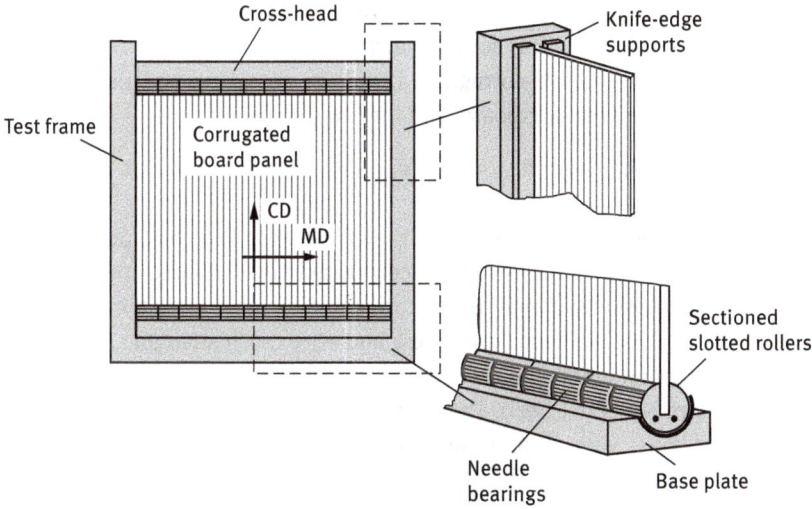

Fig. 3.17: Panel compression test rig (Nordstrand, 2004). Reproduced with permission from the author.

3.4.2 Empirical models for static box strength

Models of box strength provide guidance for package design and optimization. Although corrugated board has been used for more than a century, the design process of packages is still mostly empirical or semi-empirical and based on box testing, which is time consuming and expensive. Finite element simulations of box loading offer a modelling strategy that requires only the dimensions of the box and the mechanical properties of the face sheets and core.

Already before World War II, engineers at the British Air Ministry had established an empirical relationship between the collapse load of a panel and its buckling load and material compression strength. This formula is often called the Cox formula (Cox, 1933). McKee et al. (1963) developed a design formula (3.1) for corrugated board boxes based on the Cox formula and published the classical article "Compression Strength Formula for Corrugated Boxes". This formula incorporates an empirical relation between the bending stiffness of the corrugated board, its geometry, and the material properties of the fluting and the linerboard, and the box perimeter,

$$\text{BCT} = a \cdot \text{ECT}^b (\sqrt{S_{\text{corr, MD}} S_{\text{corr, CD}}})^{1-b} Z^{2b-1}. \tag{3.1}$$

In eq. (3.1), a and b are empirical constants, Z is the perimeter of the box, and $S_{\text{corr, MD}}$ and $S_{\text{corr, CD}}$ are the bending stiffnesses of the corrugated board in machine direction (MD) and CD. The compressive strength of the corrugated board in CD is the ECT value (edge crush strength). Both the bending stiffness and edge crush strength can be estimated from the fluting and linerboard properties and the corrugated board dimensions (Patel, 1996) and (Carlsson et al., 1985). The model is simple to use and often gives satisfying accuracy for boxes with all four panels of similar dimensions. The McKee formula has several limitations. It underestimates BCT of shallow boxes and other boxes that do not buckle. It is inaccurate for boxes with panels of very different size, or when one linerboard is significantly heavier than the other. Extensive box testing is required for calibrating the parameters a and b for new board types, and the model cannot predict the optimum balance between linerboard and fluting. The McKee formula for corrugated board, eq. (3.1), is limited to boxes with panels without any cut-outs.

For carton board boxes, it has been shown that the compression strength correlates well with the strength of its panels. Based on this observation and the Timoshenko theory for post-buckling collapse of edge-loaded simply supported plates, Grangård (1970) derives a formula for box strength

$$P_{\text{collapse}} = k \sqrt{\text{SCT} \sqrt{S_{\text{b, MD}} S_{\text{b, CD}}}} \tag{3.2}$$

where P_{collapse} is the ultimate strength of the panel, SCT is the strength value obtained from the short-span compression test, $S_{\text{b, MD}}$ and $S_{\text{b, CD}}$ are the bending stiffnesses in the MD and CD, and k is a parameter that depends on the panel dimensions and geometry. Thus, k is usually determined by a fit to experimental data (Fig. 3.18).

3.4.3 Finite element models

Finite element models of corrugated board boxes may be conducted in a number of different ways (Gilchrist et al., 1999; Nordstrand, 2004; Nyman, 2004; Patel, 1996; Pommier et al., 1991; Rahman, 1997; Urbanik and Saliklis, 2003). Most of the approaches focus on a panel of the box. In this section, we compare a highly detailed and refined model of a corrugated board panel with a homogenized model that has fewer elements and therefore is computationally much less expensive (Allansson

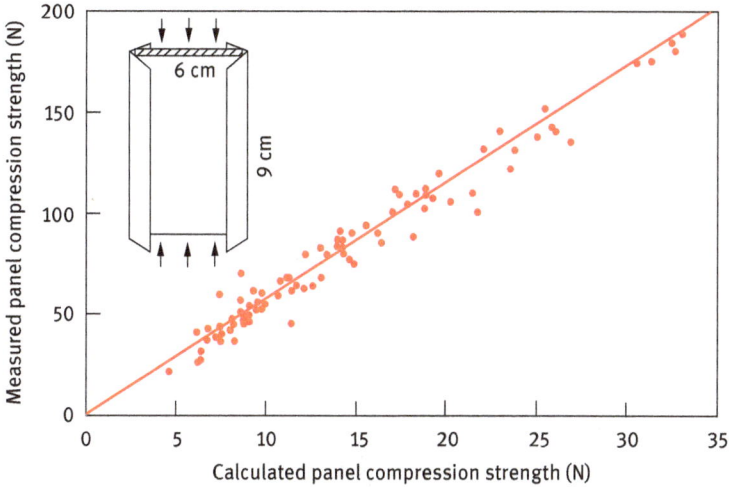

Fig. 3.18: Measured panel compression strength against predictions from eq. (3.2) with $k = 5.7$. Reproduced from Fellers et al. (1983) with permission from Innventia AB.

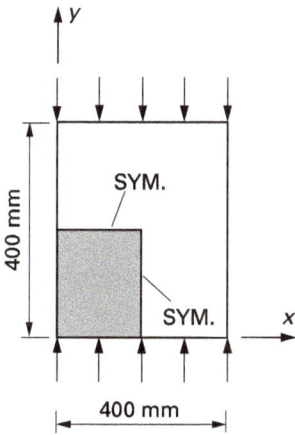

Fig. 3.19: Geometry of the corrugated board panel analysed using both the refined model and the homogenized model of Allansson and Svärd (2001). Only one-quarter of the panel is considered due to symmetry. For each symmetry section, one translation and two rotations are allowed. Reproduced with permission from the first author.

and Svärd, 2001). The corrugated board considered is a 400×400 mm^2 C-flute that is simply supported and subject to edge-wise compressive loading (Fig. 3.19).

The refined model describes the corrugated structure of the board in detail. Each of the fluting and linerboard layers is modelled by four-node quadrilateral rectangular isoparametric shell elements (Bathe, 1982). *Isoparametric* means that the element can take the form of a distorted rectangle. The elements are flat, which makes the corrugated fluting core piece-wise linear in a side view (Fig. 3.20). The facings and core are connected by common nodes at the fluting crests. Allansson and Svärd (2001) found that a mesh with evenly distributed elements in all layers with six elements per half-wave length is the coarsest mesh that can produce reliable results. This model has 73,008 elements. A non-linear solution scheme for large deformations and rotations is necessary for accurate analysis of stress state and deformation. Besides geometric parameters, all the orthotropic elastic constants of the constituent paper layers are required (Section 2.2). Several of these are not straightforward to measure because of the small thickness of paper. Sometimes parameters can be approximated using engineering estimations (Baum et al., 1981; Mann et al., 1980).

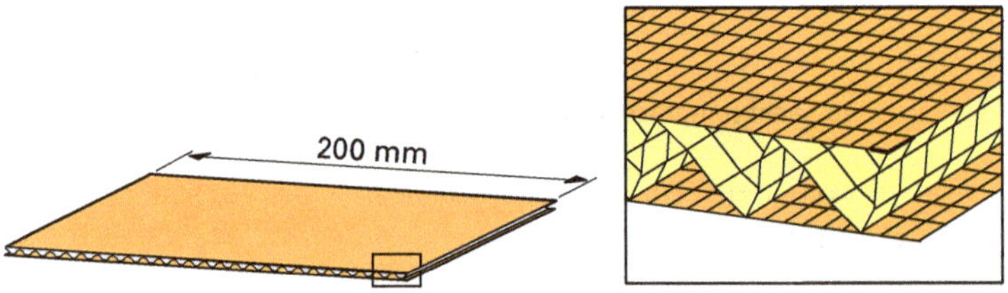

Fig. 3.20: Details of the refined model of Allansson and Svärd (2001). Reproduced with permission from the first author.

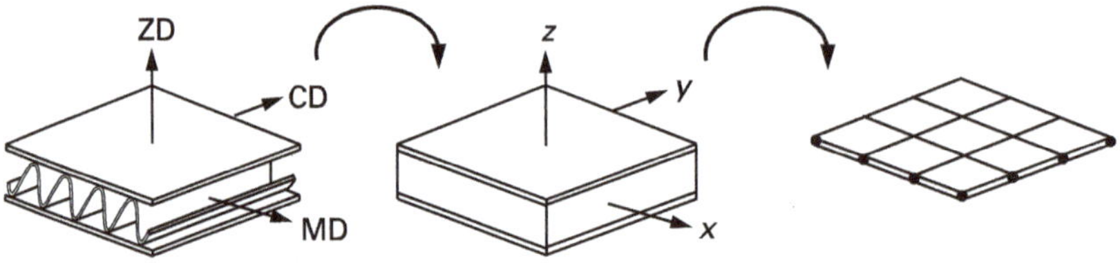

Fig. 3.21: Homogenization of the corrugated core sandwich.

To simplify modelling of corrugated board packaging, the corrugated board may be transformed into an equivalent homogenous layered structure (Fig. 3.21) using homogenization (Carlsson et al., 2001) (Patel, 1996) and laminate theory (Agarwal and Broutman, 1990). The homogenous model of the sandwich consists of three layers, where each layer has its own effective elastic moduli, Poisson's ratios, and shear moduli. The effective elastic constants of the core layer may be estimated as

$$E_x^{\text{eff}} \cong 0, \quad E_y^{\text{eff}} \cong \alpha_{\text{take-up}} \frac{d_{\text{liner}}}{h_{\text{core}}} E_y^{\text{fluting}}, \quad \nu_{xy} \cong 0, \quad G_{xy} \cong 0, \tag{3.3}$$

where $\alpha_{\text{take-up}}$ is a geometric take-up factor (the length ratio of the fluting paper web before and after fluting), E_y^{fluting} is the elastic modulus of the fluting material in CD, d_{liner} is the thickness of the linerboard, and h_{core} is the height of the fluting waves (Fig. 3.2).

Allansson and Svärd (2001) use both structural and homogenized models in their analysis of a compression-loaded corrugated board panel. Both models predict global buckling of the panel. Figure 3.22 shows the load versus out-of-plane deformation response of the models along with experimental results. It is assumed that failure of the corrugated board panel is triggered by material failure of the linerboard facing on the concave side. The Tsai–Wu failure criterion (Tsai and Wu, 1971; eq. (2.8)) is used to predict failure. Figure 3.23 shows the resulting Tsai–Wu stress index f^{TW} for both

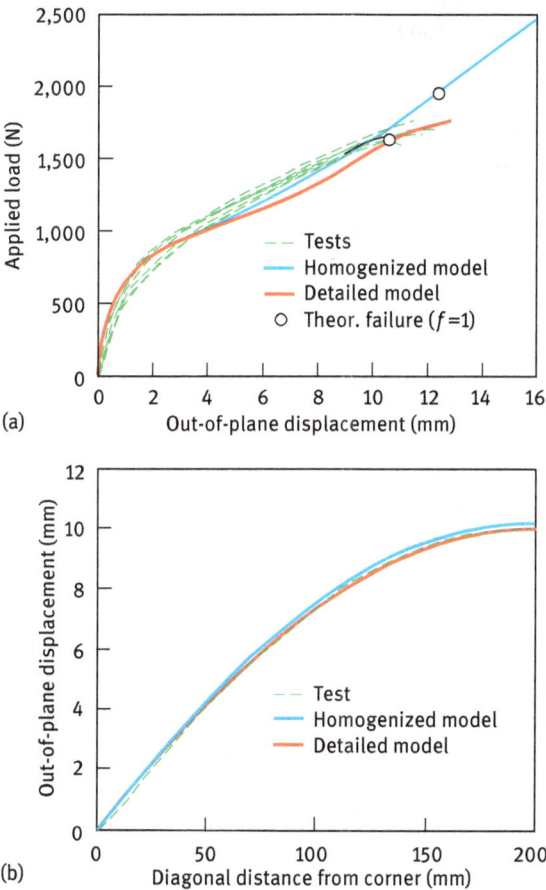

Fig. 3.22: Applied load against out-of-plane displacement of the panel in Fig. 3.19 (Allansson and Svärd, 2001). Reproduced with permission from the first author.

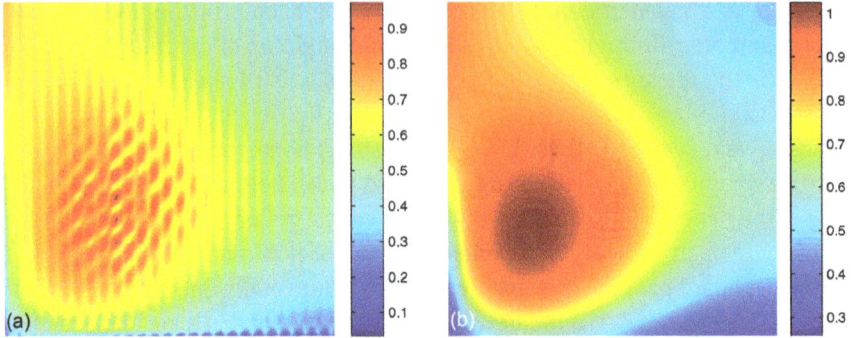

Fig. 3.23: Tsai–Wu stress index f_{TW} on the concave side of the panel in Fig. 3.19 according to the structural model (a) and homogenized model (b) (Allansson and Svärd, 2001). Failure is indicated by $f_{TW}=1$. The applied load and the out-of-plane displacement in the top right-hand corner (centre of the panel) are 1,630 N and 10.6 mm in (a) and 1,950 N and 12 mm in (b). Reproduced with permission from the first author.

the structural and homogenized models. One can see that both models give similar stress distributions. However, the structural model reveals local buckling of the face sheets between the fluting crests. Failure was predicted to occur at a corner of the panel, consistent with the experimental result in Fig. 3.14.

Hence, a structural model is required for a detailed analysis of complex failure mechanisms that typically occur when the constituent papers are thin. It can capture local buckling and other failure mechanisms. The homogenized model is an effective alternative when the amount of local buckling is limited. Such a model predicts the ultimate failure with satisfying accuracy (Nordstrand, 2004), but does not capture local buckling. The Tsai–Wu stress index plot (Fig. 3.23) is useful for the box design. For example, cut-outs like handholds and ventilation holes should be positioned in areas of the panel where the Tsai–Wu stress index is small.

Carton board has a simpler structure than corrugated board and is well-suited for finite element analysis (FEA). Beldie et al. (2001) analysed a three-layer carton board box under compressive loading using FEA. The outer layers of the board were thin and stiff, while the middle layer was thick and of low stiffness. The material of each layer was assumed to be elastic–plastic, orthotropic, and described by Hill's orthotropic yield criterion (Hill, 1950). The resulting stress distribution (Fig. 3.24) shows that failure is predicted to occur at the highly stressed corners, consistent with experimental observations, but the finite element model overestimates both the stiffness and ultimate strength of the box (Fig. 3.25). This is most likely due to the creases in the actual box, which tend to cause local softening and eccentric loading.

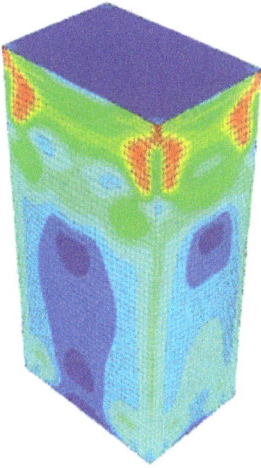

Fig. 3.24: Stress distribution in a carton board package under top-to-bottom compressive load (Beldie et al., 2001). Reproduced with permission from John Wiley & Sons.

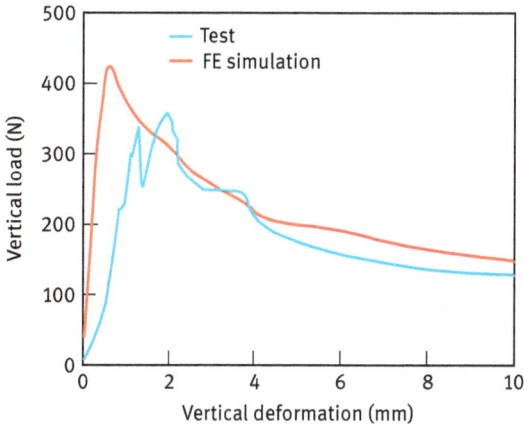

Fig. 3.25: Vertical load versus vertical deformation of the carton board package in Fig. 3.24, according a measurement (blue) and finite element calculation (red; Beldie et al., 2001). Reproduced with permission from John Wiley & Sons.

3.4.4 Long-term loading

One of the most significant environmental factors that make paper packaging weaker is exposure to high humidity or water. This is due to the hygroscopic nature of paper. In corrugated packaging, water not only makes paper softer (Section 2.3.2) but also weakens the water-based glue joints keeping the box panels together. An increase in ambient humidity has a dramatic effect on the strength of a box. Exposure to 85% RH can reduce the strength of a corrugated board box by 40% compared to exposure to

50% RH. Box performance may be enhanced by using moisture-resistant papers, waterproof adhesives, and wax or plastic coatings that make the board surface water repellent. The effects of moisture are especially important when loading occurs over long times, as it is typical in transportation and storage. Carton board packaging sometimes fails during transportation and storage.

The gradual deformation of paper when loaded over long times is called creep. If loading is maintained, creep continues until the package breaks. The worst case is when humidity varies cyclically (Bronkhurst et al., 1994). In this case, the creep is accelerated (Fig. 3.26). Consequently, if the ambient conditions vary in a warehouse or during transportation, stacked boxes may collapse prematurely. For engineering purposes, it is thus important to be able to determine the allowable load that does not cause collapse within the foreseen storage time. To characterize this box behaviour, one uses the concept of lifetime, the time that it takes for a box to fail under relatively high static loads. In Chapter 7, the creep properties of paper materials and the lifetime evaluation of boxes are discussed.

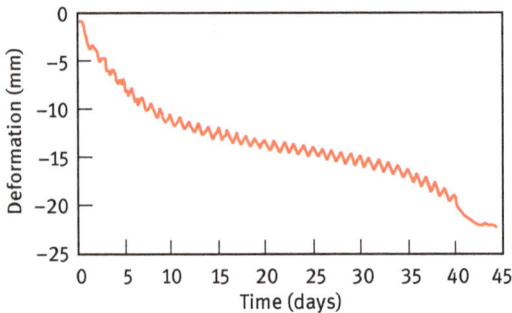

Fig. 3.26: Example of creep deformation vs. time curve of a corrugated box under constant load when humidity varies cyclically between 50 and 90% RH.

3.5 Curved corrugated board structures

This section presents a new class of corrugated board packages designed with curved panels (see Fig. 3.27). This type of corrugated board material is especially favourable for use, for example, in packaging applications where curved shapes are required, such as cylinders or packaging having rounded corners. It has been shown that such structures provide improved material utilization in compressive loading compared to flat panels (Patel et al., 1997). This can be explained by fundamental difference in buckling behaviour between curved shell and flat panel sections, which is reflected in the load carrying capability. Curved panels are also used to create a visual appearance more attractive for the consumer.

Fig. 3.27: Curved corrugated board products.

3.5.1 Manufacturing of flexible corrugated material

To enable smooth bending of the board, properties of the layers must be designed according to a certain scheme. The outer layer (convex), with reference to Fig. 3.28, consists of a paper having relatively high thickness while the inner layer (concave) has a small thickness. The thin paper will exhibit local buckling upon bending and the local buckles propagate along the curved region. This behaviour explains the smooth bending behaviour of this type of board.

Fig. 3.28: Bendable corrugated board.

3.5.2 Compression strength of corrugated board cylinders

This section presents an experimental investigation of the compression strength of corrugated board tubes. Tubes of different dimensions has been tested in axial compression. The dimensions of a tube are given by its radius (R) and its height (h) (see Fig. 3.29).

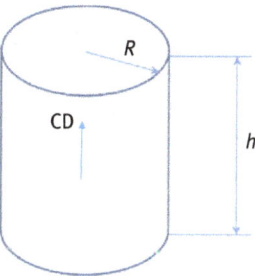

Fig. 3.29: Dimensions of cylinder. The cylinder axis is aligned with the CD of the board.

The board material used in this study consists of a 200 g/m² coated kraftliner as outer layer, a 127 g/m² semi-chemical web core (fluting) and a 65 g/m² lightweight coated inner facing. The material was produced as flat panels in a conventional full-scale corrugator. The cylinders were made by bending rectangular panels cut from large panels using a digital cutting table. In order to minimize the influence of the vertical glue joint, 19 mm of the inner facing and fluting layer were removed prior to gluing (see Fig. 3.30a). To achieve a perfect circular shape of the cylinders, circular pieces of corrugated board were placed inside the tube at both ends (see Fig. 3.30b).

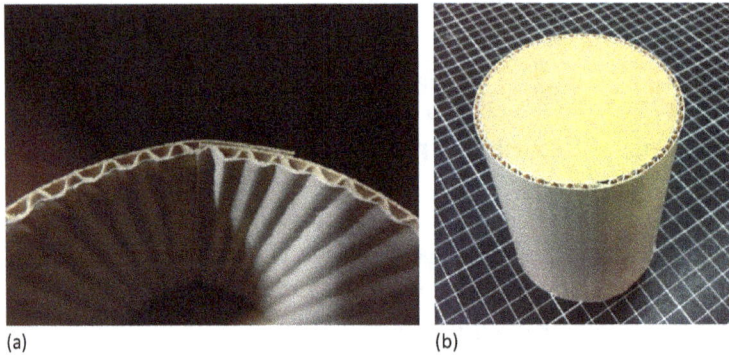

(a)   (b)

Fig. 3.30: The vertical glue joint (a) and cylinder with end supports (b).

Two test series were conducted: one where the radius, R, of the cylinders, was varied, while the height, h, was held constant at 100 mm, and another, where cylinders of radii, $R = 75$ and 340 mm, with heights, h, in the range from 100 to 800 mm. Ten replicate test specimens were prepared. To enable compressive load application without crushing the edges of the board cylinders, loading end caps made from aluminium (see Fig. 3.31) were attached to the ends of the tubes.

Testing was conducted in a general-purpose servo-hydraulic MTS test machine of 20 kN load capacity. The specimens were placed between flat metal plates. The upper loading plate was attached to a load cell mounted in the moving cross-head of the test machine. The test climate was approximately 23 °C and 50% RH. The cylinders were loaded in axial compression at a cross-head speed of 1 mm/min. The test was interrupted when global collapse of the cylinders occurred. This corresponds to a maximum in the load–displacement curve.

In the first test series, cylinders of a constant height of 100 mm were tested. The radius varied between 20 and 157.5 mm. The collapse load is normalized with respect to the edge-wise compression strength of the material, denoted ECT. The ECT value is 6.46 kN/m for this board. Figure 3.32 shows the collapse load test results versus the radius of the cylinders. The data is compared with results from earlier

Fig. 3.31: Corrugated paperboard cylinder compressed between two flat metal plates.

studies on compression strength of large corrugated board cylinders (Patel, 1996) and edge-wise-loaded flat panels study (Nordstrand, 2004).

The results shown in Fig. 3.32 reveal that the collapse loads for cylinders are significantly higher than for flat panels. The cylinders with a radius of 25 mm fail at a load about four times that of the flat panel. The cylinders of radius 37.5 mm exhibit wrinkling near the loading plates upon progressive loading where the outer facing folds in

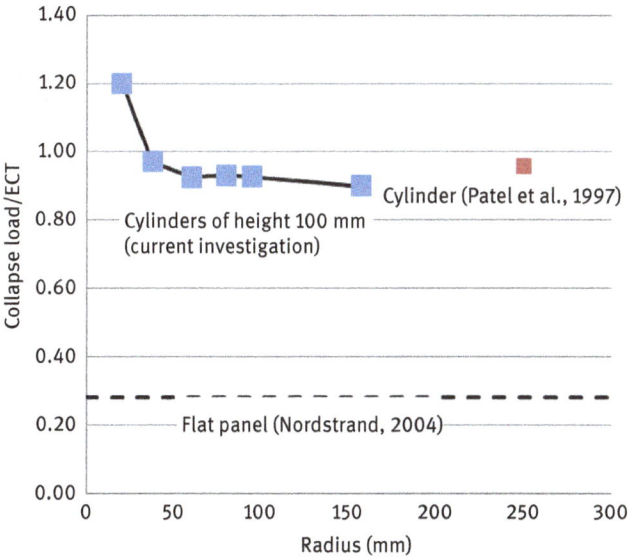

Fig. 3.32: Normalized collapse load versus radius of curved corrugated board cylinders. Data are compared with results from large corrugated board cylinders (Patel, 1996) and edge-wise-loaded flat panels (Nordstrand, 2004).

Fig. 3.33: Failure of corrugated cylinder with a radius of 37.5 mm at progressive loading.

a manner of an accordion, without any global buckling of the cylinder (Fig. 3.33). It is noteworthy that the collapse load of cylinders with a large radius is in the same order of magnitude as the previous investigation by Patel (1996).

In contrast to the cylinders of radius 37.5 mm, the cylinders of radius 157.5 mm exhibited significant global buckling at failure (Fig. 3.34). The reason for the

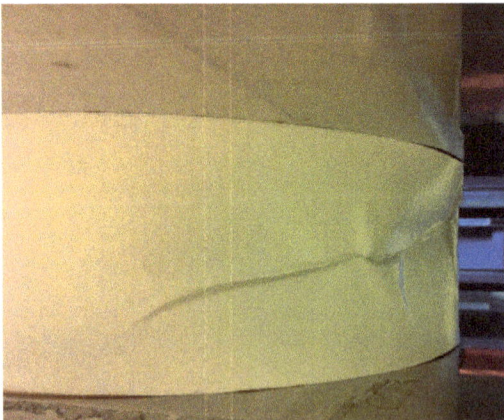

Fig. 3.34: Failure of corrugated cylinder with a radius of 157.5 mm.

relatively high compression strength is due to the high buckling loads of curved shells, which is well established in literature (Timoshenko and Gere, 1961). The reason why the collapse load is higher than the ECT at the smallest radius is believed to be due to the suppression of local buckling of the layers due to the curved shape of the cylinders.

In the second test series, the influence of height on the collapse load of the cylinders was studied. Two radii where considered, 75 and 340 mm. The height was varied between 100 and 800 mm. The normalized collapse load is shown in Fig. 3.35. Test results shown in Fig. 3.35 reveal that the collapse load for the cylinders with 340 mm radius decreases significantly as the height increases, while the collapse load is almost constant for the cylinders with 75 mm radius. The failure mode of the cylinders with large radius (340 mm) shifted from local face wrinkling to global buckling when the height increased. Cylinders with small radius failed essentially near the supporting steel plates.

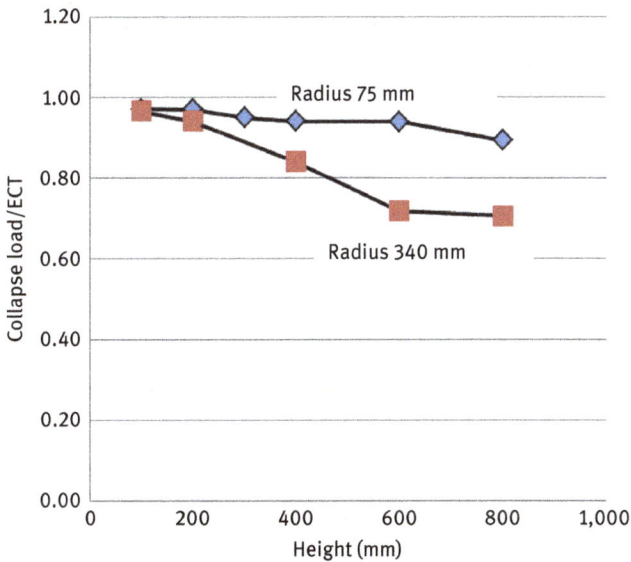

Fig. 3.35: Normalized collapse load versus height for corrugated cylinders of radii 75 and 340 mm, respectively.

3.6 Concluding remarks

Packaging is an important application area for paper materials. Corrugated board is the most popular material for secondary packaging of a wide variety of products. The performance of a corrugated board box is governed by its dimensions and the mechanical properties of its constituent papers. Carton board is used mainly for primary packaging.

Compressively loaded corrugated board boxes with flat panels usually fail after the panels have buckled. The final failure is caused by local buckles that develop in the face sheets on the concave side of a panel and coalesce into wrinkles that form diagonally from the corners towards the centre of a panel. The semi-empirical McKee formula can be used to estimate the compression strength of RSC. The formula does not apply to shallow boxes, asymmetric board, boxes of different length and width, or boxes containing holes or perforations, where other analysis methods such as FEA are needed. Carton board boxes loaded in compression also exhibit global buckling under compressive loading, but the failure is controlled by the strength of the corners, discussed in Chapter 4. Curved corrugated board panels provide packaging top-to-bottom compressive strength much higher than for flat panel packaging, due to the high buckling loads of curved shell structures.

In practice, the most critical situation is a packaging that fails during transportation or storage. The long-time behaviour of packages is difficult to study and relatively little work has been published. Long-time loading is different from the short-time strength properties because the deformations of paper gradually accumulate over time. The deformations are strongly influenced by variations in humidity. The long-time creep behaviour of paper materials and paper-based packages is discussed in Chapter 7.

Literature references

Agarwal, B.D. and Broutman, L.J. (1990). Analysis and performance of fiber composites, New York, NY, USA: John Wiley & Sons.

Allansson, A. and Svärd, B. (2001). Stability and collapse of corrugated board. Master's thesis. Report TVSM-5102, Division of Structural Mechanics, Sweden: Lund University.

Bathe, K.J. (1982). Finite element procedures in engineering analysis, Englewood Cliffs NJ, USA: Prentice-Hall.

Baum, G.A., Brennan, D.C. and Habeger, C.C. (1981). Orthotropic elastic constants of paper. Tappi J. 64(8), 97–101.

Beldie, L., Sandberg, G. and Sandberg, L. (2001). Paperboard packages exposed to static loads – finite element modelling and experiments. Packag. Tech. Sci. 14, 171–178.

Bronkhurst, C.A. and Riedemann, J.R. (1994). The creep deformation behaviour of corrugated containers in a cyclic moisture environment. In: Proceedings of the Symposium on Moisture Induced Creep Behaviour of Paper and Board, Stockholm, Sweden, December 5–7, 249–273.

Carlsson, L., Fellers, C. and Jonsson, P. (1985). Bending stiffness of corrugated board with special reference to asymmetrical and multi-wall constructions. Das Papier (39), 149–156.

Carlsson, L.A., Nordstrand, T. and Westerlind, B. (2001). On the elastic stiffnesses of corrugated core sandwich. J. Sandw. Struct. Mater. 3, 253–267.

Cavlin, S.I. and Edholm, B. (1988). Converting cracks in corrugated board: Effect of liner and fluting properties. Packag. Tech. Sci. 1, 25–34.

Confederation of European Paper Industries (CEPI). (2011). 250 Belgium: Avenue Louise Brussels.

Cox, H.L. (1933). The buckling of thin plates in compression. Technical Report of the Aeronautical Research Committee Report 1554.

European Carton Makers Association (ECMA) (2011). ECMA code of folding carton design styles, The Hague, Netherlands: ECMA. http://www.ecma.org.

Federation of Corrugated Board Manufacturers (FEFCO). (2011). FEFCO-Code. Brussels, Belgium: FEFCO. http://www.fefco.org.

Fellers, C., de Ruvo, A., Htun, M., Carlsson, L., Engman, C. and Lundberg, R. (1983). Carton board – profitable use of pulps and processes, Stockholm, Sweden: Swedish Forest Products Research Laboratory.

Gilchrist, A.C., Suhling, J.C. and Urbanik, T.J. (1999). Nonlinear finite element modeling of corrugated board. Mechanics of cellulose materials. ASME AMD 231, 101–106.

Grangård, H. (1970). Compression of board cartons part 1: Correlation between actual tests and empirical equations. Sv. Papperstidn. 73, 16.

Hägglund, R. and Isaksson, P. (2008). Mechanical analysis of folding induced failure in corrugated board: A theoretical and experimental comparison. J. Compos. Mater. 4(9), 889–908.

Hill, R. (1950). Mathematical theory of plasticity, Oxford, UK: Clarendon Press.

Isaksson, P. and Hägglund, R. (2005). A mechanical model of damage and delamination in corrugated board during folding. Eng. Fract. Mech. 72(15), 2299–2315.

Kellicutt, K.Q. (1963). Effect of contents and load bearing surface on compressive strength and stacking life of corrugated containers. Tappi J. 46(1), 151–154.

Kuusipalo, J. (2008). Paper and Paperboard Converting, 2nd Helsinki: Finnish Paper Engineers' Association.

Maltenfort, G.G. (1956). Compression strength of corrugated containers. Fibre Contain 41, 7.

Mann, R.W., Baum, G.A. and Habeger, C.C. (1980). Determination of all nine orthotropic elastic constants for machine-made paper. Tappi J. 63(2), 163–166.

McKee, R.C. and Gander, J.W. (1957). Top-load compression. Tappi J. 40(1), 57–64.

McKee, R.C., Gander, J.W. and Wachuta, J.R. (1963). Compression strength formula for corrugated boxes. Paperboard Packaging 48, 149–159.

Nordstrand, T. (2004). Analysis and testing of corrugated board panels into the post-buckling regime. Compos. Struct. 63(2), 189–199.

Nyman, U. (2004). Continuum mechanics modeling of corrugated board. Doctoral Thesis, Sweden: Lund University.

Patel, P. (1996). Biaxial failure of corrugated board. Licentiate thesis. Division of Engineering Logistics, Sweden: Lund University.

Patel, P., Nordstrand, T. and Carlsson, L.A. (1997). Instability and failure of corrugated core sandwich cylinders under combined stresses. Kalluri, S. and Bonacuse, P.J. Eds. Multiaxial fatigue and deformation testing techniques. 264–289 ASTM STP 1280.

Pommier, J., Poustis, C., Bending, J., Fourcade, J. and Morlier, P. (1991). Determination of the critical load of a corrugated box subjected to vertical compression by finite element method. In: Proceedings of the 1991 International Paper Physics Conference, Kona, Hawaii, 437–447.

Rahman, A. (1997). Finite element buckling analysis of corrugated fiberboard panels. In: Proceedings of the 1997 joint ASME/ASCE/SES summer meeting, Mechanics of Cellulosic Materials, June 29–July 2, Evanston, Illinois, 87–92.

Thakkar, B., Gooren, L.G.J., Peerlings, R.H.J. and Geers, M.G.D. (2008). Experimental and numerical investigation of creasing in corrugated paperboard. Phil. Mag. 88, 3299–3310.

Timoshenko, S. and Gere, J.M. (1961). Theory of elastic stability, New York, NY: Mc Graw-Hill. 457–519.

Tsai, S.W. and Wu, E.M. (1971). A general theory of strength for anisotropic materials. J. Compos. Mater. 5, 58–80.

Urbanik, T.J. and Saliklis, E.P. (2003). Finite element corroboration of buckling phenomena observed in corrugated boxes. Wood Fiber Sci. 35(3), 322–333.

Westerlind, B.S. and Carlsson, L.A. (1992). Compressive response of corrugated board. Tappi J. 75 (7), 145–154.

Mikael Nygårds
4 Behaviour of corners in carton board boxes

4.1 Introduction

The preceding chapter discussed the manufacture and performance of boxes, with focus on corrugated board that is used in secondary packaging. In this chapter, we consider the mechanics related to carton board boxes used as primary packaging. The converting process of a carton board into packages is in essence the same as that used with corrugated boards, but the difference in the board structure means that the critical properties of a carton board are somewhat different from those of a corrugated board. A typical corrugated board box fails through a coalescence of small buckles that form between the fluting crests after one or several of the box faces have buckled. This mechanism is absent in a carton board box because the board is less heterogeneous (there are no flutes). Instead, the strength of corners is critical for the strength of carton board boxes (Fig. 3.24; Beldie, 2001).

Carton board packages are often used to protect and carry liquids, frozen food, cigarettes, and other products that are sensitive to moisture, temperature, and air. Cracks in the carton board must be avoided because they can lead to package failure, leakage, or oxidization of the product. In addition, the converting and filling of carton board packages are fast processes, and it is important that the board material behaves uniformly and performs in the same way all the time.

In converting of carton board, the making of corners is the critical phase. The corners must be sharp and, if folded by 90°, no cracks should be formed. The generally poor folding behaviour of uncreased carton boards is easy to observe if one tries to fold a carton board sheet by hand. It is practically impossible to achieve a straight and sharp fold line. In contrast, it is easy to fold thin paper sheets sharply. In an untreated board, the tensile stresses at the outer edge of the corner may break the board there (Fig. 4.1a). However, if the board is delaminated into layers before folding, the material on the inside can buckle inwards (Fig. 4.1b). This reduces the thickness of the layer that is bent and decreases the tensile stress on the outer edges. This is how a controlled amount of delamination in the corner area can be used to improve the converting behaviour of a carton board.

The desired deformation of a carton board in the box-making process is shown schematically in Fig. 4.2. The creasing phase creates delamination of the carton board structure. Figure 4.3 demonstrates the effect of creasing in reality. In this example, the uncreased board forms a sharp bend and rupture at the inside of the corner. A carton board box made in this way would not protect the contents from external conditions. Even the creasing operation can rupture the board if the crease

Mikael Nygårds, KTH Royal Institute of Technology, Stockholm, Sweden

https://doi.org/10.1515/9783110619386-004

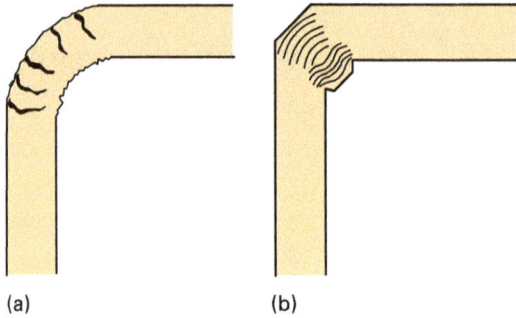

Fig. 4.1: Cracking in a folded corner of uncreased (a) and creased (b) carton board. Uncreased carton board tends to crack in the outer surface. In a creased carton board, the inner surface bulges out after planar cracking or delamination.

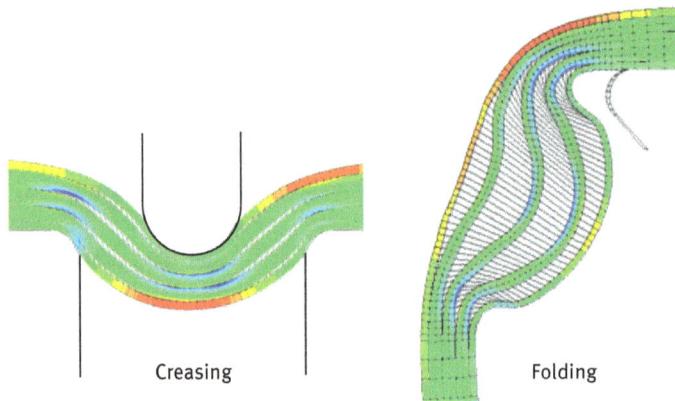

Fig. 4.2: Principle of creasing and folding.

is too deep. In the optimal case, fibres are not pulled apart in the in-plane direction but only in the out-of-plane direction. This is important for preserving the mechanical strength of the corner.

There are alternative ways to achieve good delamination in the creasing phase by designing the structure of the carton board (Fig. 4.4). Carton boards are usually made by making several plies simultaneously in the forming section of the paper machine, before the wet web enters the press section (Fig. 2.23). If high thickness is needed, up to five plies can be used. Commonly, the middle ply has lower density than the outer plies. This gives an "I-beam" type of structure that has high bending stiffness and delaminates easily in the middle ply. By using the I-beam structure, high bending stiffness can be achieved with a minimum of fibres. Alternatively, all the plies can be similar in properties, but the interfaces between them are made weak. This gives a carton board that has very precisely set delamination planes.

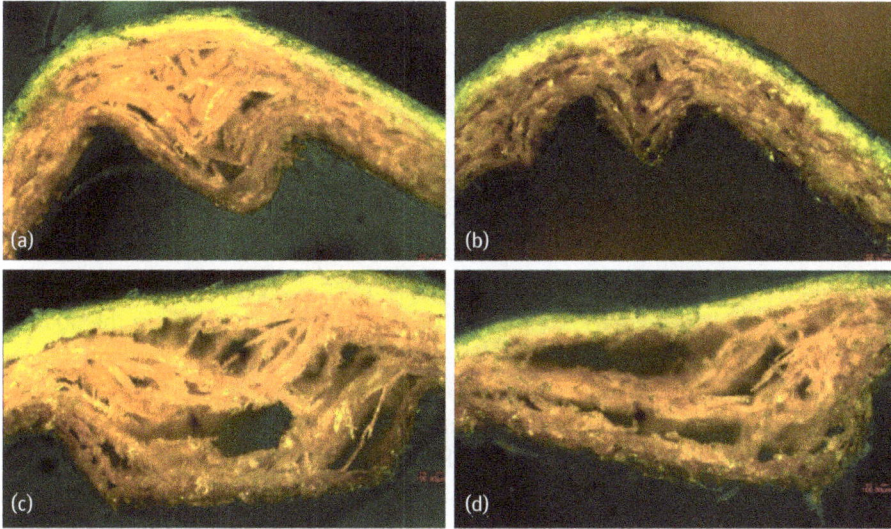

Fig. 4.3: Cross sections of folded uncreased carton board in MD (top left) and CD (top right) and creased carton board in MD (bottom left) and CD (bottom right); photos courtesy of Hui Huang.

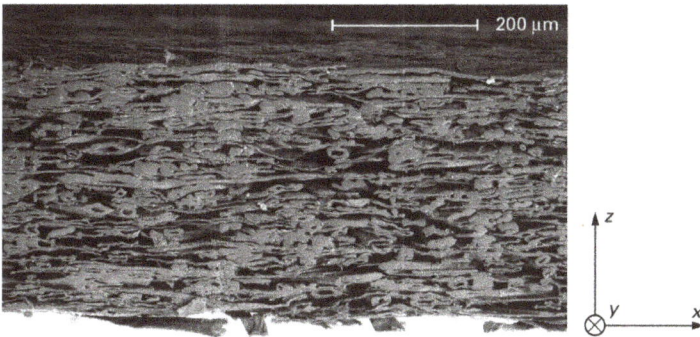

Fig. 4.4: Cross section of a carton board, showing the fibre network; photo courtesy of Christer Fellers.

However, both these examples are idealized. In reality, it is difficult to precisely engineer the z-direction (ZD) structure since carton boards have a porous structure (Fig. 4.4). It should be noticed that there is only about 20–40 layers of fibres in a carton board with thickness of 0.4 mm. Apart from the fibres, there are also fines that are small material residues from the refining and chemicals in the paper. It should also be recalled that the carton board was formed with 99% water (see Section 2.5). This water needs to be removed in the machines, which generate property gradients in the ZD; fines and chemicals tend to follow the water during dewatering. Good control of the papermaking process, the fibre sources, and chemical additives used is crucial to achieve high-quality boards. Typically, boards with high quality have low variability, and hence behave the same, independently of where it has been loaded or converted. Structures with an actual I-beam or with weak interfaces should therefore be considered as design criteria for different board grades. For applications where high bending stiffness and low weight is important, it is good to aim for an I-beam type of structure. This is typically the case for most folding box board qualities, where the bending stiffness is the most critical parameter. For other board qualities, such as liquid boards and tobacco board, the convertibility at high speed with no paper jams in the machines is crucial. Then weak interfaces are a better design criterion. This contributes to fewer variations in the boards, since delamination then can occur at the same location in all folds.

In summary, the mechanical properties related to creasing and folding are crucial for the performance of a carton board in packaging. The stacking strength of the filled carton board package is actually of lesser importance compared to the performance in the packaging process. Usually, it is the corrugated secondary package where strength is needed during transportation and storage (see Section 3.3). In this chapter, we discuss the mechanisms that control the creasing and folding behaviour of a carton board.

4.2 Folding of a multi-ply carton board

The folding performance of carton boards can be tested using the set-up in Fig. 4.5. One end of the board is fixed on a rotating pneumatic clamp, while a load cell measures the force P needed to bend the sample. Since one end of the beam has a rigid support, while the other is loaded by a point contact load, the stress state inside the specimen varies along the free length. The highest stresses occur at the rigid clamps. According to Timoshenko (1934), the in-plane stress is

$$\sigma_x = \frac{PL}{I}z = \frac{12PL}{wd^3}z. \tag{4.1}$$

In addition, the transverse shear stress is

$$\tau_{xz} = \frac{P}{2I}\left(\frac{d^2}{4} - z^2\right) = \frac{3P}{2wd}\left(1 - 4\frac{z^2}{d^2}\right). \tag{4.2}$$

Here, I is the surface moment of inertia of the specimen,

$$I = \frac{wd^3}{12}, \tag{4.3}$$

L is the length, d is the thickness, w is the width, and z is the distance in ZD from the centre line of the specimen. Usually the transverse shear stress is small and can be ignored. However, paper materials are highly anisotropic and the shear stresses are important to consider.

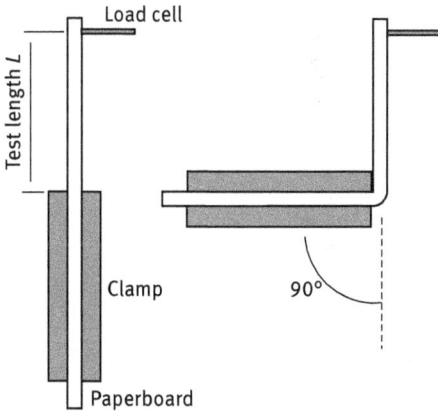

Fig. 4.5: Principle of a folding test (Lorentzen and Wettre, Sweden). One end of the carton board is fixed in a rotating pneumatic clamp, while a load cell measures the force needed to fold the sample.

Figure 4.6 shows the measurement result when a three-ply uncreased carton board is folded by 90° and released both in machine direction (MD) and in cross-machine direction (CD). The result is expressed as the bending moment $M_b = PL$. There is an initial linear region that can be used to calculate the bending stiffness $S_b = EI$. The difference between the behaviour in MD and CD arises from the anisotropy of paper (Section 2.2.2). In both directions, the bending moment is linear almost all the way to a maximum value that is followed by a regime of fairly constant bending moment. The irregular curve shape in this region is due to slippage between the carton board and load cell and entanglement of fibres in the fold.

The initial unloading stiffnesses in Fig. 4.6 are close to the initial loading stiffnesses, indicating that the in-plane properties of the carton board are not affected even though a permanent fold of approximately 40° is created. This means that the average degree of inter-fibre bonding is not changed enough to reduce the in-plane elastic modulus. However, some inter-fibre bonds are probably broken by the shear stresses. To see this, consider a simple experiment. Take a copy paper sheet, fold it by 180° and then unfold. This creates a local damage along a straight line. If you make a tensile test, the sheet will most likely not fail along the fold line. However, if you instead tear the paper, it is easiest to tear it along the fold line. This shows that the paper is weaker along the fold line since bonds between the fibres have been broken.

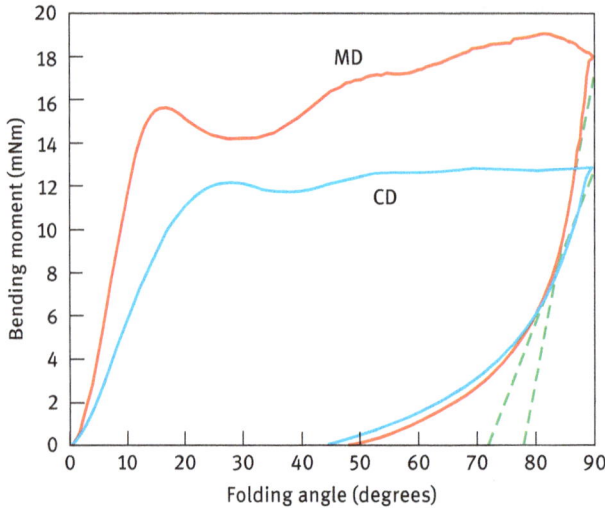

Fig. 4.6: Bending moment against folding angle in MD and CD for a folding–unfolding cycle of uncreased 0.4 mm carton board. The straight lines in the unloading phase indicate the slope detected in the initial bending phase.

Consider again the elastic stresses (4.1) and (4.2) during folding. The maximum bending moment in Fig. 4.6 gives the stress distributions shown in Fig. 4.7(a). The normal stresses are much larger than the shear stresses. If the stress values in different plies are divided by the measured strength values of the carton board (Table 4.1), the renormalized normal and shear stresses are closer to each other (Fig. 4.7b). The renormalized values of the axial stresses are still larger than values of the shear stresses. Furthermore, the compressive stress values could be divided by the compressive strength, not the tensile strength. As illustrated in Fig. 2.14, the compressive strength values are much smaller than the tensile strength values, but it should be noted that compressive failure also involves delamination within the board as discussed further in Chapter 5. One infers that the uncreased carton board fails in compression in the inside edge of the fold line and activates bulging of the ply. This is qualitatively consistent with Fig. 4.3. There is a small caveat in the argument in that the stress distributions in Fig. 4.7 were estimated assuming an ideal linear elastic behaviour. In reality, plastic deformations in the different plies will influence the stress distributions and may influence the mode of failure that would be caused by the folding.

Stress distributions through the thickness of a paper board depend on the board thickness. From eqs. (4.1) and (4.2), one can see that the shear stress component increases relative to the axial stress component if thickness d is increased. The ratio $\tau_{xz}^{max}/\sigma_x^{max} = 0$, if $d = 0$ mm, and increases linearly to $\tau_{xz}^{max}/\sigma_x^{max} = 1.5$ for $d = 1$ mm. This shows that the deformations caused by folding are different in thin

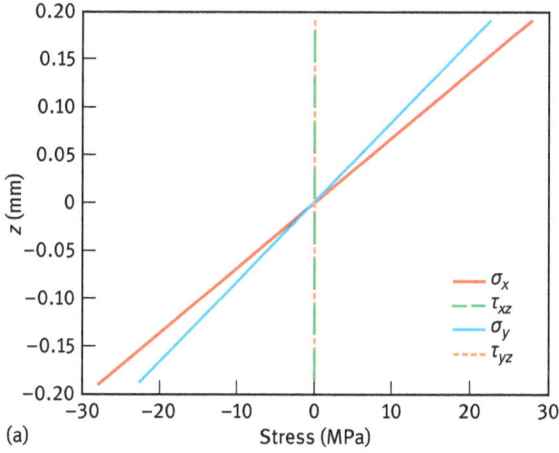

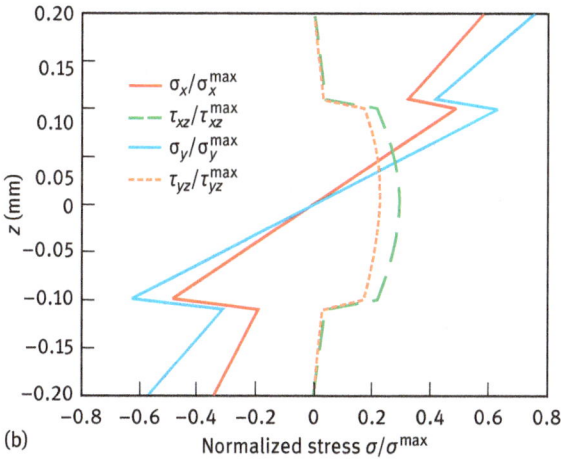

Fig. 4.7: Profile of tensile and shear stresses across board thickness (a) and the same stresses when divided by tensile strength and shear strength (Table 4.1) measured separately for different plies of the carton board. The stress profiles have been calculated for the maximum bending moment in Fig. 4.6.

paper grades (newsprint, office paper) and thick carton boards. Thin grades can be easily folded with no visible cracking (Barbier et al., 2002). Thick carton boards usually break when folded. Breaking usually occurs in the inside edge of the fold, and it is practically impossible to achieve straight folding line. A box corner of this type would have low strength because an irregular folding line can easily trigger cracking, and the folding cracks themselves reduce the strength of a box corner.

Table 4.1: Experimentally determined properties of different plies in a three-ply carton board (Nygårds, 2008).

Ply	Thickness (mm)	Elastic modulus		Tensile strength		Shear modulus		Shear strength
		E_x (MPa)	E_{ye} (MPa)	σ_x^{max} (MPa)	σ_y^{max} (MPa)	G_{xz} (MPa)	G_{yz} (MPa)	$\tau_{xz}^{max}, \tau_{yz}^{max}$ (MPa)
Bottom	0.090	8,760	3,030	85	40	50	78	5
Middle	0.210	3,110	1,210	30	18	23	21	1
Top	0.090	5,920	2,670	50	30	73	85	5

4.3 Creasing

Creasing is used to prevent folding cracks and irregular fold line. The resulting folding performance depends greatly on the creasing operation. In the creasing operation, the ruler is pressed into the carton board that rests against the die (Fig. 4.8), and a permanent indentation is created. High shear stresses and compressive out-of-plane stresses arise in the carton board. The shear stresses break bonds between the board plies and between the fibres and deform fibres so that several delamination planes are created in the carton board. However, delamination occurs only in a narrow zone, where the bending stiffness is therefore dramatically reduced.

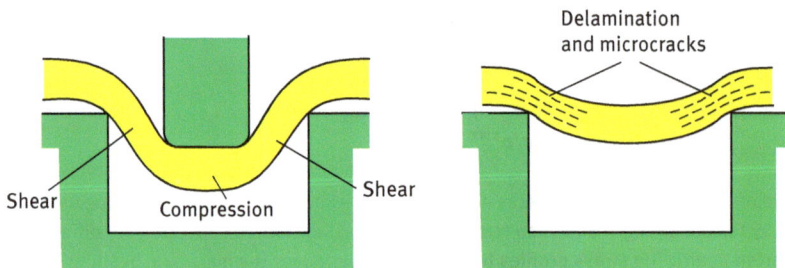

Fig. 4.8: Generation of shear-induced delamination and microcracks in the creasing operation.

The force–displacement curve of a creasing operation in Fig. 4.9 is calculated from a finite element model. The vertical displacement of the ruler is defined to be zero when it is in plane with the die surface. The area under the curve gives the amount of energy consumed to deform the board irreversibly. Different deformation mechanisms act during the operation. At first, a uniform shear stress field is created between the ruler and the die. The shear stress breaks inter-fibre and inter-ply bonds. Next, the carton board is compressed under the ruler. The compression prevents

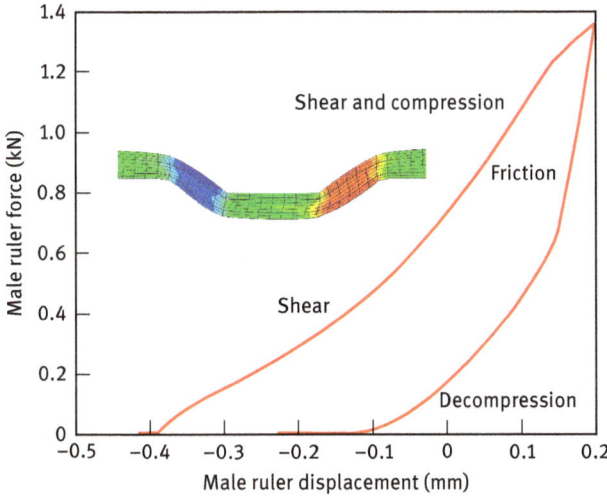

Fig. 4.9: Force–displacement curve calculated using FEM for the creasing of a 32 mm wide carton board specimen. The inset shows the shear strain field at peak load.

delamination from propagating under the ruler because friction between fibres within the board increases. After the elastic recovery, one can see that the carton board is permanently deformed as force reaches zero before the ruler has returned to its original position. The shape of the creasing curve, and hence the amount of energy consumed, depends on the structure of the carton board and on the friction between it and the surfaces of the ruler and the die.

The shear-induced delamination of the carton board reduces the bending stiffness of the creased carton board and enhances buckling of the carton board during folding. Due to the local damage in the creased area, the damaged region will behave similar to a hinge, which reduces the folding resistance. The folding properties do depend on the properties of the carton board, the geometry of the male ruler and female die, and on the creasing depth.

The influence of creasing on folding behaviour (bending moment as a function of folding angle) is shown in Fig. 4.10 for the same 0.4 mm thick carton board that was used in Fig. 4.6. Different crease depths from $d_{crease} = -0.15$ to 0.15 mm are compared. One can see that the initial bending stiffness S_b and the maximum bending moment M_b decreased when the crease depth d_{crease} was increased. In other words, a deeper crease makes a box corner easier to fold. However, it can also be seen from Fig. 4.10 that roughly half of the 90° folding angle recovers when the bending moment is removed. This "spring-back" is unaffected by creasing. If box corners are folded by only 90° in the box forming, then the box tends to be rounded.

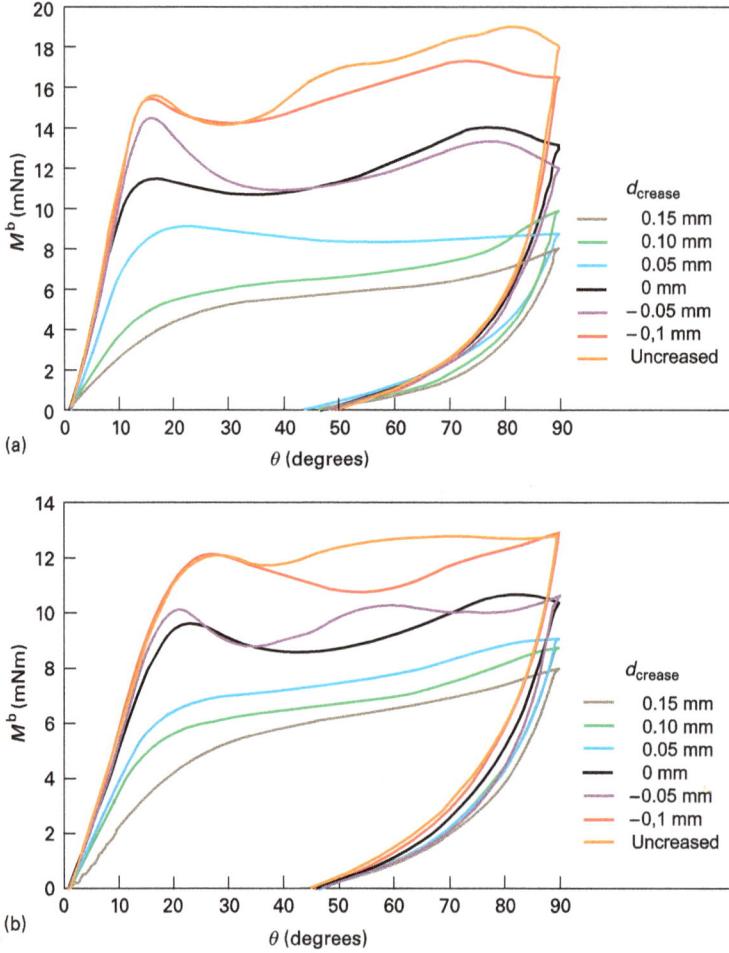

Fig. 4.10: Bending moment against folding angle for the same 0.4 mm thick carton board as in Fig. 4.6 in MD (a) and CD (b) for a range of crease depths d_{crease} using a ruler of width 0.6 mm and die of width 1.2 mm (Nygårds, 2008).

In order to facilitate a comparison of the different crease depths, we define a damage parameter D through both the reduction in the initial bending stiffness S_b,

$$D = \Delta S_b^{creased} / \Delta S_b^{uncreased}, \tag{4.4}$$

and the maximum bending moment, M_b,

$$D = \Delta M_b^{creased} / \Delta M_b^{uncreased}. \tag{4.5}$$

In an uncreased and hence undamaged material, $D = 0$, and in a completely failed material, $D = 1$. As seen in Fig. 4.11, the two definitions of D give the same result.

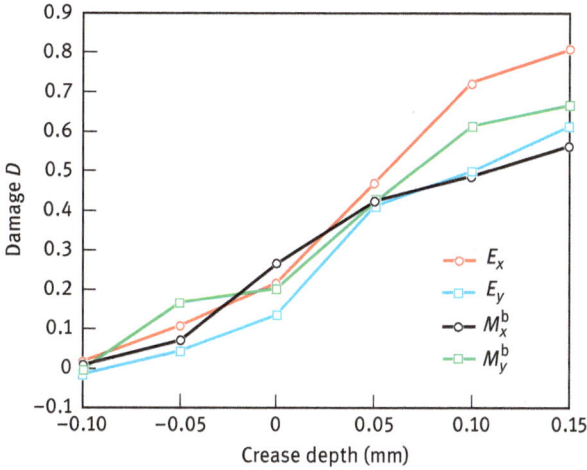

Fig. 4.11: Damage against crease depth, calculated from eqs. (4.4) and (4.5).

Also, there is essentially no difference in the damage evolution between MD and CD folding behaviour. Hence, the degree of damage creased in a creasing process seems to be independent of the anisotropy in board properties and controlled mainly by the creasing process.

Although the folding behaviour is reduced due to creasing, the in-plane strength is almost unaffected. The shear loading will break the fibre–fibre bonds, but it will not tear the fibres out of the fibre network. The fibre length is in the range 0.7–2.5 mm, and the creasing region between the ruler and die is about 0.2–0.6 mm, hence the creasing will not break all bonds along a fibre, only the ones in the creased region. However, if the strength of the crease is drastically reduced, then it has been creased too deep such that cutting has been initiated. The crease depth then needs to be reduced.

4.4 Important material properties

Carton board that is converted into boxes should ideally fold easily, but still maintain strength and stiffness along the folded corner. This is necessary for good package performance because box corners carry a great deal of compressive loads imposed on the box. Ideally, the delamination in the creasing operation should be confined accurately and reproducibly along the crease line. In-plane cracking is not allowed in the folding operation. The desired behaviour can be achieved and optimized by adjusting the material properties. Especially the ZD properties govern the bending stiffness and performance in converting operations such as creasing, folding, and printing. Understanding how these can be changed is therefore essential in successful product development.

The out-of-plane mechanical properties of a carton board can be characterized with respect to compression, tension, and shear. The transverse shear failure mode is not only the most important increasing and folding operations, but also ZD tension is important. The most dominant characteristic during shear and ZD tension loading is the softening behaviour (Fig. 2.15). It is caused by a local cohesive failure in planes where inter-fibre bonds break and fibres bridge the planar crack surfaces (Fig. 4.3). The ZD stress–strain curve in compression is generally described by an exponential curve since the material is porous.

From the preceding discussion of creasing, we can conclude that good creasing behaviour is controlled by two factors. First, low shear strength is needed so that delamination occurs instead of in-plane cracking. Second, the delamination needs to occur at well-defined planes without having fibres that bridge the crack surfaces. This can be achieved with multi-ply sheet structures. If delamination takes place, either in the middle ply or in the interfaces, then shear damage occurs in controlled positions. At the same time, the outer plies remain undamaged and can contribute to high in-plane tensile properties. From a board perspective, it is hence wise to design the middle ply and interfaces to be weak to enable shear damage, while the outer plies are strong such that they contribute to high tensile properties and bending stiffness.

The out-of-plane shear behaviour, in the middle ply and in the vicinity of the interfaces, of a carton board can be determined using the notched shear test (Fig. 4.12; Nygårds et al., 2009a). Characterization is done with a specimen with two notches; shear failure occurs along the plane at which the two notches meet. If the position of this plane is varied, one can determine the variation of shear strength through the material thickness. In the example shown in Fig. 4.12, the carton board has weak interfaces close to the interfaces, indicated by the low shear strength values. It should be favourable to have both low average shear strength and well-defined minima in the profile. In Fig. 4.12, four samples have been tested; it should be noted that the variation between the samples are smaller than the variation in ZD. Hence, this carton board should have well-defined delamination points and perform well in converting. Carton boards normally perform badly in converting if the delamination planes are scattered at random and thus folding is not reproducible.

To optimize the ZD profile, it should comply with the creasing operation. The stress states during creasing has been analysed by Nygårds et al. (2009b) and from the simulations, it can be seen that there will be higher stresses in the paperboard closer to the male ruler. Since the male ruler penetrates the paperboard from the top surface, it will load these regions more. The crease depth does however need to be deep enough to also cause delamination close to the bottom. If the board is creased too much, it will cause crease cracks to cut the paperboard from the top surface; hence, creasing needs to be optimized. Tools and crease depths should therefore be chosen to generate as much shear deformation as possible.

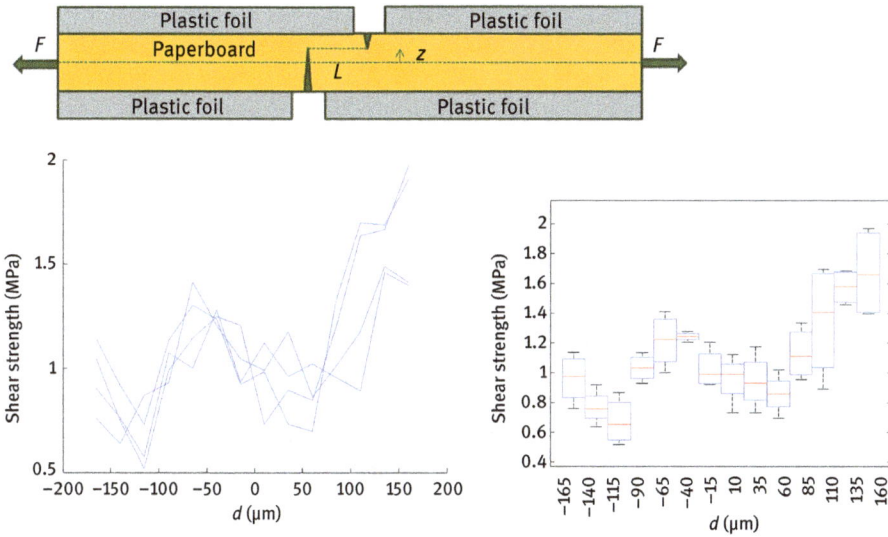

Fig. 4.12: Shear strength profiles in a paperboard measured by the DNS specimen. Four samples were tested, and both the raw data and evaluated box plots are shown.

Folding box boards (FBB) do often have chemi-thermomechanical pulp (CTMP)-based middle plies. These are bulkier and weaker than the kraft pulp plies that often are used as outer plies. The ZD shear profile of an FBB will often have a characteristic shape where the outer plies are strong, and the middle ply is weak and uniform. This paperboard structure will easily delaminate in the whole middle ply. The uncertainty with it is, however, that there is no well-defined weakest position in the structure. Delamination can therefore occur at different positions in ZD. This results in variability from crease to crease and also a possible variation along a crease line. As a consequence, crease lines from CTMP-based boards can be uneven when they are studied from bulged inside. A machine trial was done on a carton board machine, where the effect of weak interfaces is illustrated. First, a board with weak middle ply was made; thereafter, the middle layer was strengthened (see Fig. 4.13). The samples were creased to different crease depths and folded, and the resulting maximum bending moment shows that the board with the reinforced middle ply at all crease depths had lower maximum bending moment. This was because it delaminated along the interfaces, and less fibres entangled the delamination cracks.

If carton boards are creased deeply, the maximum bending moment, similarly, becomes independent of board design, since severe delamination has been activated. In Fig. 4.13, the board with two weak interfaces has been evaluated against a board with strong bottom interface and weak top interface. The latter board has much higher bending moments at low crease depths. Hence, during folding delamination cracks will open up between the top and middle plies. The middle and

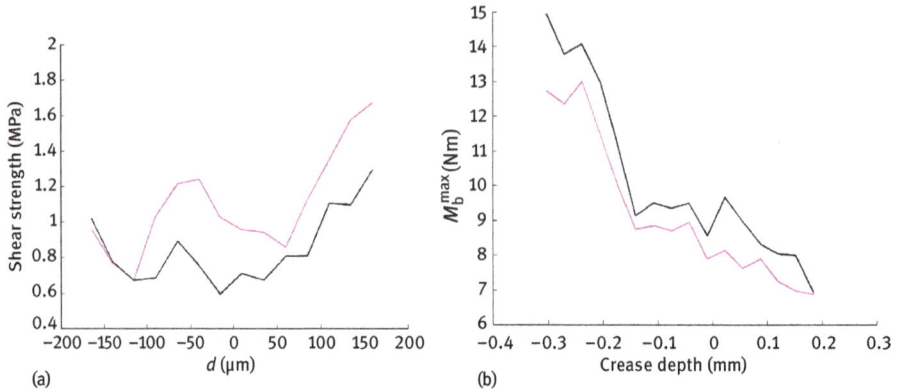

Fig. 4.13: (a) Average shear strength profiles for two boards: one with weak middle ply and one with strengthen middle ply. (b) Evaluated maximum bending moment after creasing to different crease depths.

bottom plies then need to bulge out which result in a high bending moment. When delamination has been initiated further down in the board structure the bending moment decreases.

Figure 4.14 shows a typical shear strength profile for a solid bleached board. If the same pulp is used in the whole paperboard structure, then the shear strength should be uniform in ZD, with a slight increase towards the ends. Therefore, the aim is to create a weak middle point in the middle of the structure. This can be achieved by surface sizing or a multi-ply structure. In Fig. 4.15, two boards with different shear strength levels and V-shaped profiles are shown. Folding of the boards show that the boards will delaminate mainly in the middle, where shear strength is

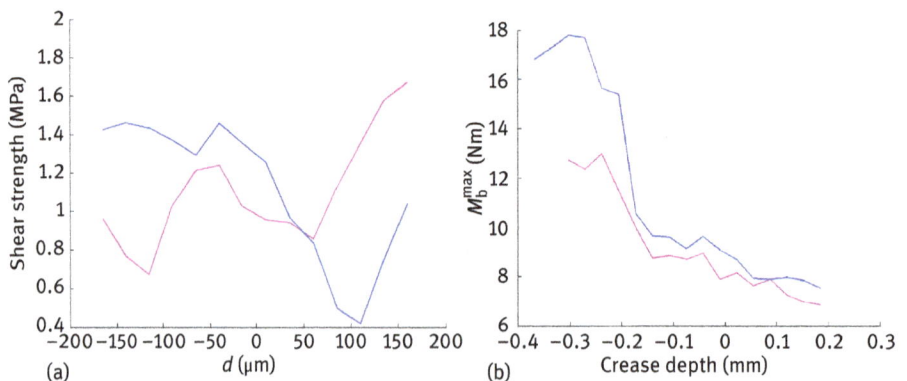

Fig. 4.14: (a) Average shear strength profiles for two boards: one with weak top and bottom interface and one with strong bottom interface and weak top interface. (b) Evaluated maximum bending moment after creasing to different crease depths.

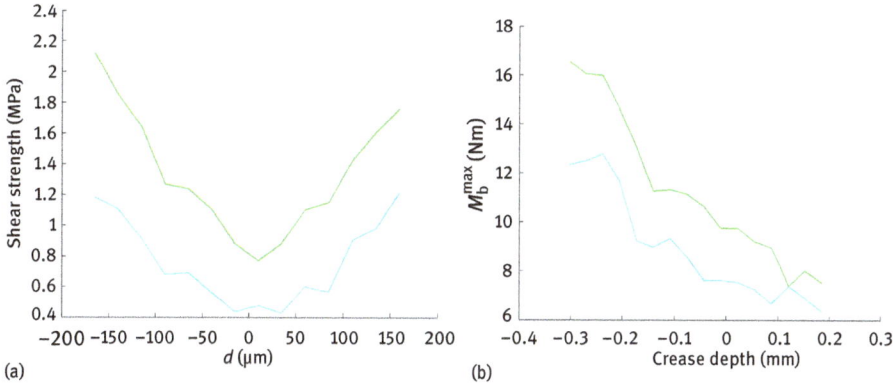

Fig. 4.15: (a) Average shear strength profiles for two boards with V-shaped profile, but different shear strength levels. (b) Evaluated maximum bending moment after creasing to different crease depths.

lowest. The resulting bending moments will become lower for the boards with lowest shear strength; basically, because it delaminates more easily.

4.5 Concluding remarks

We have discussed the behaviour of box corners, which relies on creasing and folding of carton board. This is a fairly well-defined boundary value problem, but also involves many different deformation and fracture mechanisms. We have presented a qualitative description of the main mechanisms of creasing and folding, but we have not considered the effect of corners on the box strength. A thorough study of the strength of a box with a given geometry would require knowledge of not only the panel properties (Section 3.4.3; Beldie et al., 2001) but also the strength properties of the corners, which in turn are sensitive to the delamination of the creases. In principle, the latter effect can be studied using fracture mechanics (Chapter 5), although little work has been published on this.

The mechanical properties of a carton board and the behaviour of box corners also depend on the loading rate, moisture content, and temperature of the carton board. The moisture content also influences the thickness of carton boards, which in turn influences the creasing, folding, and buckling behaviours. The general effects of moisture content, temperature, and loading rate will be discussed in Chapters 7 and 9.

Literature references

Barbier, C., Larsson, P.-L. and Östlund, S. (2002). Experimental observation of damage at folding of coated papers. Nord. Pulp Paper Res. J. 17(1), 34–38.

Beldie, L., Sandberg, G. and Sandberg, L. (2001). Carton board Packages Exposed to Static Loads – Finite Element Modelling and Experiments, Packag. Tech. Sci. 14, 171–178.

Nygårds et al. 2009b. Experimental and numerical studies of creasing of paperboard. Int. J. Solids Struct. 46, 2009, 2493–2505.

Nygårds, M. (2008). Experimental techniques for characterization of elastic-plastic material properties in carton board, Nord. Pulp Paper Res. J. 23(4), 432–437.

Nygårds, M., Fellers, C. and Östlund, S. (2009a). Development of the notched shear test, Advances in pulp and paper research, transactions of the 14th fundamental research symposium, l'Anson, S.J. ed. Oxford UK, PPFRC, 877–897.

Timoshenko, S. (1934). Theory of elasticity, engineering societies monographs, Ed Craver, H.W.. New York: McGraw-Hill Book Company Inc..

Sören Östlund, Petri Mäkelä

5 Fracture properties

5.1 Introduction

The presence of defects may reduce the strength of paper structures considerably. In a typical office paper sheet, the introduction of a 10 mm long cut at the sheet edge reduces the failure load by half and the failure elongation to a quarter, as compared to the undamaged sheet (Fig. 5.1). Although somewhat artificial, this example demonstrates the impact that defects have on the strength of paper products. The exemplified strength reduction comes from the fact that when a notched sheet is subjected to mechanical loading, stress concentrations are induced in the vicinity of the defect. Sharp defects, denoted cracks, give rise to more severe stress concentrations than non-sharp defects (Fig. 5.2). Consequently, cracks are the most critical defect type for the load-carrying capacity of structures.

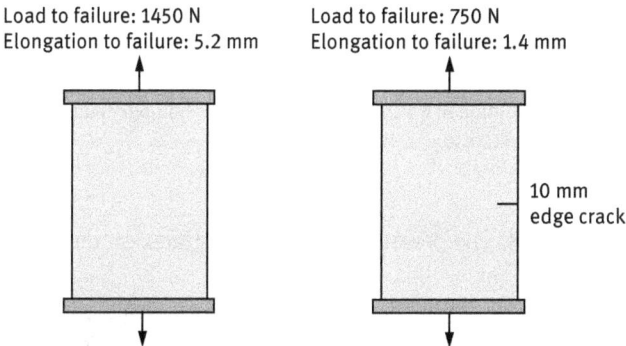

Load to failure: 1450 N
Elongation to failure: 5.2 mm

Load to failure: 750 N
Elongation to failure: 1.4 mm

10 mm
edge crack

Fig. 5.1: Comparison of the tensile load and elongation to failure of ordinary A4 office paper sheets (210 mm wide and 297 mm long). The left sheet is unnotched, while the right sheet contains a 10 mm long edge cut.

At small external loading, a defect remains stable and the structure does not break macroscopically. As the external loading is monotonically increased, damage will start accumulating in the neighbourhood of the defect, eventually followed by rupture of the structure. Accordingly, defects will affect the performance of paper products. It is also possible that ruptures are triggered by factors other than macroscopic defects, such as the non-uniformity of the paper material at different length scales,

Sören Östlund and Petri Mäkelä, KTH Royal Institute of Technology, Stockholm, Sweden

https://doi.org/10.1515/9783110619386-005

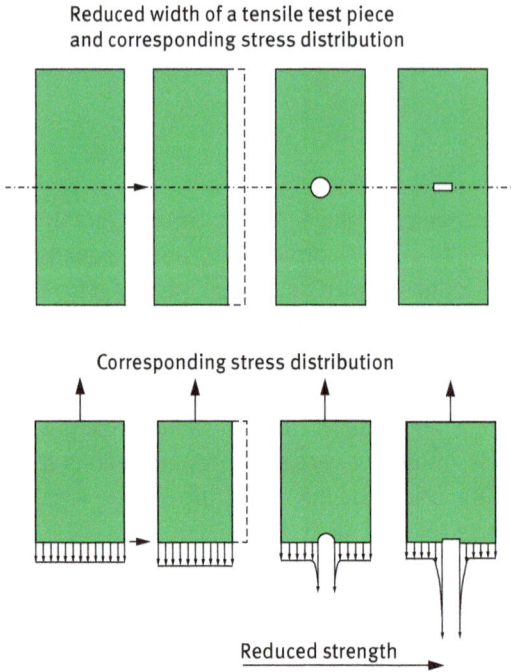

Reduced width of a tensile test piece
and corresponding stress distribution

Corresponding stress distribution

Reduced strength

Fig. 5.2: Change in the tensile stress distribution in a paper sheet caused by a homogeneous width reduction, a circular hole, and a sharp defect, respectively.

which will be discussed in Chapter 8. The present chapter concentrates on the effect of macroscopic crack-like defects.

Fracture mechanics is a discipline within solid mechanics that deals with the strength of structures containing defects. Knowledge of fracture mechanics is fundamental for the ability to prevent undesired paper failures in converting and end-use situations, as well as for controlling desired paper failures, such as delamination during creasing and folding operations (Chapter 4) and rupture performance of perforated tear lines in paper products.

Three principal factors influence the strength of a notched structure made of a specific material (Fig. 5.3): the external loading, the geometry of the structure (including the size and shape of defects), and the fracture toughness, that is, the capability of the material to sustain stress concentrations. The objective of fracture mechanics is to relate these three factors so that one of them can be calculated when the other two are known.

The general deformation of a defect under loading is described by the displacements of the two crack surfaces relative to one another. The relative displacement can be separated into the principal fracture modes I, II, and III defined in Fig. 5.4. In practice, defects influence the integrity of structures first, and foremost, under tensile loading, resulting in fracture mode I, which has been the focus in the discipline of fracture mechanics. However, the stability of structures under compressive loading is also sensitive to the existence of defects. Hagman et al. (2013) show that crack-like defects deteriorate the transverse shear failure strength under short-span compression loading. The creasing quality (Chapter 4) of carton board has large impact on the top-to-bottom compression strength of carton board packaging (Chapter 3), i.e. the delamination cracks that are created in the creasing operation are essential for the structural package performance.

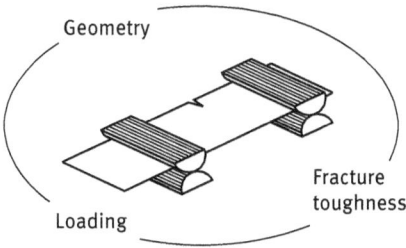

Fig. 5.3: Summary of the principal factors that influence the strength of notched structures. The size of the notch is part of the geometry definition. Reproduced from Mäkelä (2002) with permission from Svenska Pappers- och Cellulosaingeniörsfüreningen.

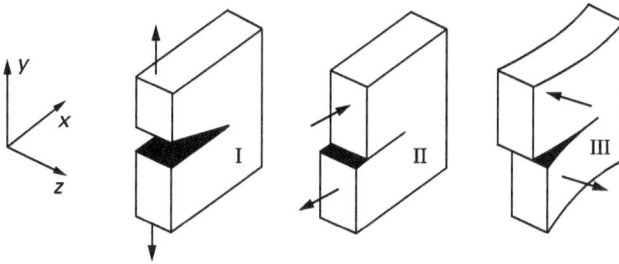

Fig. 5.4: Principal fracture modes I, II, and III in terms of the relative displacement of the crack surfaces.

5.2 Examples of practical applications of fracture mechanics

Before entering into the specific theories of fracture mechanics for paper materials, we briefly discuss a few applications where fracture mechanics are important for conceptual understanding and problem solving. It is appropriate to consider in-plane failure and out-of-plane failure (also known as delamination) separately.

5.2.1 Mode I failure under in-plane tension

Mode I failure under in-plane tension is the dominating mode of failure in many paper applications and has been a subject of considerable interest in the scientific literature. Web breaks (Fig. 5.5) are commonly caused by macroscopic defects, such as rigid fibre bundles, dirt spots, wrinkles, holes in the web, or edge cuts caused by careless handling of paper rolls. Reducing the size and frequency of defects, reducing the external loading level (including improved loading uniformity), and increasing the fracture toughness of the paper material (improving its resistance to stress concentrations) are three actions that will undoubtedly lead to less web breaks caused by the presence of macroscopic defects (Fig. 5.3). The discipline of fracture mechanics offers a framework for analysing the effect of macroscopic defects on web strength.

00:53 7

Fig. 5.5: Paper web break. Photo courtesy of Albany Nordiskafilt AB.

To assess the structural integrity of a paper web, it is necessary to know the stresses close to the crack tip and the fracture toughness of the material. The state of crack tip stress depends on the crack length, crack position, and web tension (Fig. 5.6). Its explicit mathematical form is described in Section 5.3. It is important to realize that the problem statement in Fig. 5.6 contains several simplifying assumptions, such as:
 – Only uniform uniaxial loading in MD
 – Crack oriented in CD
 – No out-of-plane buckling

The validity of these assumptions should be checked when approaching the real paper web performance.

 The web is, in general, not flat, and, particularly in the crack tip region, out-of-plane buckling will affect the stress state. If the crack is not oriented exactly in the cross-machine direction (CD), the crack tip becomes deformed both in the opening mode (fracture mode I) and in a shear mode (fracture mode II) (Fig. 5.4). However, a CD crack (as in Fig. 5.6) deformed in fracture mode I is generally the most severe type of defect. When passing a pressurized turner bar, the crack tip is loaded also perpendicular

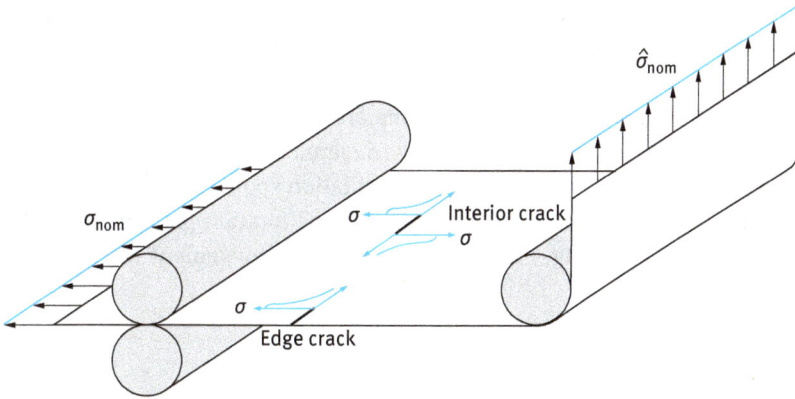

Fig. 5.6: Schematic definition of the fracture mechanics problem of web breaks caused by defects.

to the plane of the web by the air pressure (Fig. 5.7). There are also factors that can cause skew web loading or otherwise non-uniform web tension profile. Non-uniformity can arise from misaligned turner bars, frictional effects, and web tension profiles created on the machine and it can severely affect crack tip loading.

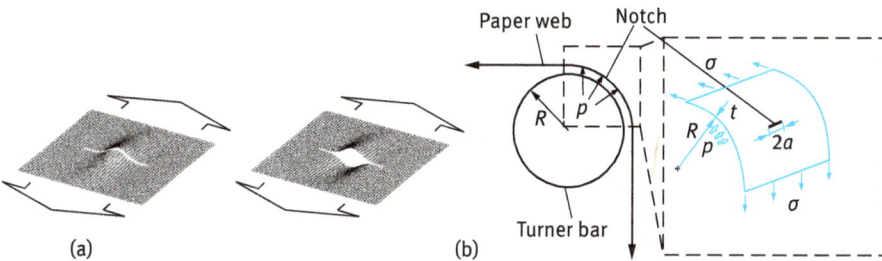

Fig. 5.7: Examples of factors that affect crack tip stresses in a paper web: tension-induced buckling (a; adapted from Skjetne, 2006) and bulging over a pressurized turner bar (b; after Nordhagen, 2009). Reproduced with permission from the authors.

The influence of loading factors on crack tip stresses in a printing press has been analysed by Nordhagen (2009). The results indicate that even in normal operating conditions, the non-ideal features may increase crack tip stress to levels that are comparable to the stresses that cause rupture in laboratory testing. Thus, it is not enough to compare the nominal web tension and the failure stress in laboratory testing. Damage tolerance methodology is a practical way to use fracture mechanics in the analysis of web breaks. One assumes the existence of a certain crack, and then uses a detailed stress analysis to investigate the integrity of the cracked structure.

In many paper products, cut-outs and perforated lines are applied to facilitate converting and end use. Examples include die cutting, ventilation holes, and hand holes in corrugated board containers, and perforated lines to enable easy opening of carton board packages. Stress concentrations are created at the tips of the cut-outs and perforations. Figure 5.8 shows a crack triggered by the die cutting of perforated lines in a corrugated container and crack initiation spots (short cracks) along the straight cut. Good performance requires that the perforation pattern should be neither too weak nor too strong. Fracture mechanics is a convenient tool for such optimization.

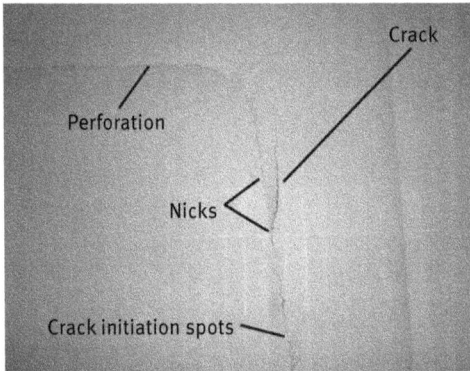

Fig. 5.8: Cracks created in the perforation operation of a corrugated board.

5.2.2 Out-of-plane delamination

Because fibres are predominantly aligned in the plane of the sheet, paper is considerably more compliant and weak in the out-of-plane direction than in the in-plane directions (Section 2.3.3). There are important cases where delamination determines the performance of paper products.

In offset printing of paper, delamination is a well-known problem that reduces production efficiency (Fig. 5.9). Transverse shear stresses induced by the loading in the printing nip (Chapter 10) can create cracks. When the web exits the nip, a crack may grow into visually unacceptable delamination due to ink tackiness. The multi-axial stress state of a material element passing through a nip varies, as shown in Fig. 5.10. Before the nip, pure in-plane tension applies (inset A in Fig. 5.10). Inside the nip, the rubber blanket causes bending, shear stresses, and compressive z-direction (ZD) stresses (inset B). Shear stresses can also arise from asymmetry. On the exit side, ink tack can cause ZD tensile stresses (inset C). The real behaviour is quite complex, and detailed mathematical analysis is needed to evaluate the stress components in the paper during its nip passage.

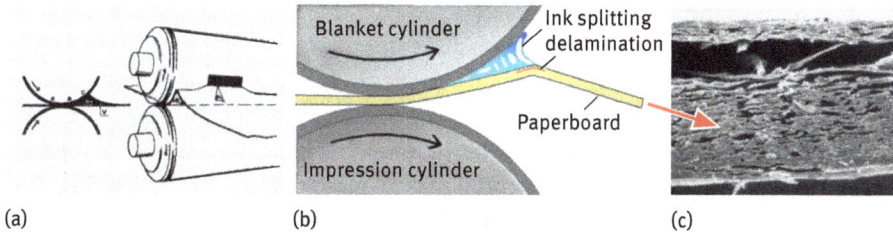

Fig. 5.9: Delamination problems in offset printing: non-uniform adhesion (a), ink tack (b), and delamination in a paperboard (c).

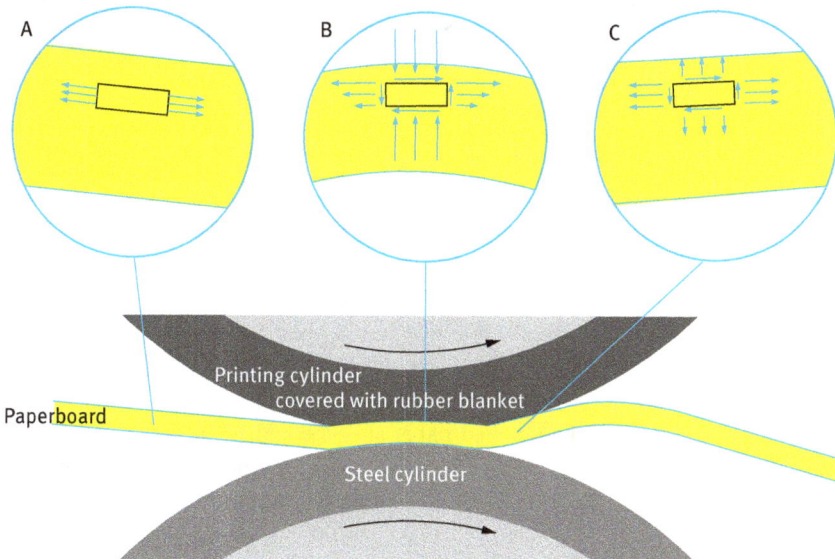

Fig. 5.10: Schematic distribution of MD and ZD stresses in a paperboard in different locations relative to an offset printing nip.

Failure analysis can be used to determine if a certain combined stress state creates macroscopic delamination or if typical existing microscopic defects can grow to unacceptable size. In the case of carton boards, the layered structure has to be accounted for because it influences the ZD variation of the normal and shear strengths. The location of an eventual delamination cannot be known in advance, although it commonly tends to occur close to the printing surface in practice. Full three-dimensional analysis is often necessary to address the mechanics of paper in a printing nip adequately, particularly when the ink tack is non-uniform as exemplified in Fig. 5.9a.

Another case where delamination is of vital importance is during the creasing and folding operations (Chapter 4). The transverse shear mechanism that creates

delamination in the creasing operation is shown in Fig. 4.8. A sharp crease line reduces the bending moment locally, which is necessary for achieving straight fold
lines and forming a box with flat panels. A successful creasing operation requires
that the material delaminates in a controlled manner, which conflicts with the requirement that delamination must be avoided in offset printing. Therefore, optimization work may be needed to avoid problems in either of these two processes.

Delamination may also play a vital role in the converting of corrugated board,
where it commonly occurs during the die-cutting process of the multi-ply material.
Delamination may also occur in the corrugation process of the fluting paper, then
causing a reduced transverse shear stiffness of the corrugated medium, resulting in
a reduced stacking strength of the box. The discipline of fracture mechanics can be
used to analyse and understand the concepts of the machine–material interaction
in the exemplified processes, allowing identification of key process parameters.

5.3 Crack tip modelling in paper materials

This section describes the theoretical foundation for crack tip modelling in paper materials. We do not cover the entire theory of continuum fracture mechanics, or review the
fracture mechanics literature for paper, but depict the most important crack tip models
for the analysis of paper materials and products. For a more detailed presentation, see
Anderson (2005) and Mäkelä (2002). Breakthrough in the application of fracture mechanics to paper materials came when Swinehart and Broek (1995) and Wellmar et al.
(2000) showed that the failure in notched paper sheets could be predicted from measured material properties. These studies shared the use of a robust fracture mechanics
theory combined with numerical methods to determine the failure point of notched
paper samples.

5.3.1 Characteristic length scale and the basis of crack tip
modelling

Most mathematical theories of fracture mechanics describe the effects of sharp cracks
where the material cohesion is lost completely. The real nature of cracks depends on
the microscopic fracture mechanisms. In paper material, that is, a cellulosic fibre network material, the failure mechanisms of the material involve mechanisms such as
inter-fibre bond failure, fibre pull-out, and fibre breakage. Among these mechanisms,
bond failure is usually the prevalent one in paper materials with requirements on
load-carrying capacity. Tens of bonds must fail to release a fibre, and the order of
hundreds of fibres must either be released or broken when a crack perpendicular to
the paper plane propagates by one fibre length. Consequently, this means that the

structural failure of paper cannot be triggered by one single microscopic event. Before macroscopic rupture can occur, microscopic damage is typically accumulated across several millimetres, a distance comparable to the length of the fibres (Fig. 5.11). The damage zone in the vicinity of the crack tip, that is, the region where irreversible damage is evolving due to the crack tip stress concentration, is referred to as the fracture process zone (FPZ).

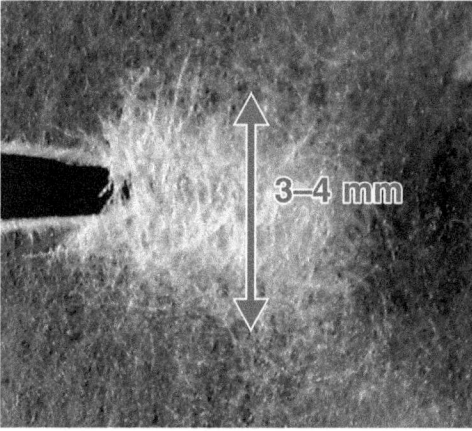

Fig. 5.11: Distribution of microscopic bond failures in a laboratory sheet made of chemical pulp, just before crack propagation starts. Damage is made visible with silicone oil impregnation (Niskanen et al., 2001). Reproduced with permission from the Pulp and Paper Fundamental Research Society (www.ppfrs.org).

As shown in Fig. 5.11, the apparent FPZ in paper can diffuse and extend well ahead of the assumed crack tip location. These features are ignored in the continuum mechanic models considered in this chapter. In similarity to other materials, the dispersion of crack boundaries does not limit the applicability of fracture mechanics in paper materials. As demonstrated later, fracture mechanics can be applied to paper even when macroscopic clearly visible cracks are absent.

Various different fracture mechanics models can be succesfully applied to paper materials and products depending on the problem and objective of the analysis, but it is best to use the simplest possible model that has predictive capability. As shown later, in certain cases a model based on elastic–plastic material properties and a point-wise fracture criterion is sufficient, while in other cases the extension of the failure process zone must be accounted for. The latter is particularly true for delamination problems. The choice of model is largely determined by the properties of the adopted continuum mechanics material model and the extension of the FPZ in comparison to the characteristic dimensions of the structure, such as the crack length and the in-plane dimensions. The point-wise fracture criterion requires that the FPZ is proportionally small, whereas this is not required if the extension of the FPZ is included in the used model. The size of the FPZ is usually not known in advance, and the accuracy of the chosen crack tip modelling approach must be verified by comparing the predictions either with experiments or with predictions from a more advanced model that has already been proven reliable.

The predictive capability of a fracture mechanics model relies on that the used fracture criterion must be transferable from laboratory testing to real structures, i.e. transferability must prevail. Transferability can be investigated by testing different test piece geometries in laboratory. It has been observed (Mäkelä and Östlund, 2007) that fracture mechanics predictions for paper can be valid, and transferability can prevail, even though the predicted stress fields are not consistent with experimental observations.

5.3.2 Linear elastic fracture mechanics

The most simple in-plane fracture mechanics model for paper materials is based on the assumption that all non-linear material behaviour, such as damage and plasticity, is confined in a failure process zone (FPZ) at the crack tip, which is small in comparison to the characteristic in-plane dimensions of the structure. Under such so called "small scale yielding" conditions, the FPZ can be mathematically treated as a point. The crack tip stress field can then be determined using a linear elastic material model. The corresponding theory is called linear elastic fracture mechanics (LEFM). Irwin (1957) showed that in mode I loading, the multi-axial stress and strain fields close to the crack tip are characterized by one single parameter, that is, the stress intensity factor K. The non-zero stress components at the crack tip are singular according to

$$\sigma_{ij} = \frac{K}{\sqrt{2\pi r}} f_{ij}(\varphi),\tag{5.1}$$

where the coordinates r and φ are defined in Fig. 5.12. Notice that here we define the x- and y-axes relative to the crack orientation, and not relative to the machine direction (MD) and cross-machine direction (CD) of the paper as elsewhere in this book (cf. Section 2.2.1). Details such as the angular functions $f_{ij}(\varphi)$, the displacement and

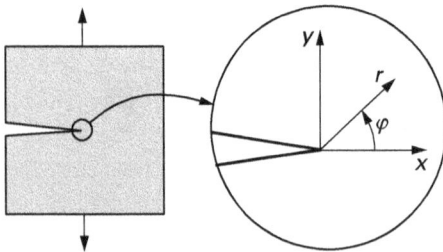

Fig. 5.12: Crack tip coordinates used in eq. (5.1).

strain fields, and the effects of orthotropic material properties (of particular importance in the case of paper) are available in the fracture mechanics literature (see, e.g. Sih and Liebowitz, 1968), but eq. (5.1) is sufficient for the present purpose.

Equation (5.1) is the first term in a series expansion, and its validity is therefore restricted to the so-called singularity-dominated zone near the crack tip. Further away from the crack tip, higher order terms should be included. In reality, singular stresses do not exist because no real material can carry infinite stresses, and eq. (5.1) is, therefore, invalid in the FPZ. Figure 5.13 gives a schematic picture of real crack opening stress σ_{yy} in a linear elastic material.

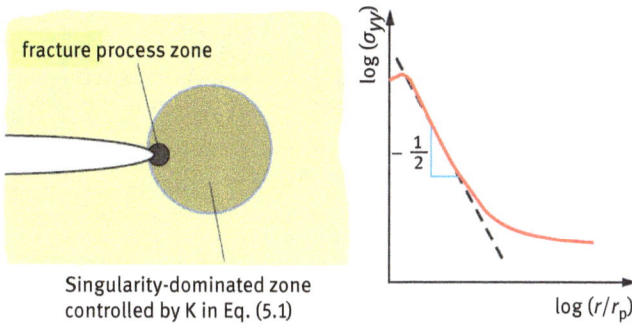

Fig. 5.13: Qualitative distribution of the crack opening stress σ_{yy} in a linear elastic material. The characteristic length scale r_p may be chosen to be the size of the fracture process zone.

The stress intensity factor K is useful because it can be determined from tensile testing of notched specimens. Closed-form solutions for K are available for some simple geometries and loading types, but in most cases, numerical methods are required for evaluation of K. The numerically obtained relations between K and external stresses for finite specimens are usually expressed in the form

$$K = \sigma\sqrt{\pi a}\, f(\text{geometrical parameters}),\qquad(5.2)$$

where σ is the remote stress due to external loading, a is the crack size (e.g. diameter or length after symmetry considerations), and f is a function of the geometry and characteristic dimensions of the structure. The factor $\sigma\sqrt{\pi a}$ originates from the closed-form solution of a mode I centre crack in an infinite plate. Tensile test data for notched specimens can be used to evaluate the critical value of K at which crack growth begins. In LEFM, this critical value is denoted K_c and is taken to be a material property called *fracture toughness*. If K_c is known, one can calculate from eq. (5.2) the critical stress state that initiates crack growth. Because K_c is a material property in LEFM, the theory predicts that, independent of geometric dimensions and load type, crack growth and, therefore, macroscopic fracture begin when K calculated from eq. (5.2) reaches K_c. Thus, the LEFM fracture criterion is

$$K = K_c. \tag{5.3}$$

The first studies on the defect sensitivity of paper materials were based on the LEFM theory, but later investigations have shown that LEFM works excellent for predominantly liner elastic and brittle materials, such as glass and ceramics. LEFM has been shown to apply also to non-linear and tough materials, such as paper materials, but only when restricted to analysis of large cracks in large structures. In the case of paper materials, LEFM does generally not apply to problems of commercial interest.

Several attempts have been made to introduce corrections to LEFM, aiming to broaden the application of the theory to materials that exhibit non-linear material behaviour. The first and most known attempts, which were suggested by Dugdale (1960) and Barenblatt (1959), result in successful application of LEFM, to certain problems involving construction materials, such as metals. A more recent attempt to correct LEFM, which originates from the paper physics community, was suggested by Coffin et al. (2013).

5.3.3 Non-linear fracture mechanics using J-integral

If the material behaviour in the crack tip region deviates distinctly from linear elasticity in a large region, more advanced fracture mechanics models are needed. This is in general the case for paper materials. The most well-known crack tip model for non-linear materials assumes a parabolic non-linear elastic material model. The uniaxial version of the stress–strain expression is

$$\varepsilon = \frac{\sigma}{E} + \left(\frac{\sigma}{E_0}\right)^n, \tag{5.4}$$

where n is a strain-hardening exponent and E_0 is called the strain-hardening modulus. This leads to the following stress field in the vicinity of the crack tip:

$$\sigma_{ij} = \alpha \left(\frac{J}{r}\right)^{\frac{1}{n+1}} g_{ij}(n, \varphi). \tag{5.5}$$

The scalar multiplier α depends on the material and stress state (plane strain or plane stress). This crack tip solution is known as the Hutchinson–Rice–Rosengren (HRR) field, after Hutchinson (1968) and Rice and Rosengren (1968). They showed that the crack tip conditions in a non-linear elastic material are characterized by a path-independent contour integral known as the J-integral that was introduced by Rice (1968).

The HRR field predicts that (a) the stresses in the vicinity of a crack tip are singular and governed by the stress–strain behaviour of the material, (b) the J-integral determines the amplitude of the stresses, and (c) the stresses have a uniform distribution. Therefore, in non-linear elastic materials, the stresses close to a crack tip are characterized by J, in analogy with K in the case of linear elastic materials. In fact, LEFM is a special case of non-linear fracture mechanics. If the material is linear elastic, isotropic and plane stress conditions apply, then $J = K^2/E$, which means that the HRR model can be used to estimate the length scale outside which the LEFM model (small-scale yielding approximation) applies. A schematic illustration of this is shown in Fig. 5.14. In addition to the HRR field, other single-parameter characterizations of the crack tip stress fields that use J are possible. The necessary conditions for the existence of a J-dominated zone are that the stresses and strains scale with J/r are distributed uniformly inside this zone.

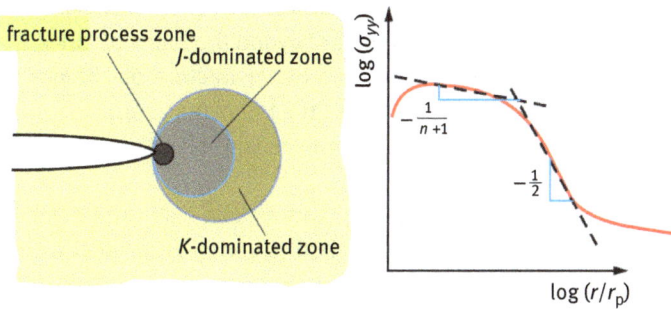

Fig. 5.14: Qualitative distribution of the crack opening stress σ_{yy} in a non-linear elastic material. The characteristic length scale r_p may be chosen to be the size of the fracture process zone.

In analogy with LEFM, tensile testing of notched specimens can be used to determine the critical value of J, i.e. the non-linear fracture toughness denoted by J_c, that is needed for crack growth to start. Hence, the criterion for crack growth is expressed as

$$J = J_c. \qquad (5.6)$$

Non-linear fracture mechanics was introduced for paper materials by Uesaka et al. (1979). Initially, the research on fracture mechanics of paper materials was focusing on the experimental methods to determine the fracture toughness J_c and its dependence on the papermaking parameters. However, J_c itself cannot be used to compare the defect sensitivity of paper materials because the relationship between J and external loading, and therefore the critical value J_c depends also on the material behaviour (n and E_0 in eq. (5.4)). Apart from a few exceptions, numerical methods are therefore needed to determine the relationship between J and the external loading. The J-integral evaluation is in fact implemented in most of the commercially available finite element software codes that are intended for non-linear fracture mechanics analyses.

Non-linear elastic materials have a unique relationship between strains and stresses, while in elastic–plastic materials, the stress state depends on both the strain state and the strain history. The latter is the case for paper (Section 2.3). However, there is no difference between the two material models if unloading does not take place. Thus, the previously discussed non-linear elastic treatment applies also to elastic–plastic materials as long as the material does not recoil because of noticeable damage creation prior to crack growth. The non-linear elastic approximation of an elastic–plastic material is called *deformation theory of plasticity*.

In the case of multi-axial stresses and deformations, the non-linear elastic treatment of elastic–plastic materials would also require proportional loading. In fact, a deformation theory relationship is generally not identical to the corresponding elastic–plastic relation, even under the special case of strictly increasing uniaxial loading, but the differences are small. Therefore, the J-integral together with the concept of deformation theory of plasticity is valid far beyond the limits of LEFM.

Figure 5.15 summarizes the practical use of non-linear fracture mechanics to evaluate the defect sensitivity of paper materials with the following steps:

1. Determine the material behaviour (eq. (5.4)) from tensile testing of standard tensile test pieces.
2. Determine the non-linear fracture toughness J_c from tensile testing of notched rectangular test pieces. The finite element method is generally needed in the evaluation.
3. Predict the failure of products or structures of interest from eq. (5.6). The calculation of J generally requires finite element analysis.

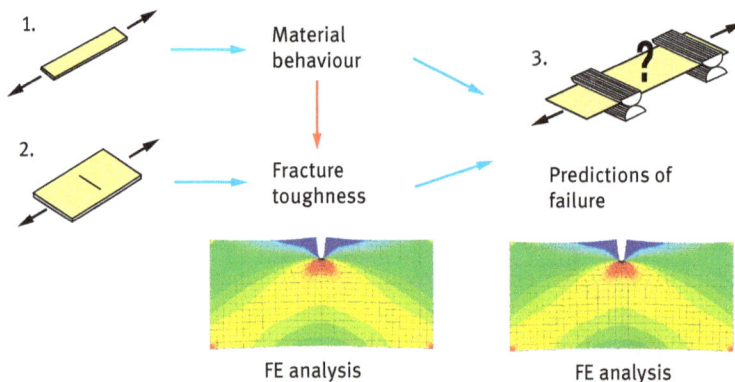

1. Material behaviour
2. Fracture toughness
3. Predictions of failure

FE analysis   FE analysis

Fig. 5.15: Summary of the use of non-linear fracture mechanics to predict failure.

5.3.4 Cohesive zone models

The cohesive zone model was originally suggested as a correction of LEFM for large-scale plasticity in metals. It was further developed for the analysis of damage evolution in concrete and later in paper (Tryding, 1996). In the cohesive zone model, the assumption is that the damage is confined to a narrow zone ahead of the crack tip. The explicit damage behaviour is replaced by a cohesive stress-widening curve. For mode I loading, this curve is a function of only the crack widening u_w (Fig. 5.16). This simple model description of damage is uniaxial so that only the material properties parallel to the external load are considered.

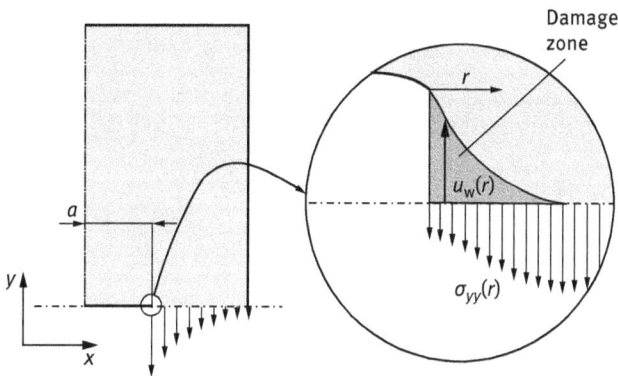

Fig. 5.16: Crack tip modelling using the cohesive zone model. The cohesive stress-widening curve is given by $\sigma(u_w)$.

The cohesive stress-widening curve describes the stress–elongation behaviour of the material after the peak stress is passed, in other words, the post-peak behaviour. Short specimens have to be used to observe this behaviour because standard test specimens of paper fail immediately and completely when the maximum stress is reached. The system becomes unstable because the recoil in elongation outside the FPZ must be equal to the elongation inside the damaged zone, and the latter easily makes up a very large strain increment if the damaged zone is narrow compared to specimen length. Thus, a short enough specimen length removes the instability, and the post-peak stress-elongation curve can be detected (Hillerborg et al., 1976). In order to determine the cohesive stress-widening curve, one assumes that the total elongation of a test piece (Fig. 5.17) can be decomposed into a sum of uniform elongation in the undamaged material and widening u_w in the damaged zone

$$u = \varepsilon L + u_w, \tag{5.7}$$

where ε denotes the strain in the undamaged material and L is the unstrained length of the test piece. In the post-peak region ε can be expressed as

$$\varepsilon = \frac{u_0}{L} - \frac{\sigma_0 - \sigma}{E}. \tag{5.8}$$

Here σ_0 and u_0 are the peak-stress (also referred to as the cohesive stress) and the corresponding elongation (Fig. 5.18), and E is the elastic modulus. Combining eqs. (5.7) and (5.8), one obtains the cohesive stress-widening relation

$$u_w = u - u_0 + \frac{\sigma - \sigma_0}{E} L. \tag{5.9}$$

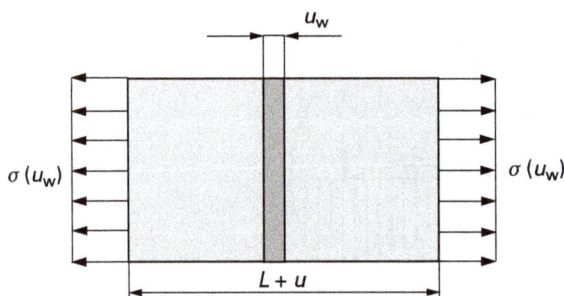

Fig. 5.17: Elongation of a test piece with a narrow damage zone, used to determine the post-peak cohesive stress. Reproduced from Mäkelä (2002) with permission from Svenska Pappers- och Cellulosaingeniörsfüreningen (SPCI).

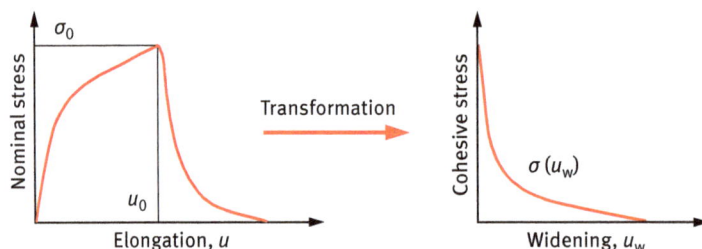

Fig. 5.18: Transformation of a stable load-elongation curve into a cohesive stress-widening curve. Reproduced from Mäkelä (2002) with permission from Svenska Pappers- och Cellulosaingeniörsfüreningen (SPCI).

The third term of this expression accounts for the elastic recovery (unloading) of the structure outside the damaged zone. The main assumptions behind eq. (5.9) are that this recovery occurs far from the damaged zone and that the irreversible elongation in the damage zone consists only of post-peak damage growth. Observe that the measurement of the cohesive stress-widening relation of a material does not require notched structures.

Measured cohesive stress-widening relations can be included in a material model to evaluate the defect sensitivity of paper structures. Then, a separate fracture criterion is no longer needed. When the calculated local deformation exceeds the peak value u_0, the softening of the material starts, given by eq. (5.9). When the softening has reached such an extent that the overall load-carrying capacity of the structure starts to decrease, instability and rapid failure will occur as response to further increase of the external loading, while further increase in the external elongation may cause the structure to became instable or remain stable. In the latter case, one has to determine if the system is stable or unstable against further infinitesimal increases in deformation.

Usually the cohesive zone is not fully developed before a notched paper structure reaches instability. This means that some of the theoretical fracture energy,

$$W_f = \int_0^{u_w^{max}} \sigma(u_w) du_w, \tag{5.10}$$

is *not* consumed before the instability occurs.

The cohesive crack model is not limited to situations where a macroscopic defect exists. Any situation that brings the local stress above σ_0 will create a damaged region that may subsequently transform into a growing crack due to the localized softening of the material. Practical use of the cohesive crack model in fracture analysis of paper, as exemplified by Mäkelä and Östlund (2012), requires the following three steps:

1. Determine the stress–strain properties of the material from tensile testing of standard tensile test pieces.
2. Determine the cohesive stress-widening relation of the material from tensile testing of short tensile test pieces.
3. Analyse the real problem (geometry and loading situation) using a cohesive zone model, involving the measured stress–strain properties and the cohesive stress-widening relation, and evaluate the load or elongation that causes instability. This step requires finite element analysis.

In Fig. 5.19, the results from three different crack tip models are compared in the case of MD and CD loading of centre-cracked copy paper specimens. The analysis includes LEFM, the J-integral method, and cohesive zone modelling with an incremental elastic–plastic material behaviour. The comparison with experiments shows that, in general, the linear elasticity assumption yields poor accuracy. The nonlinear J-integral method gives accurate predictions for medium- and large-size cracks. Cohesive zone modelling with parabolic strain hardening plasticity gives accurate predictions for the load and strain at failure for all the crack sizes.

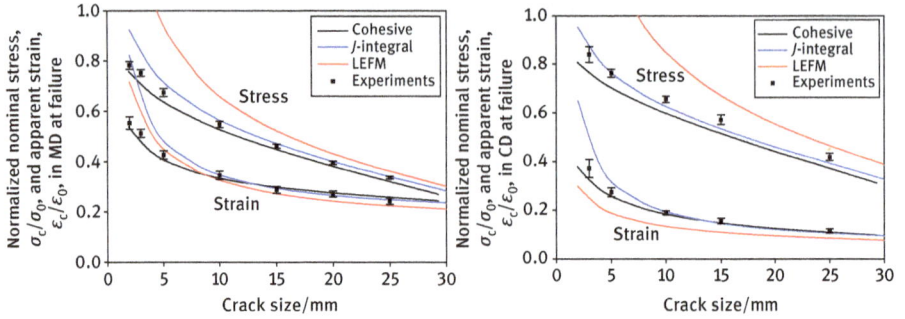

Fig. 5.19: Comparison between numerically predicted and experimentally determined critical stress and critical strain at failure in MD (right) and CD (left) for 100 mm long and 50 mm wide centre-cracked copy paper test pieces. The predictions were performed using three different fracture mechanics models, namely, a cohesive crack model, the *J*-integral, and LEFM. The results are normalized with respect to the stress and strain at break for a non-cracked paper test piece of the same dimension.

Recent work on the in-plane cohesive properties of paperboard has focused on a deeper understanding of the relations between the shape of the cohesive stress-widening curve, the fracture energy, and the paperboard tensile strength properties. (Tryding et al., 2017).

5.3.5 Continuum damage-mechanics modelling of paper

In paper materials, the microscopic fracture mechanisms are fibre–fibre bond failure and fibre breakage. Continuum damage mechanics describes the effect of microscopic fractures at the macroscopic level without making any assumptions about the nature of the microscopic processes (Kachanov, 1958). The foundation of this theory is a damage parameter that characterizes the average degradation of the material. In the simplest case, a scalar damage parameter D ranges from $D = 0$ (undamaged material) to $D = 1$ (completely ruptured material). Several damage mechanisms could also be treated simultaneously, and anisotropy in the damage development could be included by treating D as a higher order tensor, but that will not be considered here. There are different theories regarding the effect of D on the constitutive parameters, such as the elastic energy equivalence used by Isaksson et al. (2005). In their analysis of isotropic paper sheets, the change in elastic energy is described by a change in the stress tensor σ_{ij},

$$\sigma_{ij} = (1 - D)\hat{\sigma}_{ij}, \qquad (5.11)$$

where the effective stress tensor $\hat{\sigma}_{ij}$ represents the behaviour of the undamaged material. The elastic stiffness tensor C_{ijkl} of the damaged material is related to the stiffness tensor C^0_{ijkl} of the undamaged material by

$$C_{ijkl} = (1-D)^2 C^0_{ijkl}. \tag{5.12}$$

The driving force for damage evolution at a point is the damage energy release rate Y given by $Y = \partial\psi/\partial D$, where ψ is the complementary elastic stress energy density. If the material is linear elastic in its undamaged state, then

$$Y = (1-D)^{-3} \sigma_{ij} \left[C^0_{ijkl} \right]^{-1} \sigma_{kl}. \tag{5.13}$$

Therefore, when Y reaches a critical value Y_0 at a material point, damage will start to grow at that point.

Because damage in paper materials is intrinsically non-local (Fig. 5.12), the damage accumulation at a point is affected not only by the local value of Y, but also by stresses in the neighbourhood. To address this, Isaksson et al. (2005) introduced a non-local damage driving force characterized by a length scale that turns out to be of the order of the mean fibre length. To complete the model, Isaksson et al. (2005) proposed the damage evolution law

$$D = 1 - e^{-k(Y-Y_0)^m} \quad \text{for} \quad Y = H, \tag{5.14}$$

where k and m are parameters that can be determined from experiments, and H is a threshold value that Y has reached during the loading history representing a damage hardening behaviour. Figure 5.20 shows a typical comparison of the model with a measured stress–strain curve and damage evolution in a short and wide paper

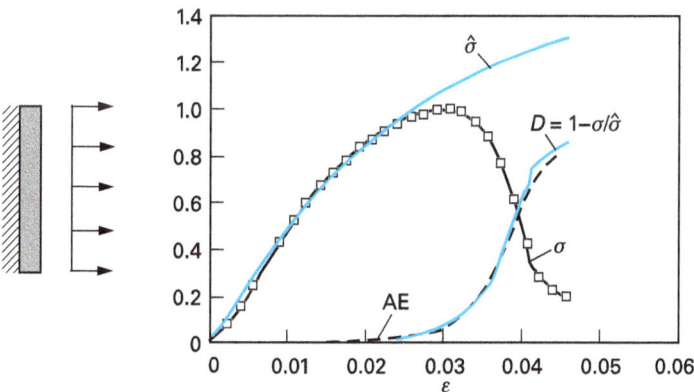

Fig. 5.20: Measured and modelled stress–strain and damage curves of a short and wide paper specimen, shown on left. Acoustic emission (AE) events are used to measure damage accumulation. Reprinted from Isaksson et al. (2005) with permission from Elsevier.

specimen. The cumulative number of acoustic emission events is used as the experimental measure of damage. In the calibration and application of continuum damage mechanics models, it is necessary to use the finite element method.

5.3.6 Delamination of paper materials

The importance of out-of-plane mechanical properties was discussed in Section 5.2.2. Considerable efforts have been put into the development of robust and reliable methods for experimental determination of the stiffness and strength under ZD normal and shear loading. Such measurements tend to be extremely complex. Fracture mechanics-based test methods, such as the double cantilever beam (Anderson, 2005), have been suggested for the purpose. However, Girlanda et al. (2005) have shown that crack tip stress fields cannot fully develop in the ZD because of the heterogeneous structure and low thickness of paper materials.

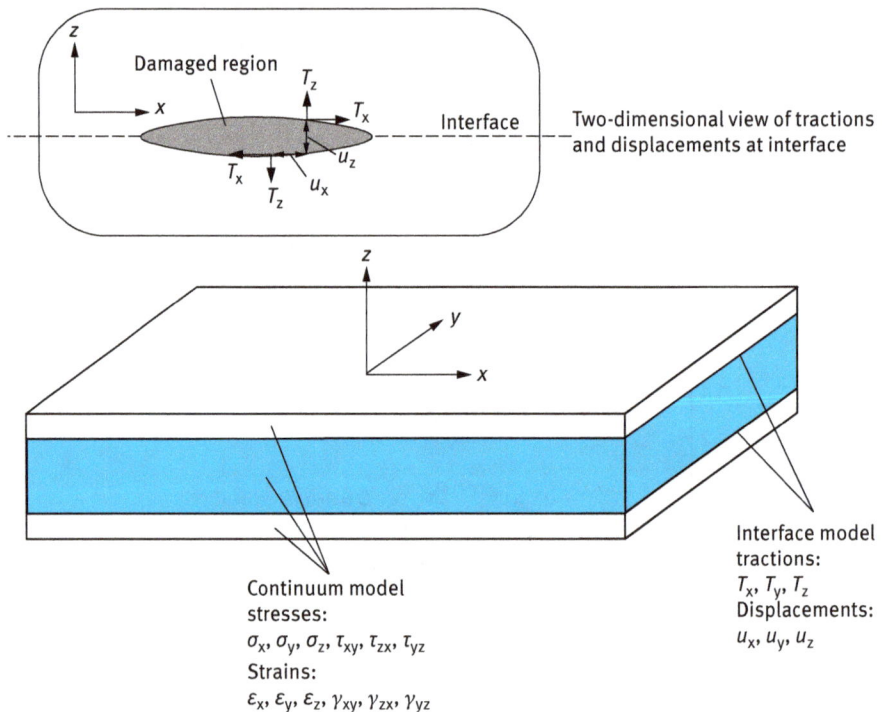

Fig. 5.21: Combination of a continuum model and an interface model for a three-ply carton board.

Figure 5.21 shows one approach to model the ZD deformation behaviour of a multi-ply carton board. The undamaged material is described with a continuum material model and the interfaces between plies with a traction–displacement relation similar to the cohesive zone model in Section 5.3.4. An advanced anisotropic elastic–plastic model is needed to describe the undamaged material behaviour. The interface model works without macroscopic defects, which is useful in the present case. A structural integrity analysis based on the existence of defects could also be used. The interface model described here could be applied at any ZD position, not only at the interfaces between the carton board plies.

The interface model describes how two material surfaces are separated, or delaminated, during ZD loading. Delamination is assumed to initiate due to a combination of out-of-plane normal loading (mode I) and shear loadings (modes II and III). The displacements of the two opposing surfaces relative to one another are assumed to consist of elastic and damage components,

$$u_i = u_i^e + u_i^d, \tag{5.15}$$

where i represents the three directions x, y, and z in the local coordinate system. This assumption means that plastic deformations can happen only inside the undamaged plies. The relevant stress components are σ_z, τ_{zx}, and τ_{yz}. The total displacement grows in increments

$$\Delta u_i = \Delta u_i^e + \Delta u_i^d, \tag{5.16}$$

and we assume that the corresponding traction increments are proportional to the change in the elastic displacement, or

$$\Delta T_i = C_i \left(\Delta u_i - \Delta u_i^d \right) \quad \text{(no summation)}. \tag{5.17}$$

Here C_i is the component of the instantaneous interface stiffness in the i-direction. Experimental data (Stenberg, 2002) demonstrate that damage affects the interface properties. In the model, this is accounted for by *reductions* in the interface strength components $\sigma_i^{\max}$ and stiffnesses C_i. Interface failure caused by the damage is then described by a failure criterion analogous to the Tsai–Wu criterion (Section 2.4), stating that local failure occurs when

$$f^{\text{interface}} \left(\sigma_i^{\max}, C_i \right) = 0. \tag{5.18}$$

In the analysis of the creasing of a carton board, Nygårds et al. (2009) used the expression

$$f^{\text{interface}} = \left(\frac{T_x}{\sigma_x^{\text{max}}}\right)^2 + \left(\frac{T_y}{\sigma_y^{\text{max}}}\right)^2 + \frac{T_z}{\sigma_z^{\text{max}}} - 1. \tag{5.19}$$

No initial crack is needed in the model. Initiation of damage is modelled in a manner equivalent to plastic deformation in metals, and the integration of the previous equations follows a procedure analogous to the one used in the theory of plasticity (Ottosen and Ristinmaa, 2005). In the case of pure shear stresses ($T_z = 0$), an associated damage rule will result in normal dilatation, that is, an increase in paper thickness. This has also been observed in experiments. For details on the application of an interface model of the type described here, see Nygårds et al. (2009).

Although the principle of the model is straightforward, its experimental calibration is challenging. First, low thickness means that high precision is needed when measuring the ZD displacements. Second, there is no well-established method for separation of the non-linear part of the ZD deformation into the continuum and damage components required by the model.

5.4 Compressive failure

Compression properties are important in the performance of boxes under stacking load, as discussed in Chapter 3. The measurement of the compressive failure of paper materials is challenging because thin sheets buckle easily. Macroscopic buckling is the most important compressive failure mechanism in many products (see Section 3.4). Different approaches to avoid buckling during compression testing have been proposed (Fellers, 1986), including short-span testing, geometrical stiffening (e.g. by using tubular specimens), or the use of anti-buckling guides to prevent out-of-plane deformations (e.g. by using narrow support blades).

As demonstrated in Fig. 2.14, the compressive stress–strain behaviour of paper is different from the tensile behaviour. The compressive strength is only about 30% of the tensile strength. Compressive loading of a paper specimen to levels close to failure will not affect the strength obtained in a subsequent tensile test with certainty, whereas tensile loading to levels close to failure is expected to lead to a pronounced reduction in the compressive strength. Thus, the microscopic mechanisms of failure in compression are different from the mechanisms in tension.

The microscopic failure of paper in compression is caused by a structural instability (not rupture) of the fibre network. This happens through either buckling of free fibre segments or shear dislocations in fibre walls (Fig. 5.22). The first happens primarily in low-density sheets, and the latter in medium- and high-density sheets. The

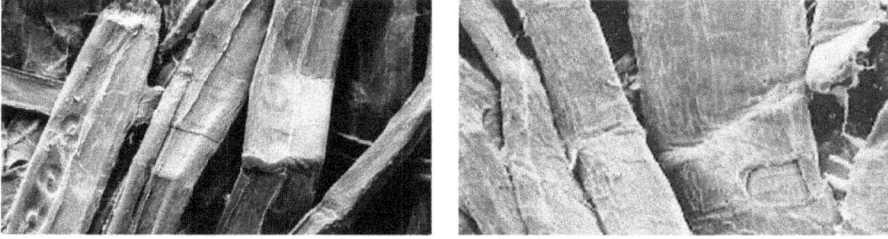

Fig. 5.22: Microscopic compressive failures in paper: buckling of fibre segments (left) and shear dislocations in fibre walls (right). Courtesy of Christer Fellers.

microscopic buckling and shear dislocations in the fibre network change the distribution of compressive stresses. When compressive load is increased, more fibre segments buckle and shear dislocations increase until the whole sheet becomes structurally unstable. The macroscopic compressive failure is often associated with a shear slip dislocation that shows also delamination (Fig. 5.23).

Fig. 5.23: Shear band slip failure in compression, also showing delamination. Courtesy of Christer Fellers.

The microscopic failure mechanism in compression explains why prior tensile loading affects compressive strength but prior compressive loading does not affect tensile strength. Inter-fibre bond failures in tensile loading increase the mean length of free fibre segments. As a result, the buckling threshold of fibre segments decreases. In the same way, if the creasing or folding of a carton board creates excessive delamination outside the fold area, then the compressive strength of a box decreases.

For example, in carton board packaging applications, the box compression strength is highly affected by the in-plane compression strength of the carton board materials. As discussed in Section 3.4.1, the short-span compression test (SCT) is for this purpose used to determine the compression strength, and papermakers strive to increase its value. Due to its importance and recent developments in numerical analysis methods, this method has been subjected to detailed analyses. (Borgqvist et al., 2016; Hagman et al., 2013). An important observation is that the SCT value is strongly affected by the in-plane stiffness properties and the transverse shear strength of the carton board.

5.5 Concluding remarks

Let us return to the problem of tensile web breaks caused by defects, and the possibility to predict such failures, by taking a closer look at a study by Mäkelä et al. (2009). The study comprises a series of tensile experiments, performed for large edge-notched paper webs, and predictions of the encountered failures using a non-linear fracture mechanics model. The large-web experiments were performed for 1880 mm long and 800–1000 mm wide paper webs with man-made edge cracks of various different sizes. The used fracture mechanics model, which was calibrated solely by laboratory tensile testing of 100 mm long test pieces, was based on isotropic deformation theory of plasticity (see Section 5.3.3) and the J-integral. The failure experiments and predictions were repeated for six different paper grades, comprising fluting paper, sack paper, newsprint, linerboard, medium-weight coated paper and supercalendered paper.

The results of the study are summarized in Fig. 5.24, showing reasonable agreement between the experimentally measured and predicted force and elongation at failure of the large webs. Consequently, the study shows that a non-linear fracture mechanics model, calibrated by laboratory testing, can successfully predict failures in notched large paper webs.

As previously stated, non-linear fracture mechanics analysis generally relies on finite element analysis (see Fig. 5.15), that is, a numerical complexity that constitutes an obstacle to engineering application of fracture mechanics. Thus, an analytic method for determination of the fracture toughness of paper materials (Mäkelä and Fellers, 2012) and an analytic fracture mechanics analysis procedure (Mäkelä, 2012) were developed. These analytic procedures, which enable engineering fracture mechanics analysis without the need for finite element analysis, were standardized (ISO/TS 17958). The standard method was further implemented in a commercially available testing apparatus (L&W Tensile Tester, code 066), which automatically determines the fracture toughness from material test data. Additionally, the testing apparatus automatically provides predictions of the fracture strength and fracture strain of a notched standard web geometry, that is, two predicted parameters that allow ranking of paper materials in terms of their fracture properties.

The goal of this chapter was to describe the use of continuum fracture mechanics for analysis of paper materials and products. The choice of the appropriate model depends on the size of the FPZ in relation to the structural dimensions. If it is assumed that the FPZ is proportionately small, then stress and strain fields can be modelled as

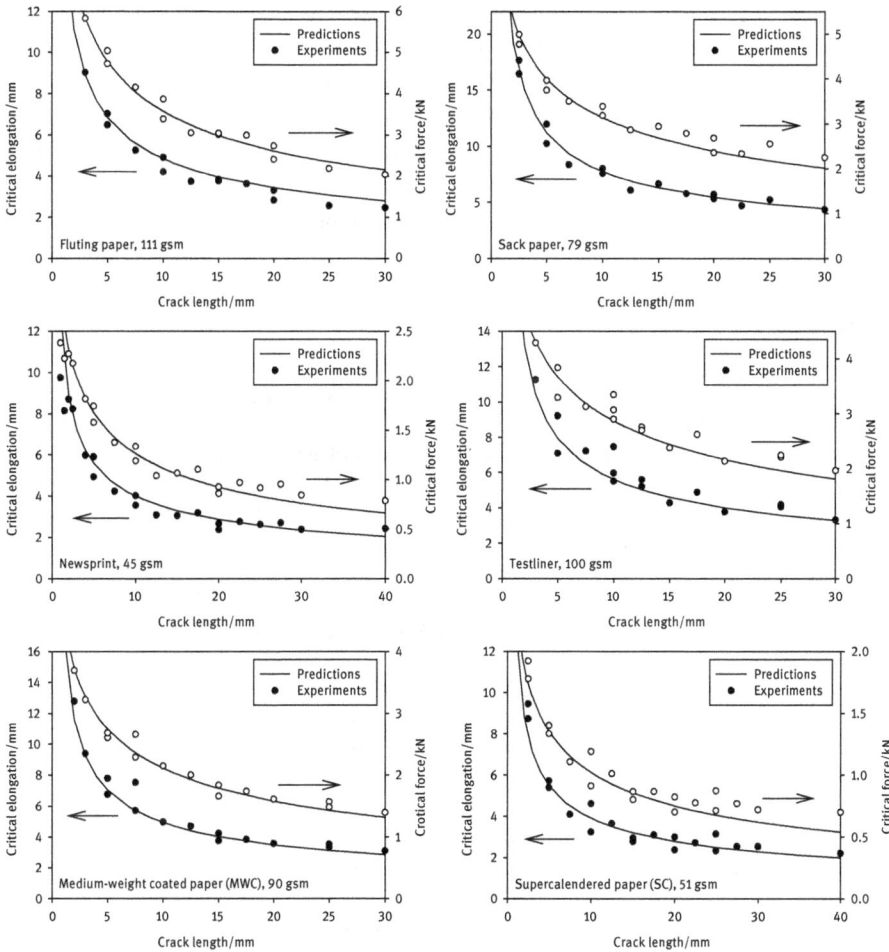

Fig. 5.24: Comparison of measured and calculated load (unfilled marks, upper curves) and elongation (filled marks, lower curves) at failure versus edge crack length in 1,800 mm long and 800–1,000 mm wide paper webs of fluting paper, sack paper, newsprint, linerboard, MWC paper, and SC paper. The predictions were performed using non-linear fracture mechanics based on isotropic deformation theory of plasticity (Mäkelä et al., 2009).

singular at the crack tip. The results in Fig. 5.24 demonstrate that, in this case, non-linear fracture mechanics based on the *J*-integral succesfully predicts the mode I in-plane failure of notched paper structures, indicating that a zero-sized FPZ constitutes a reasonable assumption. If the non-zero size of the FPZ cannot be ignored, then the cohesive zone model based on the cohesive stress-widening curve gives excellent predictions for mode I in-plane failure (see Fig. 5.19). Models based on the cohesive stress-widening curve are also applicable for out-of-plane failure of paper materials where the concept of crack tip singularity is not necessarily applicable.

Fracture mechanics modelling can thus be used to predict the failure point of notched structures as well as for damage tolerance analysis of structures that contain defects. The latter approach has not been fully exploited in the performance analysis of paper products, in contrast to its widespread use in other branches of engineering. Special complications with paper materials arise from the effects of moisture, creep and relaxation, loading dynamics, and the heterogeneity of paper properties. These aspects of paper materials are further explored in the next chapters.

Literature references

Anderson, T.L. (2005). Fracture Mechanics: Fundamentals and Applications, 3rd Boca Raton, FL, USA: CRC Press, Taylor & Francis Group.

Barenblatt, G.I. (1959). The formation of equilibrium cracks during brittle fracture. General ideas and hypotheses. Axially-symmetric cracks. J. Appl. Math. Mech. 23, 622–636.

Borgqvist, E., Wallin, M., Tryding, J., Ristinmaa, M. and Tudisco, E. (2016). Localized deformation in compression and folding of paperboard, Packag. Technol. Sci. 29, 397–414.

Coffin, D.W., Li, K. and Li, J. (2013). Utilization of modified linear elastic fracture mechanics to characterize the fracture resistance of paper. In: Advances in Pulp and Paper Research, Cambridge 2013. Proceedings of the 15th Fund. Res. Symp. Cambridge, 2013, I'Anson, S.J., (ed.), 637–672.

Dugdale, D.C. (1960). Yielding of steel sheets containing slits. J. Mech. Phys. Solids 8, 100–104.

Fellers, C. (1986). The significance of structure for the compression behavior of paperboard, Paper – Structures and Properties, Bristow, J.A. and Kolseth, P. eds. New York, NY, USA: Marcel Dekker, 281–310.

Girlanda, O., Hallbäck, N., Östlund, S. and Tryding, J. (2005). Defect sensitivity and strength of paperboard in out-of-plane tension and shear. J. Pulp Pap. Sci. 31, 100–104.

Hagman, A., Huang, H. and Nygårds, M. (2013). Investigation of shear induced failure during SCT loading of paperboards. Nord. Pulp Paper Res. J. 28, 415–429.

Hillerborg, A., Modéer, M. and Petersson, P.-E. (1976). Analysis of crack formation and crack growth in concrete by means of fracture mechanics and finite elements. Cement Concrete Res 6, 773–782.

Hutchinson, J.W. (1968). Singular behavior at the end of a tensile crack tip in a hardening material. J. Mech. Phys. Solids 16, 13–31.

Irwin, G.R. (1957). Analysis of stresses and strains near the end of crack traversing a plate. J. Appl. Mech. 24, 361–364.

Isaksson, P., Gradin, P.A. and Kulachenko, A. (2005). The onset and progression of damage in isotropic paper sheets. Int. J. Solids Struct. 43, 713–726.

ISO/TS 17958:2013 Paper and board – Determination of fracture toughness – Constant rate of elongation method (1.7 mm/s)

Kachanov, L.M. (1958). Time of the rupture process under creep conditions. Izv. Akad. Nauk. SSSR. Otd Tekhn Nauk (In Russian) 8, 26–31.

Mäkelä, P. (2002). On the fracture mechanics of paper. Nord. Pulp Paper Res. J 17, 254–274.

Mäkelä, P. (2012). Engineering fracture mechanics analysis of paper materials. Nord. Pulp Paper Res. J 27, 361–369.

Mäkelä, P. and Fellers, C. (2012). An analytic procedure for determination of fracture toughness of paper materials. Nord. Pulp Paper Res. J 27, 352–360.

Mäkelä, P., Nordhagen, H. and Gregersen, Ø.W. (2009). Validation of isotropic deformation theory of plasticity for fracture mechanics analysis of paper materials. Nord. Pulp Paper Res. J 24, 388–394.

Mäkelä, P. and Östlund, S. (2007). Cohesive crack modelling of paper materials. In: Proceedings of the 61st Appita Annual Conference and Exhibition, Gold Coast, Australia, pp. 357–364.

Mäkelä, P. and Östlund, S. (2012). Cohesive crack modelling of thin sheet material exhibiting anisotropy, plasticity and large-scale damage evolution. Eng. Fract. Mech. 79, 50–60.

Niskanen, K., Kettunen, H. and Yu, Y. (2001). Damage width: A measure of the size of fracture process zone, The Science of Papermaking, Trans. of the 12th Fund. Res. Symp., Oxford, September 2001 Baker, C.F. ed.UK: The Pulp and Paper Fundamental Research Society, 1467–1482.

Nordhagen, H. (2009). Development of fracture mechanics to study end use of paper webs, Doctoral thesis Trondheim, Norway: Norwegian University of Science and Technology.

Nygårds, M., Just, M. and Tryding, J. (2009). Experimental and numerical studies of creasing of paperboard. Int. J. Solids Struct. 46, 2493–2505.

Ottosen, N.S. and Ristinmaa, M. (2005). The Mechanics of Constitutive Modeling, Oxford, UK: Elsevier.

Rice, J.R. (1968). A path independent integral and the approximate analysis of strain concentration by notches and cracks. J. Appl. Mech. 35, 379–386.

Rice, J.R. and Rosengren, G.F. (1968). Plane strain deformation near a crack tip in a power-hardening material. J. Mech. Phys. Solids 16, 1–12.

Sih, G.C. and Liebowitz, H. (1968). Mathematical theories of brittle fracture, Fracture, Vol. 2, Liebowitz, H. ed. New York, NY, USA: Academic Press, 67–190.

Skjetne, B. (2006). Numerical studies of brittle-elastic fracture in random media, Doctoral thesis Trondheim, Norway: Norwegian University of Science and Technology.

Stenberg, N. (2002). Out-of-plane shear of paperboard under high compressive loads. J. Pulp Paper Sci. 17, 387–394.

Swinehart, D. and Broek, D. (1995). Tenacity and fracture toughness of paper and board. J. Pulp Paper Sci. 21, J389–J397.

Tryding, J. (1996). In-plane fracture of paper, Doctoral thesis. Division of Structural Mechanics Lund, Sweden: Lund University.

Tryding, J., Marin, G., Nygårds, M., Mäkelä, P. and Ferrari, G. (2017). Experimental and theoretical analysis of in-plane cohesive testing of paperboard. Int. J. Dam. Mech. 26, 895–918.

Uesaka, T., Okaniwa, H., Murakami, K. and Imamura, R. (1979). Tearing resistance of paper and its characterization. J. Japan Tappi 33, 403–409.

Wellmar, P., Gregersen, Ø.W. and Fellers, C. (2000). Predictions of crack growth initiation in paper structures using a J-integral criterion. Nord. Pulp Paper Res. J 15, 4–11.

Part II: **Dynamic stability**

Tetsu Uesaka

6 Web dynamics in paper transport systems

6.1 Introduction

Paper and board are manufactured and converted mostly in a web form. The smooth, trouble-free transport of the paper web is essential in papermaking, printing, and converting. Examples of web transport systems may be seen in paper machines, corrugators, coaters, printers, label machines, and others. For example, Fig. 6.1 shows a typical web transport system used for a web heatset offset printing press. Figure 6.2 shows an example of the system in the wet end of a paper machine. Another process is creping used in tissue production to impart softness to the web (by micro-folding and structure explosion) (Fig. 6.3).

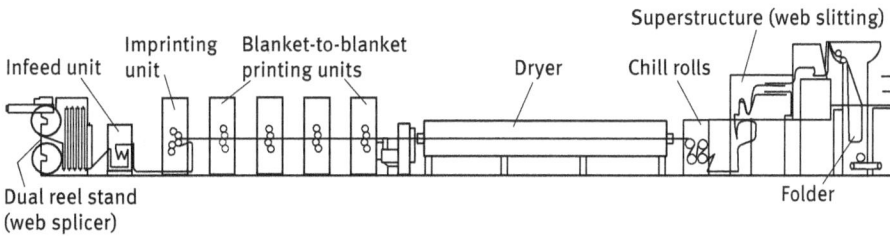

Fig. 6.1: Web transport system in commercial web heatset offset press (Source: © Helmut Kipphan, Handbook of Print Media, Springer, 2001). Reproduced with permission.

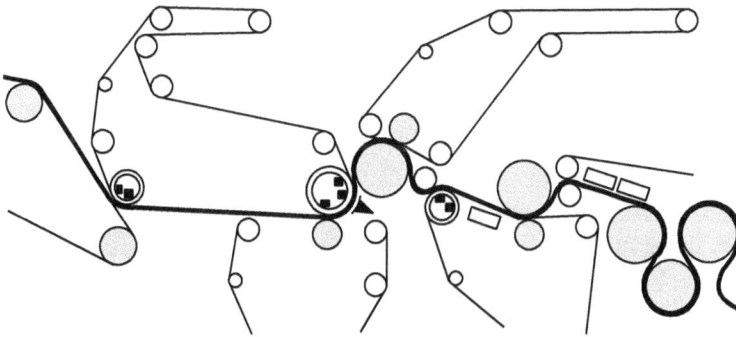

Fig. 6.2: Web transport system in the press section of paper machine.

Tetsu Uesaka, Mid Sweden University, Sundsvall, Sweden

https://doi.org/10.1515/9783110619386-006

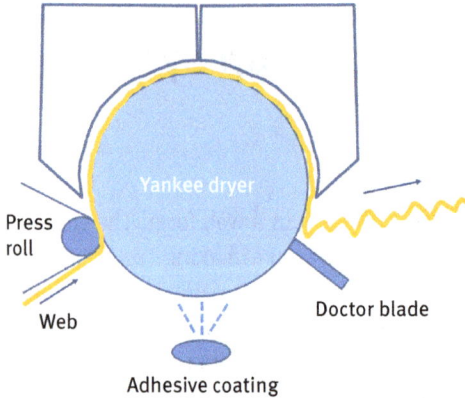

Fig. 6.3: Creping process in tissue manufacturing.

A web transport system typically consists of (1) the section called open draw, where the web is running without any supporting fabric, (2) the section called closed draw, where the web is supported by one fabric (or two fabrics) or metal surface, and (3) the nip or doctor blade where a specific force is applied to drive the web or to change its moving direction. The web is transported through a number of rolls, among which some rolls are placed to drive other rolls, and some are set simply to change the direction of the running web. The relative speed difference between the two sets of nipped rolls is called draw, and it is an important parameter to describe web transport problems, as we will discuss in more details later.

When considering practical problems in web transport systems, one may need to pay attention to the phenomena related to three length scales: (1) web behaviour between rolls (metre or sub-metre scale), (2) the phenomena happening within the nip (sub-centimetre scale), and (3) dynamic phenomena happening at the doctor blade in creping process (sub-millimetre scale). In this chapter, we focus on the first and the third length scale of the problem. Phenomena related to the nip are discussed separately in Chapter 10.

The primary motivation for investigating the dynamics of web transport system comes from practical issues, such as web breaks on paper machines and printing presses and web instabilities (e.g. wrinkles, leading to creases and breaks). The same issues also exist on board machines. The literature from early days has already shown that the way in which the web is run in the transport system has enormous impacts on web breaks and wrinkles (Grant, 1967; Larsson, 1984; Page and Bruce, 1985; Uesaka and Ferahi, 1999; Uesaka et al., 2001). In the field of printing, operators long recognized that there are difficult and easy feeds depending on the press units. It is also not overstating that the history of paper machine development is filled with a series of attempts to stabilize web transport systems (Paulapuro, 2000).

Typical problems encountered in practical situations are (1) tension variations due to draw changes; (2) large transverse vibrations of web due to aerodynamic effects and system resonances; (3) time-varying and/or non-uniform adhesion on press rolls and dryer rolls; (4) instabilities in peeling and creping, as associated with web property variations and hygroexpansion or shrinkage; and (5) web tension non-uniformity in the cross-machine direction (CD) (often called "baggy edges"). Although the entire area contains the whole web handling issues, we will specifically focus, in this chapter, on the web running in the open draw section of printing machines and the wet end of paper machines. In other words, we will leave the discussions on winding/reeling mechanics and paper roll deformation in the finishing section of paper machine operation to other appropriate textbooks. There are already excellent textbooks on winding mechanics and finishing (Roisum, 1994, 1998; Jokio, 1999; Good, 2007; Paulapuro, 2008).

6.2 Dynamics of web transport

6.2.1 Basic formulation of web transport problems

When analysing the dynamics of web transport, one needs the conservation equations for mass and momentum. These equations are later approximated to specific situations and solved either analytically or numerically. In this section, we introduce the basic formulation of the problems as a preparation for the subsequent analyses.

Figure 6.4 shows a portion of the web moving in a web transport system. For generality, we consider the web as a three-dimensional (3D) object and consider a small (infinitesimal) volume element inside the web.

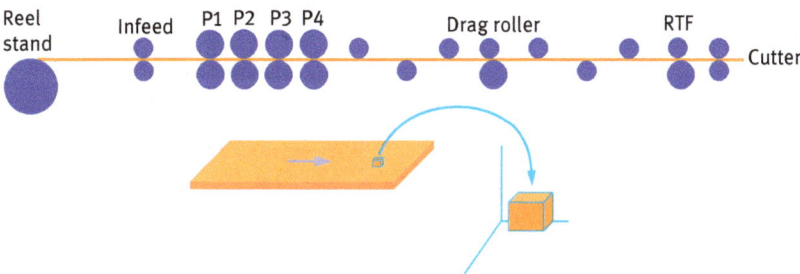

Fig. 6.4: Portion of the web moving through a web transport system (e.g. printing press).

The conservation law of the mass states that the rate of change in mass in this volume element is equal to the mass entering and leaving the volume element (Fung and Tong, 2001),

$$\frac{\partial \rho}{\partial t} + \frac{\partial \rho v_i}{\partial x_i} = 0, \tag{6.1}$$

where ρ is density and v_i is velocity in the x_i-direction. The first term represents the rate of density change within the volume element, and the second term represents the net mass flow leaving the volume element (per unit volume). It should be noted that both density ρ and velocity v_i are generally dependent on the position and time. For example, in web dynamics, density changes are mainly associated with the web strain, which is non-uniform along the web length (or width) and also varies with time, for example, a sudden tension surge.

The second important law is the conservation of momentum (or Newton's second law) (Fung and Tong, 2001),

$$\rho\left(\frac{\partial v_i}{\partial t} + v_j \frac{\partial v_i}{\partial x_j}\right) = \frac{\partial \sigma_{ij}}{\partial x_j} + B_i, \tag{6.2}$$

where σ_{ij} is a stress tensor component, and B_i is a body force component (e.g. gravity). The left-hand side represents the rate of momentum change for the moving volume element, and the right-hand side represents the total forces applied to the element (per unit volume), consisting of the surface forces on the element (stresses) and the forces applied on the element (body force).

Equations (6.1) and (6.2) are the general conservation laws and must be solved, together with boundary conditions and a stress–strain relationship, to obtain web motions in 3D space and stress/strain distributions in the web. The boundary conditions typically encountered in web transport systems are (1) the conditions applied at the contact points with rolls, such as velocity, contact forces, and web-roll adhesion forces, and (2) the conditions on the web surface, such as forces due to interactions with the surrounding air (drag force, air friction, and inertial force due to air motion). In the following sections, we specialize the earlier equations to extract some of the important characteristics of the web transport systems in a printing press and paper machine.

6.2.2 Axially moving web problem

The case of an axially moving web is shown in Fig. 6.5.

Web section 1

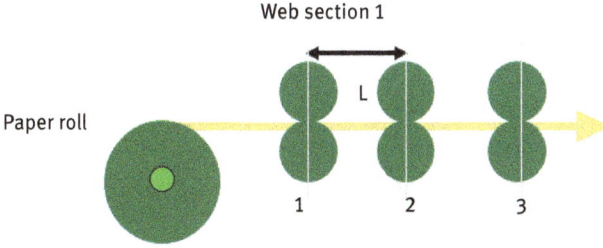

Paper roll

L

1   2   3

Fig. 6.5: Web section defined for an axially moving web, where L is the length of the span.

In this simplified case, all the field variables are a function of only one coordinate (say, x-direction taken in the machine direction (MD) of the web). We also assume that the stress–strain relationship is linear,

$$\left.\begin{array}{l} \mathbf{v} = (v_{MD}, 0, 0) \\ \sigma_x = C_x(\varepsilon_x - \varepsilon_{x0}) \end{array}\right\},\tag{6.3}$$

where v_{MD} is the speed and ε_x is the strain in the x-direction, ε_{x0} is the hygrostrain or residual strain, and C_x is the corresponding elastic constant. (Note that, when considering an isolated section, such as shown in Fig. 6.5, the strain ε_x is actually the strain increment created in this section and the stress σ_x is the stress increment.) By considering the case where there is no hygro- or residual strain, the previous mass conservation and momentum conservation eqs. (6.1) and (6.2) can be written as (Hristopulos and Uesaka, 2002)

$$\frac{\partial\varepsilon}{\partial t} - \frac{\partial((1-\varepsilon)v)}{\partial x} = 0\tag{6.4}$$

and

$$\frac{\partial((1-\varepsilon)v)}{\partial t} + \frac{\partial((1-\varepsilon)v^2)}{\partial x} = c^2\frac{\partial\varepsilon}{\partial x},\tag{6.5}$$

where $v = v_{MD}$, $\varepsilon = \varepsilon_x$, $c^2 = C_x/\rho$, with c representing the speed of the longitudinal wave that propagates in the MD. The subscripts for strain and velocity are omitted for convenience.

We first consider the case of a steady state where the field variables ε and v are independent of time. Equations (6.4) and (6.5) then give

$$(c^2 - v^2)\frac{\partial\varepsilon}{\partial x} = 0,\tag{6.6}$$

where $c_0 = v(1 - \varepsilon)$ is constant (from eq. (6.4)), and

$$v^2 \ll c^2 = \frac{C_x}{\rho}. \tag{6.7}$$

Therefore, the only admissible solution for eq. (6.6) is that strain and speed are constants over the open draw section and have jumps at the nip, such as shown in Fig. 6.6.

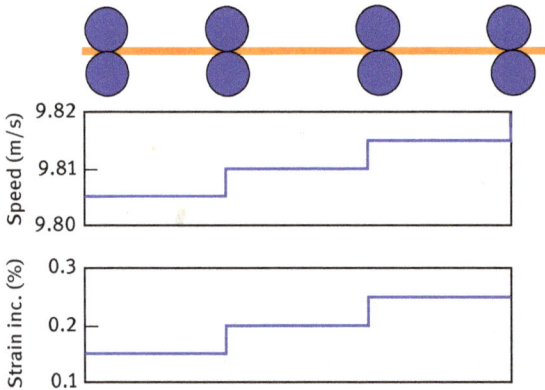

Fig. 6.6: Speed and strain changes in a steady-state web transport system, from Hristopulos and Uesaka (2002). Reproduced with permission from Pulp and Paper Technical Association of Canada (PAPTAC).

Figure 6.6 shows that speed changes stepwise at each nip and strain increment also changes stepwise, and the total strain at a certain section is the sum of all the strain increments created up to that section. Another important implication of the earlier result is that the paper is subjected to a very high strain rate at the nip, which is not seen in any of the standard strength tests. For example, suppose that the printing press is running at 10 m/s, the speed difference between downstream and upstream rolls is 0.02 m/s in the section and the nip width (or length) is 1 cm. Then, the strain rate that the paper experiences in the nip is 200%/s, which is 1,200 times higher than the strain rate used in the standard tensile test. Under this circumstance, the relevant paper property is the response to a stepwise strain increment and subsequent stress relaxation. (Stress relaxation is discussed in Chapter 7.)

In the steady state, there is an important parameter for describing the running condition of web, called *draw*. The draw, Δ_{12}, is defined as the relative speed difference between the rolls located at point 1 and point 2 (Fig. 6.5), and in the steady state, it is equal to the strain increment produced at the nip

$$\overline{\Delta_{12}} \equiv \frac{\overline{V_2} - \overline{V_1}}{\overline{V_1}} \simeq \overline{\varepsilon_2} - \overline{\varepsilon_1}, \tag{6.8}$$

where the bars over the variables are meant for those in the steady state. It is important to note that the draw is equal to the strain increment *only* when the speed is time independent (i.e. steady state). In a non-steady state, the draw has no such unique relationship with strain increment, as will be shown later.

The draw is normally determined from the rotational speeds of the two rolls. However, it can be also measured directly from the web speeds at two points by using double-beam laser Doppler sensors (Fig. 6.7). A typical draw variation measured in a US pressroom is shown in Fig. 6.8 (Deng et al., 2008).

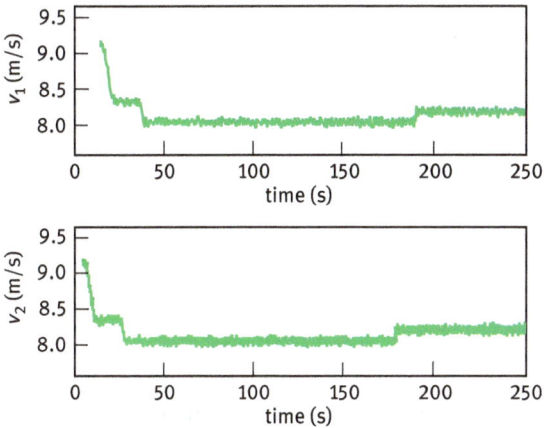

Fig. 6.7: Speeds measured at two points located at two neighbouring nip regions in a printing press using double-beam laser Doppler velocimetry, from Hristopulos and Uesaka (2002). Reproduced with permission from Pulp and Paper Technical Association of Canada (PAPTAC).

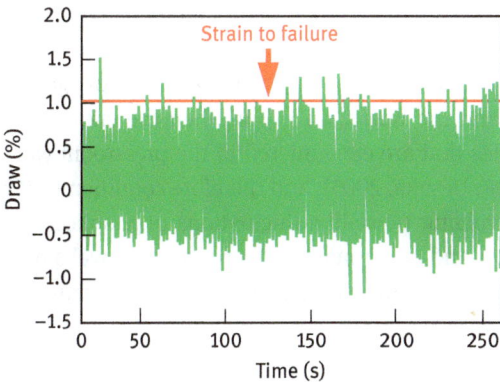

Fig. 6.8: Draw variation calculated from Fig. 6.6, from Hristopulos and Uesaka (2002). The red line indicates typical level of strain to failure for newsprint. Reproduced with permission from Pulp and Paper Technical Association of Canada (PAPTAC).

Figure 6.8 shows a very large variation in the draw during the normal operation of a printing press. The peak values sometimes exceeded the typical values of stretch (strain to failure) of the paper web. The question is whether this draw variation can be translated to actual strain (or tension) variations, and if so, why does not paper break all the time? In order to answer this question, we need to discuss a general non-steady case where velocities and draw are fluctuating.

In a general non-steady state, we need to go back to eqs. (6.4) and (6.5). By integrating eq. (6.4) on both sides in the web section, $0 < x < L$, and using the mean value theorem, we have

$$L\frac{\partial\varepsilon(\xi,t)}{\partial t} = v_2(t) - v_1(t) + \varepsilon_1(t)v_1(t) - \varepsilon_2(t)v_2(t), \quad 0 < \xi \leq L. \tag{6.9}$$

In order to specialize the earlier equation to the typical web handling conditions in printing houses, we consider low-frequency fluctuations. In this case, the wavelength of a fluctuation, say $\lambda = c/f$, where f is the fluctuation frequency, is assumed to be much larger than the span length L. Equivalently, the frequency of the fluctuation is much lower than the critical frequency, f_c, of the web section,

$$f \ll \frac{c}{L} = f_c. \tag{6.10}$$

Under this condition, the fluctuation is approximately constant over the web section with $\varepsilon(\xi,t) = \varepsilon(t)$, so that eq. (6.9) can be rewritten as (Hristopulos and Uesaka, 2002),

$$L\frac{d\varepsilon(t)}{dt} = v_2(t) - v_1(t) + \varepsilon_1(t)v_1(t) - \varepsilon_2(t)v_2(t). \tag{6.11}$$

This equation is found in the standard model of web dynamics (Reid and Shin, 1991), and also the solutions of the earlier equation were shown, numerically, to be in a close agreement with those obtained from a system of partial differential equations in the theory of elastic string (Brown, 1999). The important point is that this equation is valid under condition (6.10). For example, suppose the longitudinal wave-propagation speed of a printing paper is 3 km/s, and the web span is 1 m, then the critical frequency is 3 kHz. The typical frequencies that are encountered in the pressroom were much less than 60 Hz (Deng et al., 2008; Uesaka, 2005), and therefore condition (6.10) is well met. However, phenomena involving pulse-like fluctuations (e.g. splice and wrinkle), higher speed operation (coater and paper machine), or wet paper webs may not be captured by this equation, instead it requires a full solution of the original partial differential equations, (6.4) and (6.5), as will be described later.

Although eq. (6.11) is non-linear, we can linearize and solve it by considering weak fluctuations around the steady state (i.e. the fluctuation $\delta v \ll v$)

$$\Delta\varepsilon = \Delta + \frac{1}{L}\int_0^t \delta v(s)\exp\{-f_k(t-s)\}ds, \tag{6.12}$$

where $\Delta\varepsilon$ is strain increment, Δ is the draw for the section, δv is the fluctuation of the speed applied at the second point, and f_k is the characteristic frequency defined by

$$f_k = \frac{\overline{v_2}}{L}, \tag{6.13}$$

where $\overline{v_2}$ is the average speed at the second point as defined in Fig. 6.5. In Fig. 6.9, the frequency responses of strain are plotted together with the draw applied. For the low-frequency fluctuation (1 Hz: upper graph), the strain variation almost follows the draw variation, whereas for the higher frequency fluctuation (20 Hz: lower graph), the strain increment variation is greatly attenuated with a significant phase delay. In this specific example, the characteristic frequency of this web section is 10 Hz. This means that any disturbances of the speed with the frequency of more than 10 Hz are severely attenuated, as also expected in eq. (6.12). In other words, the web section behaves like a low-pass filter of the disturbances.

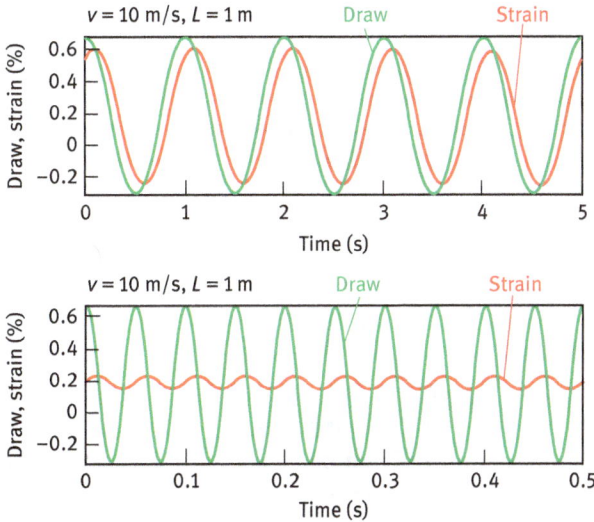

Fig. 6.9: Calculated strain response for two frequencies of draw oscillations, $f=1$ Hz (upper graph) and $f=20$ Hz (lower graph), from Hristopulos and Uesaka (2002). Reproduced with permission from Pulp and Paper Technical Association of Canada (PAPTAC).

Figure 6.10 shows the spectrum density distribution corresponding to Fig. 6.8. Except for white noises and the 8.2 Hz peak, most of the variations were in a much higher frequency range than the characteristic frequency for this section (≈10 Hz), and therefore they do not contribute to the strain variation. However, with increasing machine speed and shortening span section, the characteristic frequency could become higher so that the disturbances of higher frequencies would start contributing to strain variations and thus to tension variations. Therefore, it is important to eliminate any sources of vibrations of the rotational components in web transport systems. It was shown that such draw variations indeed contribute to web breaks in printing houses (Deng et al., 2008).

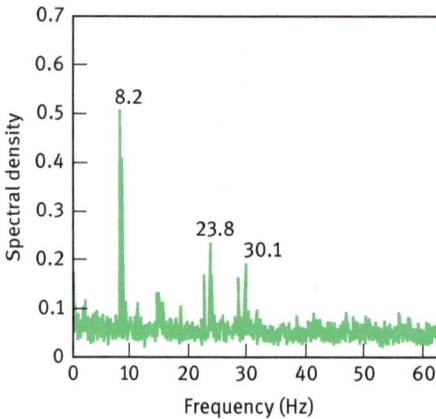

Fig. 6.10: Spectral density distribution determined from Fig. 6.8 (Hristopulos and Uesaka, 2002). Reproduced with permission from Pulp and Paper Technical Association of Canada (PAPTAC).

6.2.3 Moving thread problems

The extension of the previous *axially moving web* model can be done by adding another degree of freedom of web motion in the *transverse* direction (perpendicular to the web surface). In other words, the web moves in the x–y plane but still maintaining the one-dimensional (1D) geometry like a thread (Fig. 6.11). Although the extension seems straightforward, the conservation, eqs. (6.1) and (6.2), now becomes a set of fully coupled, non-linear differential equations that are not easy to solve analytically and even numerically. An excellent review of the 1D thread models has been given by Ahrens et al. (2004). Therefore, many researchers are forced to make assumptions on web displacements (e.g. an arc shape) and on the continuity equation (strain variations along the web length) (Kurki et al., 1995, 1997; Pakarinen et al., 1993). Another difficulty is that the boundary conditions, such as web adhesion force and contact forces on the rolls, are not known a priori, but are

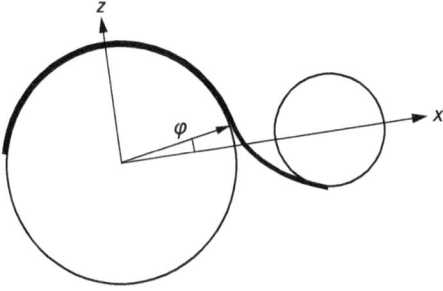

Fig. 6.11: Open draw section with web adhesion as an example of the 1D thread problem, from Edvardsson and Uesaka (2010). The web is moving from left to right, and ϕ indicates the take-off angle that represents the peeling position. Reproduced with permission from American Society of Mechanical Engineers (ASME).

actually part of the solutions. Furthermore, when considering wet web behaviour on paper machine, the stress–strain relations must represent viscoelastic–viscoplastic nature of wet paper. The time-dependent nature of wet web has been investigated intensively in the past, and found in the literature (Hauptmann and Cutshall, 1977; Kouko et al., 2006, 2007; Kouko and Sorvari, 2008; Kurki et al., 2004), as well as in Chapter 9. Nevertheless, these efforts have accumulated important insights on web dynamics of an open draw in the wet end of paper machine.

Published studies have accumulated important insights on the web dynamics in the open draw sections in the wet end of paper machine. For example, in a steady state, Pakarinen et al. (1993) estimated the contributions of different force components to total force applied on the web. The forces considered are the pressure difference between the top and bottom surfaces of the web, the gravity, the centrifugal force, and the air friction. They showed that at a 900 m/min speed, the pressure difference gave the most contribution (68%) among all forces, whereas at a 1,500 m/min, the centrifugal force became dominant (84%). In a non-steady state, tension spikes were predicted when disturbances were created either by the external pressure (Kurki et al., 1995; Pakarinen et al., 1993) or adhesion strength variations (Kurki et al., 1997).

Recently, a new numerical approach was proposed in order to overcome the earlier difficulties in solving the moving thread problem with variable boundary conditions (Edvardsson and Uesaka, 2009, 2010a). The idea is to represent the continuous thread in terms of a series of interconnected particles, such as shown in Fig. 6.12. For each particle, i, the equation of motion,

$$\frac{dp_i}{dt} = \sum_j \mathbf{F}_{ij}^{\text{int}} + \sum_j \mathbf{F}_{ij}^{\text{ext}}, \tag{6.14}$$

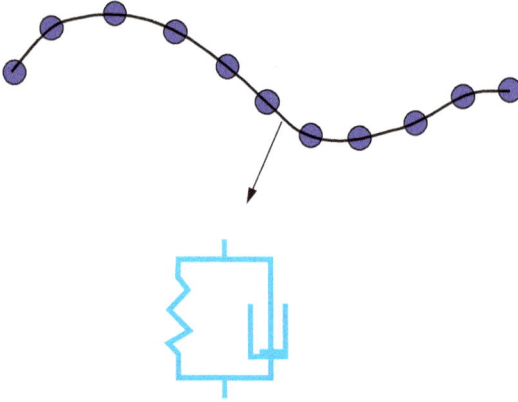

Fig. 6.12: Particle approach for the moving thread model. The thread is represented by a series of interconnected particles with interaction forces from elastic and dashpot elements.

is applied, where $\mathbf{F}_{ij}$ are the forces applied to the particle i both internally (from interconnected particles, the first term) and externally (from the pressure of the surrounding air, adhesion forces, friction, etc., the second term), p_i is the momentum of particle i, given by

$$p_i = m_i \frac{dx_i}{dt},\qquad(6.15)$$

and m_i is the mass of the ith particle, and x_i is the position of the particle. The interaction forces (the first term of the right-hand side of eq. (6.14)) can represent various stress–strain relationships (elastic, viscoelastic, and viscoplastic) both in stretching and bending modes of the thread deformation (Edvardsson and Uesaka, 2010b). This discrete representation with the particle dynamics converges well to a continuum solution of the partial differential equations as the number of particles is increased. Figure 6.13 shows such a comparison of the solutions for a web strain

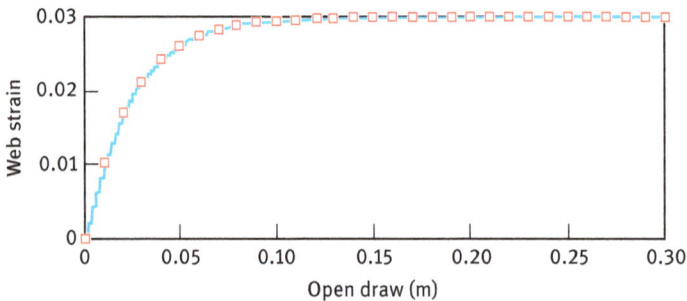

Fig. 6.13: Web strain evolution along the length of open draw section, from Edvardsson and Uesaka (2010). Reproduced with permission from American Society of Mechanical Engineers (ASME).

variation along the length of the open draw between the particle approach and continuum approach.

It is interesting to note that when a viscous term is included (Fig. 6.12), the web strain does not show the stepwise change, as shown in the elastic case (Fig. 6.6), but displays some delay depending on the assumed value of viscosity (Kurki et al., 1997). This is an important difference between the dry process (printing) and the papermaking process in the wet end.

Figure 6.14 shows the effects of machine speed on the take-off position of the web (see also Fig. 6.11 for the definition of the take-off angle ϕ). During the normal operation below a speed of 1,700 m/min, the take-off position (peeling point of the web) is constant over time. However, as the machine speed increases, this take-off point goes down, that is, the web sticks on the roll surface longer before it detaches. Above a certain speed (e.g. 1,700 m/min in this case), this open draw system loses a steady-state solution, and the take-off position continues to fall at this critical speed.

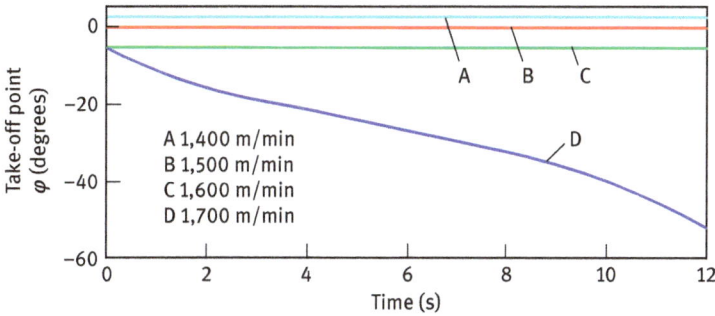

Fig. 6.14: Effect of web speed on the time evolution of the take-off position, from Edvardsson and Uesaka (2009). Reproduced with permission from the Pulp and Paper Fundamental Research Society (www.ppfrs.org).

Figure 6.15 is the corresponding web strain response. Below the critical speed, the web strain is exactly equal to the draw applied (3%), whereas over this critical speed, the web strain continues to grow without any sign of the saturation. The consequence is, obviously, a break.

The existence of such critical speed has been speculated for many years among papermakers. It is always implied in paper machine design, but this analysis showed what factors control the critical speed. It was shown that at a given draw, the tensile stiffness of the web is most influential to the critical speed among the various factors investigated (Edvardsson and Uesaka, 2009). As stiffness decreases, the critical speed decreases. This implies that in order to speed up the paper machine with an open draw in the wet end, it is important to increase the stiffness of the wet web, in other words, one needs to increase the dryness of the web.

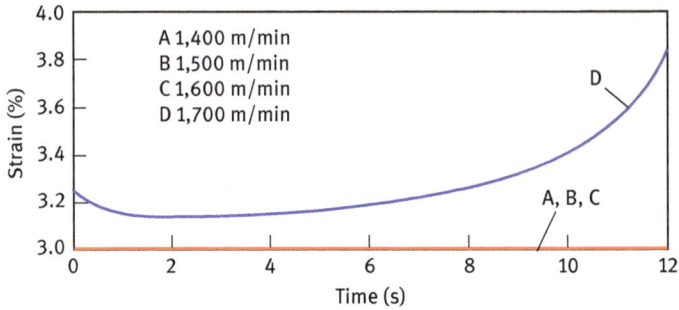

Fig. 6.15: Effect of web speed on the time evolution of the maximum strain in an open draw section, from Edvardsson and Uesaka (2009). The initial fall in the curve D is due to a numerical artefact. Reproduced with permission from the Pulp and Paper Fundamental Research Society (www.ppfrs.org).

Figure 6.16 shows the effect of the stepwise change in stiffness at a 1,400 m/min machine speed. The stiffness was decreased by 15% for the duration of 5 s. Such disturbance created strain pulses at the instants when the changes happened (Fig. 6.16). However, at an increased speed, for example, 1,600 m/min as shown in Fig. 6.17, the strain behaved in a much more complex way. For the shorter duration of the disturbance (<5 s), the web strain eventually returned to its original value, 3%

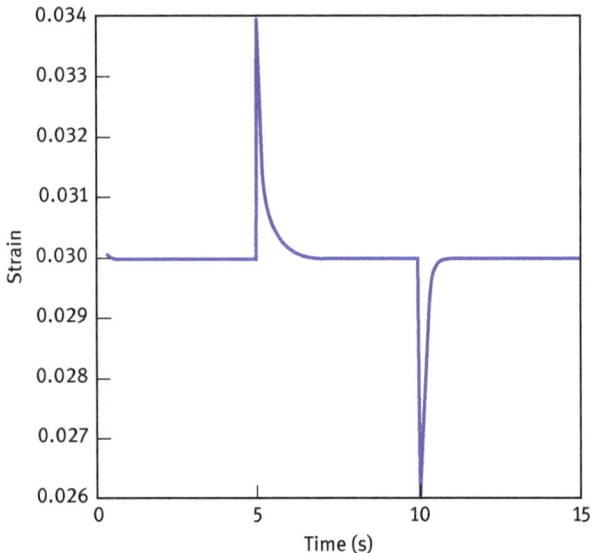

Fig. 6.16: Response of web strain to a stepwise reduction by 15% in web stiffness (at $t = 5$ s) and a stepwise return to the original level (at $t = 10$ s), from Edvardsson and Uesaka (2010). Web speed is 1,400 m/min. Reproduced with permission from American Society of Mechanical Engineers (ASME).

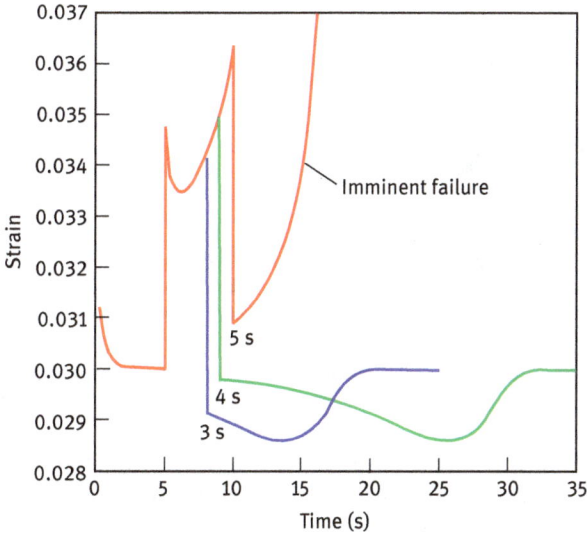

Fig. 6.17: Response of web strain to a stepwise reduction by 15% in web stiffness, lasting from 3 to 5 s, from Edvardsson and Uesaka (2010). Web speed is 1,600 m/min. Reproduced with permission from American Society of Mechanical Engineers (ASME).

(i.e. the draw applied). However, with the prolonged duration (5 s) of the disturbance, the web strain no longer returned to the steady-state value, but instead, it continued to grow.

Practically, wet-end conditions, such as wetness of the sheet coming out of the forming section, roll-paper adhesion, fibre orientation, and basis weight are always fluctuating. Therefore, it is conceivable that such fluctuations could also affect machine stability through the web stiffness, and thus affecting web breaks (Miyanishi and Shimada, 1998).

6.2.4 Creping in tissue manufacturing

Creping is a unique process for tissue manufacturing to impart desired softness to the product (Fig. 6.3). The wet web is pressed onto a large diameter dryer (Yankee dryer) by the press roll and is subjected to drying. The dryer is coated with an adhesive solution so that the wet web is glued firmly until it is scraped by the doctor blade. This scraping action by the doctor blade causes two important effects on the web: One is a zigzag, folded structure in the MD of the sheet, and the other is an "explosion" of the internal structure, that is, the breakage and delamination of fibre–fibre bonds (Fig. 6.18). The shape of the crepe structure, in terms of its amplitude and wavelength, is influenced by a large number of variables, such as basis

(a)

(b)

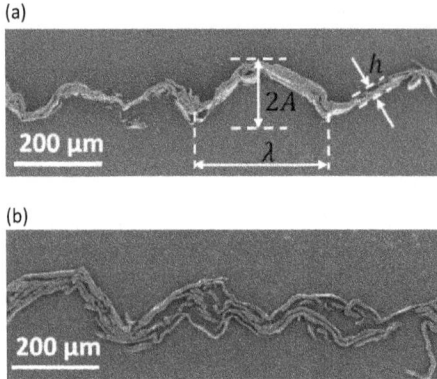

Fig. 6.18: Two major effects of creping on tissue structures. (a) Buckling of the sheet structure, and (b) Explosion of the sheet structure (debonding of fibres) (Pan, 2019). Reproduced with permission from the author.

weight, moisture content, adhesive strength, creping angle, and creping ratio. The basic relations between these variables have been established mainly through experimental work and mill trials (Hollmark, 1983).

The process is also known to have a rather small operating window (Hollmark, 1983) and is one of the sources of web instability and runnability issues. Understanding macro- and micro-dynamics of creping process is, therefore, expedient to further stabilize the process and product qualities.

Experimental creping rigs have been recently developed (Boudreau and Barbier, 2014; Hämäläinen et al., 2016; Ho et al., 2007) and were shown to be very useful to investigate impacts of the process variables. Combined with the visualization technique of the dynamic process and subsequent tensile testing (Das, 2019), one can establish the relationship between creping variables and tissue qualities.

Modelling of creping process has also been attempted (Gupta, 2013; Hossain et al., 2017; Pan, 2019, 2019; Pan et al., 2018; Ramasubramanian et al., 2011). The creping process consists of (1) the impact of the doctor blade to the web adhered to the dryer, (2) the delamination of the web from the dryer surface, (3) buckling of the web, and (4) post-buckling (Fig. 6.19) (Pan, 2019), together with the associated changes in the sheet structure. The most comprehensive model of creping at this stage is the particle-based models (Hossain et al., 2017; Pan, 2019). Similar to the model for the open draw dynamics (Edvardsson and Uesaka, 2010b), Pan and co-workers have successfully modelled the web as a 1D moving thread which consists of a number of particles (Pan et al., 2018). By incorporating a cohesive zone model into the web-adhesion layer, and assuming an elastic-plastic constitutive relation of the web (Pan, 2019), they have been able to predict typical effects of creping process variables. Among the variables investigated, the web thickness (i.e. basis weight) and creping angle showed the largest impacts on the amplitude and wavelength of the produced crepe. The creping ratio (i.e. the draw in the crepe section) and the adhesive strength also affect the crepe geometry, but the effects were still smaller than those of the former variables.

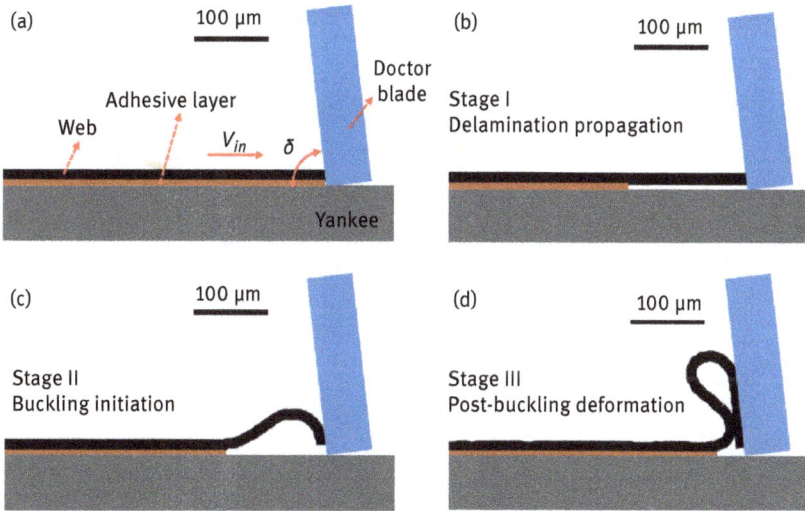

Fig. 6.19: The evolution of the web morphology during creping (the thickness of the web and adhesive layer are not scaled for visualization purpose) (Pan, 2019). Reproduced with permission from the author.

The creping dynamics is also sensitive to variabilities of the incoming web. For example, when the web has a periodic deviation from the flat shape, as typically seen as a fabric pattern, the crepe shape is distorted and shows a chaotic feature (Fig. 6.20). Basis weight variations were also shown to have significant impacts on the crepe formation (Hollmark, 1983; Pan, 2019). Such sensitivity to the variability of the incoming web properties is similar to the phenomena seen in the open draw section of a paper machine.

Modelling the web as a 3D fibre network (Hossain et al., 2017) provides the most realistic picture of the creping process. For example, breaks of fibre–fibre bonds, that is, the "explosion" of the sheet structure by creping and dusting, associated with the destruction of the structure, can be easily seen and evaluated. Although the computational cost is still high, it is a promising method to answer many questions of impacts of fibre properties and network structures on tissue qualities.

6.2.5 Fluttering of a two-dimensional web

One of the important web dynamics problems is vibrations, or what papermakers often call flutter. (In mechanics, the word "flutter" is reserved for indicating a special aeroelastic instability, but in this chapter, we follow the casual usage in papermaking.) As the machine speed increases and runnability requirements are becoming more stringent, fluttering, particularly in a dryer section, has been drawing attention as a

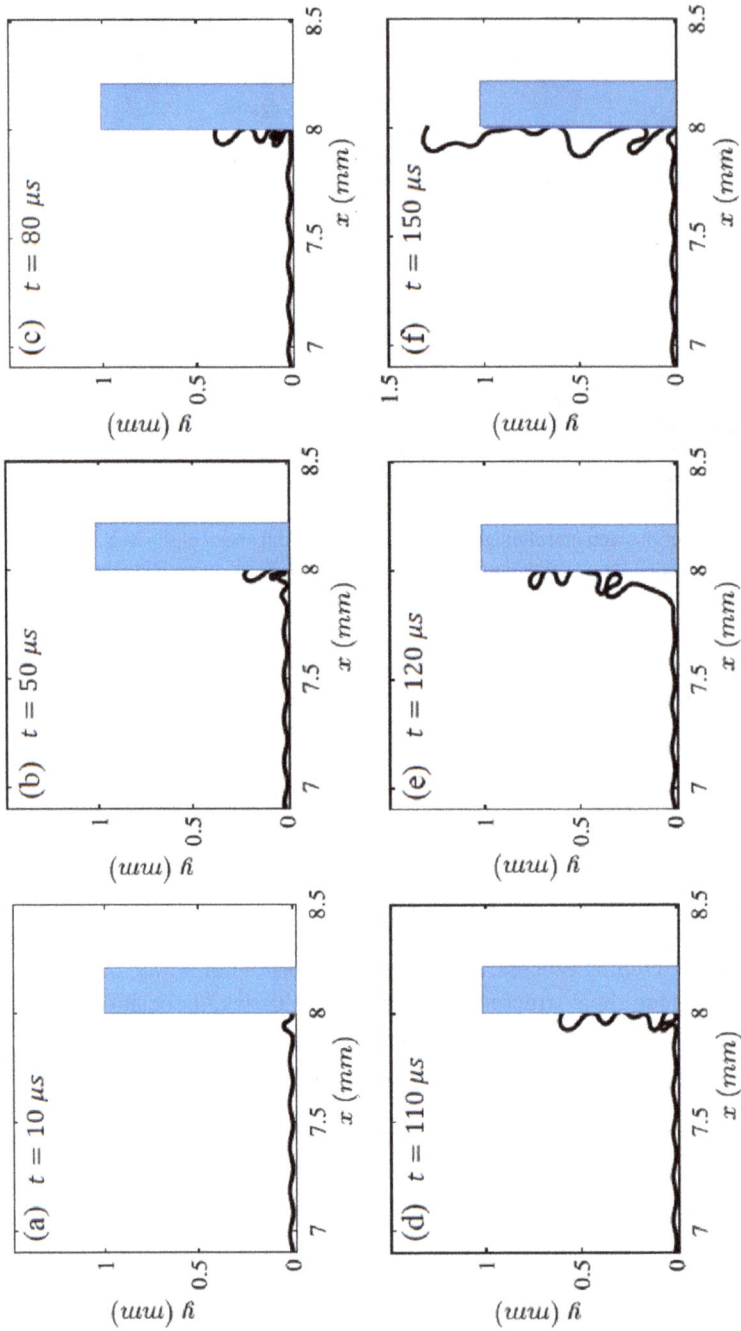

Fig. 6.20: The evolution of the web morphology with an initial geometrical imperfection (e.g. forming fabric pattern). Reproduced with permission from K. Pan.

potential culprit of web breaks. Fluttering seen in an open draw section of the double-felted dryer is a notorious example. Another example is the instability of the paper web in an air pocket in the single-felted dryer. Complex air flows inside the air pocket produce pressure at the pocket wedges that detaches the paper web from the felt, causing instabilities (Pakarinen et al., 1995).

The moving thread models (1D), as discussed in the previous section, have been used to analyse sheet flutter problems both with and without the presence of the air (Mujumdar and Douglas, 1976; Pramila, 1986). It was shown that the air surrounding the paper web has an enormous effect on the vibration characteristics. The natural frequencies predicted with air were only 15–30% of those without air. This effect of air was included by adding the mass per unit length of the surrounding air into the mass of the sheet in the equation of motion (Pramila, 1986). Another important effect which has been identified in the field is web tension non-uniformity in the CD of the paper web (Linna and Lindqvist, 1987; Linna et al., 1991, 2002; Parola and Beletski, 1999; Parola and Linna, 2000; Parola et al., 2000). In this case, the extension of the analysis to 2D is essential.

Kulachenko and co-workers proposed a non-linear finite element model for a running web that is coupled with the surrounding air through interacting air elements (Kulachenko et al., 2006). The paper web is represented by shell elements and the transport velocity is kept constant. The web was assumed to have non-uniform tension profile $T(x_2)$ in the CD (Fig. 6.21).

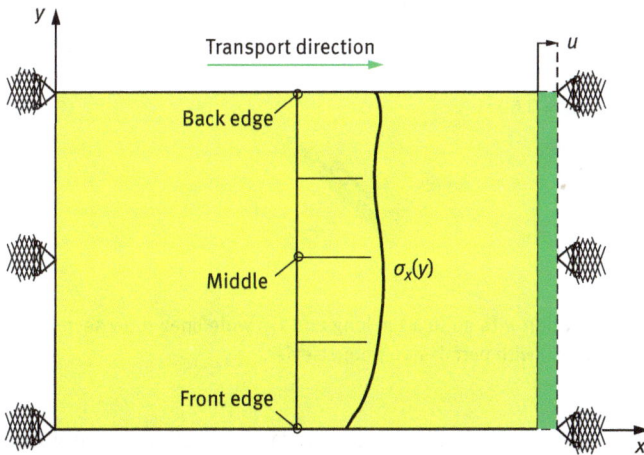

Fig. 6.21: Boundary conditions used in the study of web vibrations in the presence of web tension non-uniformity. Reprinted from Kulachenko et al. (2007) with permission from Elsevier.

Figures 6.22 and 6.23 show the predictions of natural frequencies without air (Fig. 6.22) and with air (Fig. 6.23). The presence of the air clearly affects the natural frequencies,

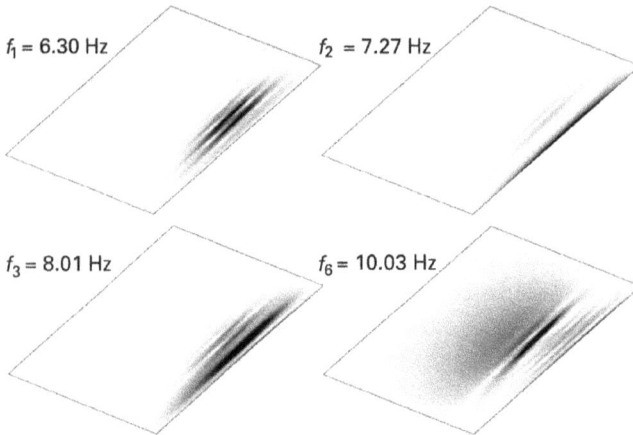

Fig. 6.22: Natural frequencies of vibration without air in a 1 m long and 1 m wide open draw section. Reprinted from Kulachenko et al. (2007) with permission from Elsevier.

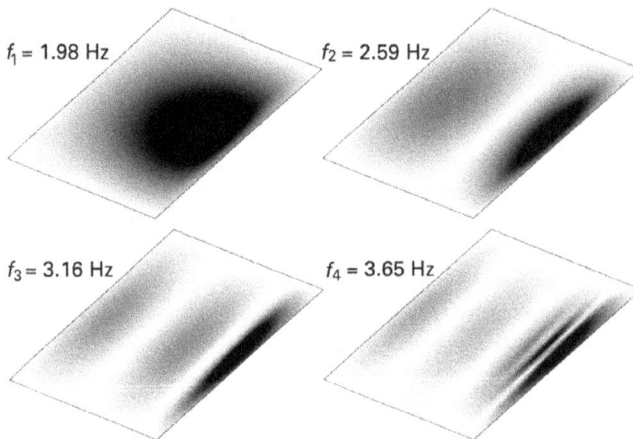

Fig. 6.23: Natural frequencies of vibration with air in a 1 m long and 1 m wide open draw section. Reprinted from Kulachenko et al. (2007) with permission from Elsevier.

depressing the frequencies drastically, as expected from the earlier discussions. In both figures, the notorious "edge flutters", as associated with the web tension non-uniformity, are observed. With the presence of air, the vibration patterns are much more complex than those without the air. In any case, the web tension non-uniformity has a dramatic effect on the edge flutters (Kulachenko et al., 2006). It was also shown that the web speed (i.e. paper machine speed) virtually had no effect on the natural frequencies.

It is a common practice that, when excessive edge flutters take place, an operator increases the web tension (through draw) to reduce the flutter and thus to try to prevent possible breaks and wrinkles. Figure 6.24 shows the effect of the tension increase on the stress variations at the front edge of the web. (Note that the stress is expressed as equivalent stress or von Mises stress, which is dominated by the tensile stress in this case.) Increasing tension indeed made the edge flutters subside, but, at the same time, the stress elevated towards more than the peak stress levels when the edge flutters occurred. Therefore, increasing tension (or draw) may not yield any better condition in terms of the stress level, except for reducing the probability of web wrinkling.

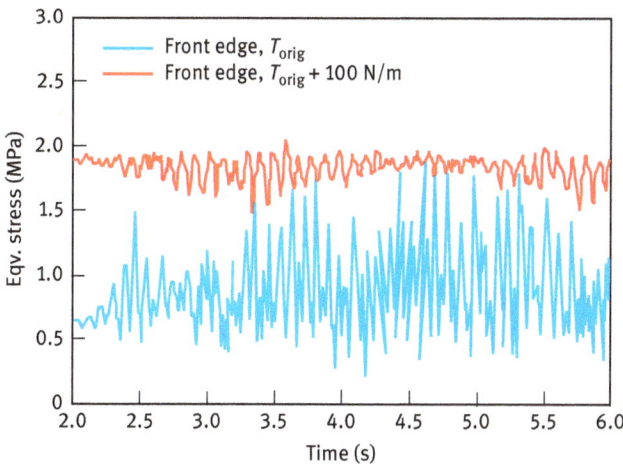

Fig. 6.24: Effect of increasing web tension on the amplitude of stress variations at one edge of a paper web. Reprinted from Kulachenko et al. (2007) with permission from Elsevier.

6.3 Concluding remarks

In this chapter, we discussed the basic characteristics of typical paper web transport systems. In a steady state, the system is simply characterized by its transport speed and draw. The draw determines strain increment (or maximum strain increment in the case of wet web), and thus the tension of the web. Such tension is balanced with various forces applied to the web, including inertial force, which is proportional to the square of the speed and adhesion force. In a non-steady state, the draw is no longer equal to the strain increment, depending on the frequency of variations. Many of the open draw sections tend to behave like a low-pass filter so that high-frequency variations are significantly attenuated. The critical frequency at which such variations are filtered depends on the length of the open span and the transport speed: the higher the speed and the shorter the span, the more the effect of higher frequency components.

The dynamic responses in typical paper web transport systems are often dominated by relatively low-frequency components. However, at nips, the paper is stretched at a very high speed which is not seen in standard laboratory tests. This raises the question of whether the current way of determining the strength properties of paper is appropriate. Another important aspect is non-uniformity of the properties (web tension, moisture, drying histories, etc.) in the CD (see also Chapter 2) and its impact on web stability. Anecdotes in the field seem to be abound, but serious analyses have only just begun. The issue of water application has not been discussed in this chapter. The state of water applied in a web transport system is complex: water can be absorbed into sheet structure, and, at the same time, a considerable amount of water can evaporate as the web moves, the latter of which is a strong function of the web transport speed (Gomér and Lindholm, 1991; Hansen, 1997). It is also well known that water applied through a fountain solution and ink in offset printing induces property changes and dimensional changes of the paper web, and therefore causes an unfavourable phenomenon in printing, called misregistration. At this stage, it is not clear whether this is a dynamic problem or the problem that can be treated in a static way. Interested readers can find a good review of the subject in the literature (Lif et al., 2005), and also consult Chapters 7 and 9.

From the point of analysing runnability performance, more and more powerful tools, including both numerical and experimental methods, are now available. Particularly, analysing across-the-machine non-uniformity problems is no longer a far-reaching goal in the web dynamics research.

Literature references

Ahrens, F., Patterson, T. and Bloom, F. (2004). Mathematical modelling of web separation and dynamics of a web adhesion and drying simulator, Int. J. Appl. Mech. Eng. 9(2), 227–271.

Boudreau, J. and Barbier, C. (2014). Laboratory creping equipment, J. Adhes. Sci. Technol. Taylor & Francis 28(6), 561–572.

Brown, J.L. (1999). Propagation of longitudinal tension in a slender moving web, Proc. Fifth Int. Conf. Web Handl., Oklahoma State University: Stillwater, Oklahoma, USA.

Das, R. (2019). Experimental studies on creping and its influence on mechanical properties of tissue paper products, The University of British Columbia, Vancouver, BC, Canada.

Deng, X., Parent, F., Manfred, T., Aspler, J.S. and Hamel, J. (2008). Pressroom draw variations and its impacts on web breaks, Part 1. Newspaper press trials, Proc. 35th Int. Res. Conf. Iarigai, Iarigai, Valencia, Spain.

Edvardsson, S. and Uesaka, T. (2009). System stability of open draw section and papermachine runnability, l'Anson, S.J. ed., Adv. Pulp Pap. Res., The Pulp and Paper Fundamental Research Society, Lancashire, UK, 557–575.

Edvardsson, S. and Uesaka, T. (2010). System dynamics of the open-draw with web adhesion: particle approach. J. Appl. Mech. 77(2), 21009.

Fung, Y.C. and Tong, P. (2001). Classical and computational solid mechanics, World Scientific Publishing Company.

Gomér, M. and Lindholm, G. (1991). Hygroexpansion of newsprint as a result of water absorption in a printing press, Proc. 43rd TAGA Annu. Conf., The Technical Association of Graphic Arts (TAGA) Rochester, New York, USA.

Good, J.K. (2007). Winding: Machines, mechanics and measurements, Destech Publications, Lancaster, PA, USA.

Grant, J.P. (1967). A decade of suffering from web breaks in a metropolitan pressroom, Pulp Pap. Mag. Canada 68(5), 143–149.

Gupta, S.S. (2013). Study of delamination and buckling of paper during the creping process using finite element method-a cohesive element approach, North Carolina State University, Raleigh, NC, USA.

Hämäläinen, P., Hallbäck, N. and Barbier, C. (2016). Development and evaluation of a high-speed creping simulator for tissue, Nord. Pulp Paper Res. J. 31(3), 448–458.

Hansen, Å. (1997). Water absorption and dimensional changes of newsprint during offset printing, Adv. Print. Sci. Technol. Pentech Press: Great Britain, 247–268.

Hauptmann, E.G. and Cutshall, K.A. (1977). Dynamic mechanical properties of wet paper webs. Tappi 60(10), 106–108.

Ho, J., Hutton, B., Proctor, J. and Batchelor, W. (2007). Development of a tissue creping test rig, Chemeca, Melbourne, Australia, 1334–1340.

Hollmark, H. (1983). Mechanical Properties of Tissue, Mark, R.E. ed., Handb. Phys. Mech. Test. Pap. Pap., New York: Marcel Dekker, 497–521.

Hossain, S., Bergström, P., Sarangi, S. and Uesaka, T. (2017). Computational Design of Fiber Network by Discrete Element Method. 16th Pulp Pap. Fundam. Res. Symp., Vol. 1, 651–668.

Hristopulos, D. and Uesaka, T. (2002). A model of machine direction tension variations in paper web with runnability applications. J. Pulp Pap. Sci. 28(12), 389–394.

Jokio, M. (1999). Papermaking Part 3, Finishing, Papermak. Sci. Technol. Gullichsen, J. and Paulapuro, H. eds., 361p, Fapet Oy, Helsinki, Finland.

Kipphan, H. (2001). Handbook of print media: Technologies and production methods, Springer, Heidelberg, Germany.

Kouko, J., Kekko, P. and Kurki, M. (2006). Effect of strain rate on the strength properties of paper, Prog. Pap. Phys. – A Seminar, Miami University, Oxford, Ohio, USA, 90–94.

Kouko, J., Salminen, K. and Kurki, M. (2007). Laboratory scale measurement procedure for the runnability of a wet web on a paper machine, part 2, Pap. ja puu 89(7–8), 424–430.

Kouko, J. and Sorvari, J. (2008). Mechanical testing and viscoelastic strength modeling of wet paper, Prog. Pap. Phys. – A Semin, Helsinki University of Technology, Otaniemi, Finland, 215–218.

Kulachenko, A., Gradin, P. and Koivurova, H. (2006). Modelling the dynamical behaviour of a paper web. Part 1. Comput. Struct. 85, 131–147.

Kurki, M., Juppi, K., Ryymin, R., Taskinen, P. and Pakarinen, P. (1995). On the web tension dynamics in an open draw, Proc. third Int. Conf. web Handl., Oklahoma State University: Stillwater, Oklahoma, USA.

Kurki, M., Kekko, P., Kouko, J. and Saari, T. (2004). Laboratory scale measurement procedure of paper machine wet web runnability Part 1, Pap. ja Puu, Suomen paperi-ja puutavaralehti oy 86(4), 256–262.

Kurki, M., Vestola, J., Martikainen, P. and Pakarinen, P. (1997). The effect of web rheology and peeling on web transfer in open draw, Proc. Fourth Int. Conf. Web Handl., Oklahoma State University: Stillwater, Oklahoma, USA.

Larsson, L.O. (1984). What happens in the press to cause web breaks?, Pulp Pap. Canada 85(9), T249–T253.

Lif, J.O., Östlund, S. and Fellers, C. (2005). In-plane hygro-viscoelasticity of paper at small deformations, Nord. Pulp Paper Res. J. 20(2), 139–149.

Linna, H.I. and Lindqvist, U. (1987). Runnability problems in web presses and paper machines, Larsson, L.O. ed., Top. Themes Newspr. – Print. Res. TFL Int. Multidiscip. Symp. Lidingö: Swedish Newsprint Research Centre (TFL), Djursholm, Sweden, 63–74.

Linna, H.I., Molianen, P. and Koskimies, J. (1991). Paper anisotropy and web tension profiles, 1991 Int. Pap. Phys. Conf., TAPPI, Atlanta, GA,, USA, Kona, Hawaii, USA, 327–339.

Linna, H.I., Parola, M.J. and Virtanen, J.O. (2002). Improving productivity by measuring web tension profiles. Appita 55(4), 317–322.

Miyanishi, T. and Shimada, H. (1998). Using neural networks to diagnose web breaks on a newsprint paper machine. Tappi J., TAPPI 81(9), 163–170.

Mujumdar, A.S. and Douglas, W.J.M. (1976). Analytical modeling of sheet flutter, Sv. Papperstidn. 79(6), 187–192.

Page, D.H. and Bruce, A. (1985). Pressroom Runnability, Newspr. Press. Conf. (Chicago, IL, USA CPPA ANPA).

Pakarinen, P., Juppi, K. and Karlsson, M. (1995). A study of air flows in single-tier pockets at high machine speeds. J. Pulp Pap. Sci. 21(2), J68–J73.

Pakarinen, P., Ryymin, R., Kurki, M. and Taskinen, P. (1993). On the dynamics of web transfer in an open draw, Proc. 2nd Int. Conf. web Handl.,, Oklahoma State University: Stillwater, Oklahoma, USA.

Pan, K. (2019). Modeling the creping process in tissue making, PhD Dissertation, The University of British Columbia, Vancouver, BC, Canada.

Pan, K., Das, R., Phani, A.S. and Green, S. (2019). An elastoplastic creping model for tissue manufacturing, Int. J. Solids Struct. 165, 23–33.

Pan, K., Phani, A.S. and Green, S. (2018). Particle Dynamics Modeling of the Creping Process in Tissue Making. J. Manuf. Sci. Eng. 140(7), 071003.

Parola, A., Sundell, H., Virtanen, J. and Lang, D. (2000). Web tension profile and gravure press runnability, Pulp Pap. Canada 101(2), T35–T39.

Parola, M.J. and Beletski, N. (1999). Tension Across the Paper web – A New Important Property, 27th EUCEPA Conf., ATIP, Paris, France, Grenoble, France, 293–298.

Parola, M.J. and Linna, H.I. (2000). Runnability and the unstable stress fields in the paper web, Graph. Arts Finl. 29(3), 9–15.

Paulapuro, H. (2000). Papermaking Part1: Stock Preparation and Wet End, Papermak, Sci. Technol.. Gullichsen, J. and Paulapuro, H. eds., 461p, Fapet Oy, Helsinki, Finland.

Pramila, A. (1986). Sheet flutter and the interaction between sheet and air. Tappi 69(7), 70–74.

Ramasubramanian, M.K., Sun, Z. and Chen, G. (2011). A Mechanics of Materials Model for the Creping Process, J. Manuf. Sci. Eng. American Society of Mechanical Engineers 133(5), 51011.

Reid, K.N. and Shin, K.H. (1991). Variable-gain control of longitudinal tension in a web transport system, Proc. First Int. Conf. Web Handl., Oklahoma State University, Stillwater, Oklahoma.

Roisum, D.R. (1994). The mechanics of winding, TAPPI, Atlanta, GA, USA.

Roisum, D.R. (1998). The mechanics of web handling, TAPPI, Atlanta, GA, USA.

Uesaka, T. (2005). Principal factors controlling web breaks in pressrooms – Quantitative evaluation. Appita J. 58(6), 425–432.

Uesaka, T. and Ferahi, M. (1999). Principal factors controlling pressroom breaks, TAPPI Int. Pap. Phys. Conf., TAPPI Press: San Diego, CA, 229–245.

Uesaka, T., Ferahi, M., Hristopulos, D., Deng, X. and Moss, C. (2001). Factors Controlling Pressroom Runnability of Paper, Baker, C.F. ed., Sci. Papermak., Lancashire, UK: The Pulp and Paper Fundamental Research Society, 1423–1440.

Douglas W. Coffin

7 Creep and relaxation

7.1 Introduction

As one delves deeper into the mechanics of paper, one gains an appreciation that the passing of time itself is a major contributing factor to the response of paper to load. This time dependence is inherently a response of the material and cannot simply be captured by incorporating changes in kinetic energy from the acceleration and deceleration of the material. The effects of time dependence on the mechanical behaviour of paper were briefly illustrated in Section 2.3.2. Sometimes these effects are negligible, and sometimes they cannot be ignored.

From an engineering standpoint, neglecting the time-dependent nature of paper will lead to reliable predictive equations and fundamental understanding. The fracture problems studied in Chapter 5 and the box corner problems presented in Chapter 4 do not require consideration of time dependence to understand the important mechanisms and behaviours. An important aspect for consideration in dealing with the mechanics of paper is the dissipation of energy, and in many situations, this could be effectively accounted for in terms of rate-independent plasticity. For the engineer, it may even be irrelevant as to the causes of energy dissipation as long as it has been accounted for in some fashion and the calculations provide a reasonable estimate for the process being modelled. However, if an approach ignoring time dependence yields unsatisfactory results, the inclusion of time dependence may be a necessity.

The time dependence of the mechanical response of paper extends across large time scales from microseconds to decades. At very short times, where relaxations have not started, paper will appear brittle and stiff. At long time scales, paper will appear more ductile and compliant. The focus of phenomena covered in Chapters 4 and 5 is on relatively short time scales, and time dependence is of less interest. In some short time scale phenomena, such as web dynamics, time dependence is sometimes of interest because stress relaxation affects the magnitude of stresses (Section 6.2.2). However, time dependence really takes centre stage in the long-time strength of stacked boxes (Section 3.4.4). Understanding the time dependence of paper over long periods of time is the focus of this chapter.

To study time dependence, we could title the chapter visco-elastic–plastic behaviour, or even rheology, but as an introduction to the time dependence of paper we focus on two phenomena: creep and stress relaxation. First, the essential elements that must be considered when treating the creep and relaxation behaviour of paper are discussed. Then, the basic observed behaviour of paper is presented and the effects of some relevant parameters are shown. Finally, a simple treatment of

Douglas W. Coffin, Miami University, Oxford, OH, USA

https://doi.org/10.1515/9783110619386-007

box lifetime is presented to underline that even in simple practical cases where creep is the main phenomenon, other factors such as material variability and environmental conditions are of equal importance.

7.2 Relaxation and creep as phenomena

Creep and relaxation are the two customary phenomena associated with time-dependent behaviour. They are opposite manifestations of the dissipation of energy from a material under some state of stress. The definitions are as follows:

- Creep: The increase in strain $\varepsilon(t)$ over time under a constant state of stress, $\sigma = \text{constant} \ \forall \ t \geq 0$.
- Relaxation: The decrease in stress $\sigma(t)$ over time under a constant state of strain, $\varepsilon = \text{constant} \ \forall \ t \geq 0$.

The state of stress or strain can be any combination of normal and shear components.

Typically, creep and relaxation testing is performed under conditions of uniaxial tension or compression. For relaxation (Fig. 7.1b), a sample of initial length L is changed to length $L + u$ and held constant for a period of time. The load P is recorded as a function of time. For creep (Fig. 7.1c), a constant load is applied, and the change in length is recorded as a function of time. From a practical standpoint, one cannot control the local state of stress and strain in every region of the material, and for testing only the applied load or applied deformation is held constant.

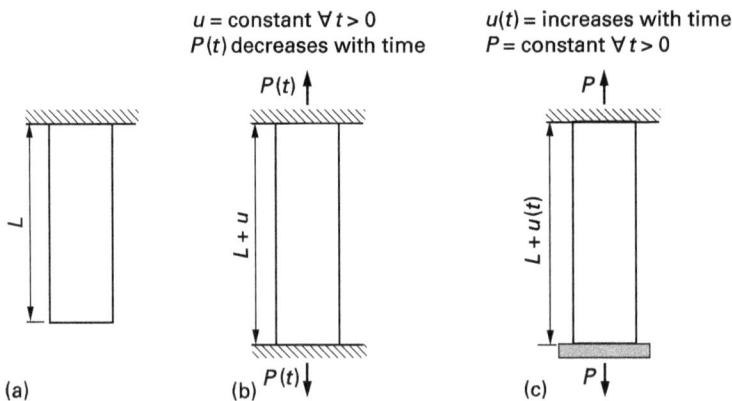

Fig. 7.1: Stress relaxation (b) and creep (c) of a strip of paper (a).

The creep and relaxation are a result of redistribution of stresses at some length scale in the material. These could be molecular, micro-, or macro-level redistributions. Here, we treat the phenomena as being valid as a continuum and ignore what leads to them. The role of structure on the global creep and relaxation response is addressed later in this chapter, but only regarding how it affects the effective continuum response.

The measured load P and displacement u are related to the stress σ and strain ε via integrating over the cross-sectional area A and the length of the sample L, or

$$P = \int_A \sigma \, dy dz \quad \text{and} \quad u = \int_0^L \varepsilon \, dx. \tag{7.1}$$

Assuming the stress and strain are uniformly distributed, we can consider the creep and relaxation test to correspond with the definition of creep and relaxation given previously. As we progress in the discussion of creep and relaxation, it is important to remember that the measured quantities are P and u. Stresses and strains are likely to be distributed unevenly through the sheet in a manner that can change with the passage of time.

Creep and relaxation tests are often carried out at various levels of load or deformation (Fig. 7.2). This yields a family of N creep curves or relaxation curves. The creep curves can be written as

$$\varepsilon(t, \sigma_i) \quad \forall t \geq 0, \quad i = 1, 2, \dots, N, \tag{7.2}$$

and relaxation curves as

$$\sigma(t, \varepsilon_i) \quad \forall t \geq 0, \quad i = 1, 2, \dots, N. \tag{7.3}$$

These curves can be used to establish material characteristics or properties. The material response is characterized as creep compliance J and relaxation modulus G. These parameters are obtained by dividing eqs. (7.2) and (7.3) with the constant stress or strain, respectively, leading to

$$J(t, \varepsilon) = \frac{\varepsilon(t, \sigma)}{\sigma}, \tag{7.4}$$

and

$$G(t, \sigma) = \frac{\sigma(t, \varepsilon)}{\varepsilon}. \tag{7.5}$$

In general, the creep compliance and relaxation modulus can be functions of the stress and strain in different directions, temperature, and moisture content. If one wishes to account for time dependence in engineering calculations, some type of model for eqs. (7.4) and (7.5) must be used. Important considerations when developing specific relations are addressed in the next section.

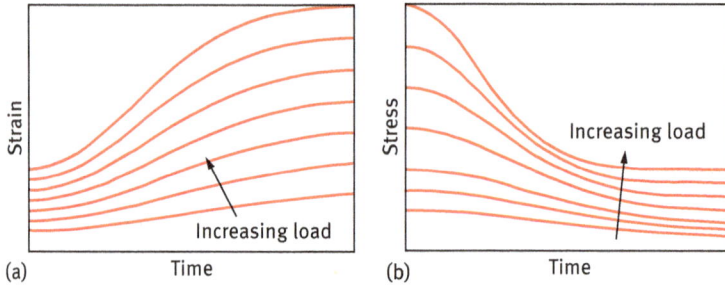

Fig. 7.2: Family of creep curves (a) and stress relaxation curves (b).

For many materials and specific time and load domains, the creep compliance and relaxation modulus are found to be independent of the level of load and strain, respectively. In other words, when normalized by the initial loading, each family of curves in Fig. 7.2 converges to one curve, either $J(t)$ or $G(t)$. At least this is an adequate approximation. The collapse implies that the material behaviour is linear in that range of time and load. If linearity is not present, it may still be possible to develop systematic shifts so that a family of curves is transformed into a so-called master curve. Even though the material response is non-linear, expressions can be developed to account for the dependence on load level. Paper materials are inherently non-linear, but it is possible to find ranges of load levels where the behaviour can be given by linear expressions or non-linear master curves.

7.3 Modelling of time dependence

Both creep and stress relaxation are manifestations of time-dependent dissipative processes. These behaviours influence the response of paper during papermaking, converting, and end use. Constitutive equations that capture this behaviour must be developed for use in models. This section discusses some of the basic features of formulating a model for paper. The discussion is centred on uni-axial loading but could be generalized to multi-axial states of stress and strain. Readers requiring more in-depth knowledge can find many sources, for example, Ward and Sweeny (2004).

7.3.1 Linear behaviour

In 1876, Boltzman proposed that creep is a function of the entire past loading history and that each step of loading makes an independent contribution to the final deformation (Ward and Sweeney, 2004). This is true for linear visco-elastic materials and allows for the description of general loading sequences using the creep compliance and stress relaxation modulus. These general equations can be written as

$$\varepsilon(t) = \int_{-\infty}^{t} J(t-t^*)\frac{d\sigma}{dt^*}dt^*,$$ (7.6)

and

$$\sigma(t) = \int_{-\infty}^{t} G(t-t^*)\frac{d\varepsilon}{dt^*}dt^*.$$ (7.7)

For both of these equations to be simultaneously valid, the creep compliance and relaxation modulus must satisfy

$$\int_{0}^{t} G(t-t^*)J(t^*)dt^* = t.$$ (7.8)

The implication is that simply fitting a convenient empirical equation to data and substituting that into eqs. (7.6) and (7.7) may not provide a good model. On the other hand, with the proper constitutive equation, time-dependent behaviour can be captured regardless of the loading conditions.

A critical factor to include in a model is the degree to which the deformation is recoverable. If a portion of the energy imparted to the sample during a test is stored in the material, the deformation may be recoverable. If the energy is dissipated (say as heat) during the process, the deformation cannot be recoverable. Figure 7.3 illustrates the need to explore time dependence beyond the simple creep test. The creep and creep recovery in the two cases have the same initial linear-elastic response and essentially the same constant-rate creep response. After a time lapse, the load is removed and the initial elastic response is recovered. In one case, there is no further recovery of deformation because all the stored energy has been dissipated. In

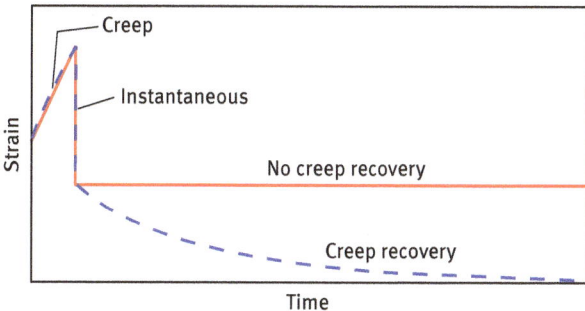

Fig. 7.3: Comparison of creep recovery for two cases with equal creep curves.

the other case, all the deformation is recovered because potential energy is stored in the material. Observe that much more time is needed for recovery than for creep. This is because the external force driving creep is greater than the internal force driving recovery.

This discussion has introduced the following three considerations that must be addressed when developing appropriate models for the time-dependent nature of paper.
1. Non-linearity
2. Recoverability
3. Time scale

7.3.2 Non-linearity

Superposition works only for linear systems. A system and process described by some operator $F[f(t)]$ is linear, if $F[f + g] = F[f] + F[g]$ holds, and non-linear otherwise. For creep compliance and stress relaxation, the response is said to be non-linear if, in addition to time, J and G are functions of stress or strain, or both stress and strain. When the material is linear, J and G are only functions of time. The degree of non-linearity can be assessed in several ways, but one must evaluate the response for a sufficient range of initial load levels to generate relevant data.

Isochronous curves provide a simple method to evaluate the degree of non-linearity (Fig. 7.4). One transforms the measured creep or stress relaxation data into separate curves that display stress and strain at one constant time, hence the term isochronous. From eqs. (7.4) and (7.5) one can see that, depending on whether creep or stress relaxation is considered, the slopes of the curves give either J or G for the time chosen. If the material is non-linear, the isochronous curves are non-linear.

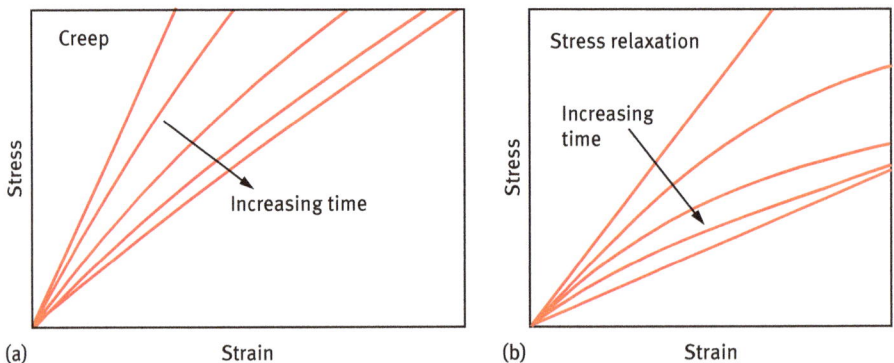

Fig. 7.4: Isochronous curves for non-linear creep (a) and stress relaxation (b).

At short times, the response is linear and corresponds to the highest slopes in Fig. 7.4. For this example, the line corresponding to the longest time also indicates a linear response. This long-time linearity results because relaxations were delayed but elastic so that at long times only linearly elastic energy is stored in the material. At intermediate times, where energy is actively being dissipated, the curves exhibit non-linear dependence such that at higher load levels a disproportional amount of energy is dissipated.

As discussed by Ward and Sweeney (2004), Leaderman was the first to generalize eqs. (7.6) and (7.7) to capture the non-linear behaviour of polymers, and then Schapery used thermodynamics of reversible systems, deriving the following constitutive equation for non-linear materials:

$$\varepsilon(t) = g_1(\sigma) \int\limits_{-\infty}^{t} D(\psi(t) - \psi(t^*)) \frac{d(g_2(\sigma)\sigma)}{dt^*} dt^*. \qquad (7.9)$$

Here

$$\psi(t) = \int\limits_{0}^{t} \frac{1}{a(\sigma(t^*))} dt^* \qquad (7.10)$$

is a stress-dependent reduced time function, and $g_1(\sigma)$, $g_2(\sigma)$, and $a(\sigma)$ are functions of stress. For a creep test, eq. (7.9) reduces to

$$\varepsilon(t) = g_1(\sigma)g_2(\sigma)D\left(\frac{t}{a(\sigma)}\right)\sigma. \qquad (7.11)$$

Similarly, one could form a master creep curve by determining appropriate shifting functions to collapse the creep compliance curves for different load levels to one curve. The shifting parameters are basically the functions g_1, g_2 and a. It could be that non-linearity does not arise from time-dependent behaviour but rather from some other mechanism, such as rate-independent plasticity. If this were the case, the isochronous curves would still be non-linear, and some shift would still be required to form a master creep curve.

7.3.3 Recoverability

Even if one were to determine all the parameters in eqs. (7.9) and (7.10), one would still have to rely on fitting curves to experimental data. The ease of using eq. (7.9) depends on whether or not an empirical equation captures the recoverability of the strain. It is reasonable to split the total strain into recoverable and non-recoverable components,

$$\varepsilon(t) = \varepsilon_{rec}(t) + \varepsilon_{non-rec}(t). \qquad (7.12)$$

The concept of recoverable and non-recoverable strain is separate from that of non-linearity. To be recoverable, the material must have potential energy stored in some manner such that when load is removed there is an internal driving force that facilitates recovery of strain. In the stress–strain discussion of Chapter 2, the permanent plastic deformation (strain at the point "∞" in Fig. 2.7) corresponds to the non-recoverable strain here, and the total elastic strain (between the points "3" and "∞" in Fig. 2.7) corresponds to the recoverable strain.

Consider now a material that is subjected to a load cycle load defined as

$$\sigma(t) = \sigma_0 H(t) - \sigma_0 H(t - t_1),\tag{7.13}$$

where $H(t)$ is the Heavyside step function, and t_1 is a sufficiently long time so that

$$\varepsilon(t > t_1) = \int_{-\infty}^{t} J(t - t^*)\frac{d\sigma}{dt^*} = \sigma_0[J(t) - J(t - t_1)].\tag{7.14}$$

Creep is recovered if the strain given by eq. (7.14) goes to zero. In other words, creep is recovered if

$$J(t) - J(t - t_1) = 0.\tag{7.15}$$

At longer times, this holds if $dJ/dt \to 0$, when $t \to \infty$. If, on the other hand,

$$\sigma_0(J(t) - J(t - t_1)) = \varepsilon(t_1),\tag{7.16}$$

then creep is not recovered. For linear materials, this occurs when $J(t) = Ct$.

One modelling approach to account for recoverable and non-recoverable behaviour would be to specify that the creep compliance consists of two components,

$$\varepsilon(t) = \varepsilon_{rec}(t) + \varepsilon_{nonrec}(t) = \int_{-\infty}^{t} \left[J_{rec}(t - t^*) + J_{non-rec}(t - t^*) \right] \frac{d\sigma}{dt^*} dt^*.\tag{7.17}$$

Similar arguments could be made for both stress relaxation and non-linear creep equations. We note that separating recoverable and non-recoverable creep requires both loading and unloading tests. The simple creep and relaxation curves by themselves are insufficient.

7.3.4 Time scales

Finally, we must consider the time scale of the loading event. When one models time-dependent behaviour, there is a minimum time scale, t^{min}, below which time dependence is ignored. This can be used as the time increment in the model, and the response at $t < t^{\min}$ can be taken to be instantaneous. Thus, the total strain consists of an instantaneous component $\varepsilon_{\text{instant}}$ and a time-delayed component $\varepsilon_{\text{delayed}}(t)$,

$$\varepsilon(t) = \varepsilon_{\text{instant}} + \varepsilon_{\text{delayed}}(t). \tag{7.18}$$

The instantaneous component of strain can be either elastic (recoverable) or plastic (non-recoverable) deformation (cf. Section 2.3.1). The same holds for the time-dependent component, $\varepsilon_{\text{delayed}}(t)$, as discussed in the preceding section. The separation point between the instantaneous and delayed responses corresponds to a minimum time, $t_{\min}$. For small, lower, cut-off times, the instantaneous elastic modulus typically increases, and the plastic deformation decreases if more of the total deformation is counted as delayed response. In a model, one could specify the $t_{\min}$ so that all history before $t_{\min}$ is lumped into an instantaneous response.

Typically, there is also an upper limit of time of interest for the problem being addressed. Events that occur at times longer than the upper limit are not important. Consideration of this fact may help simplify the analysis because one can ignore very slow relaxations. Thus, the model should capture the behaviour in the range of time scales of interest, and extrapolation to shorter or longer time scales should be applied only with the understanding that results would be more questionable.

Another way to consider time scales is the concept of a time constant τ as a material property governing a time-dependent response. The creep compliance can then be expressed as a sum of functions each active at a different time scale τ_i. For example,

$$J(t) = J_0 \delta(t) + \sum_{i=1}^{N} J_i \left(\frac{t}{\tau_i} \right), \tag{7.19}$$

where J_0 represents the material response at $t < \tau_i$.

To summarize the important characteristics of modelling time-dependent behaviour of paper, any developed equations should address the following:

1. Instantaneous elastic response to account for recoverable deformation that occurs at shorter time scales than the lower cut-off response.
2. Instantaneous plastic response to account for non-recoverable deformation that occurs at shorter time scales than the lower cut-off response.
3. Recoverable time-dependent behaviour for delayed-elastic behaviour.
4. Non-recoverable time-dependent response for a delayed-plastic or viscous response.
5. Consideration of longer time and load limits for the validity of the model.

A general expression for the creep compliance that could be used for modelling is

$$J(t) = J_0 \delta(t) + \sum_{i=1}^{N^{\text{rec}}} J_i^{\text{rec}} \left(\frac{t}{\tau_i^{\text{rec}}} \right) + \sum_{i=1}^{N^{\text{non-rec}}} J_i^{\text{non-rec}} \left(\frac{t}{\tau_i^{\text{non-rec}}} \right). \tag{7.20}$$

For creep and relaxation, there are typically time regimes where the response shows some characteristic behaviour. Consider a typical creep response that exhibits a primary, a secondary, and a tertiary creep response. Often for polymers, the primary creep is predominantly recoverable, the secondary creep is mainly non-recoverable, and the tertiary response is indicative of onset of failure or localization of stress in the material. Of course, the same mechanisms are likely to be active (to some extent) in multiple regimes. Separating out the responses is only an approximation, albeit a useful one. Figure 7.5 shows three cases where different mechanisms may contribute to the creep in different regimes.

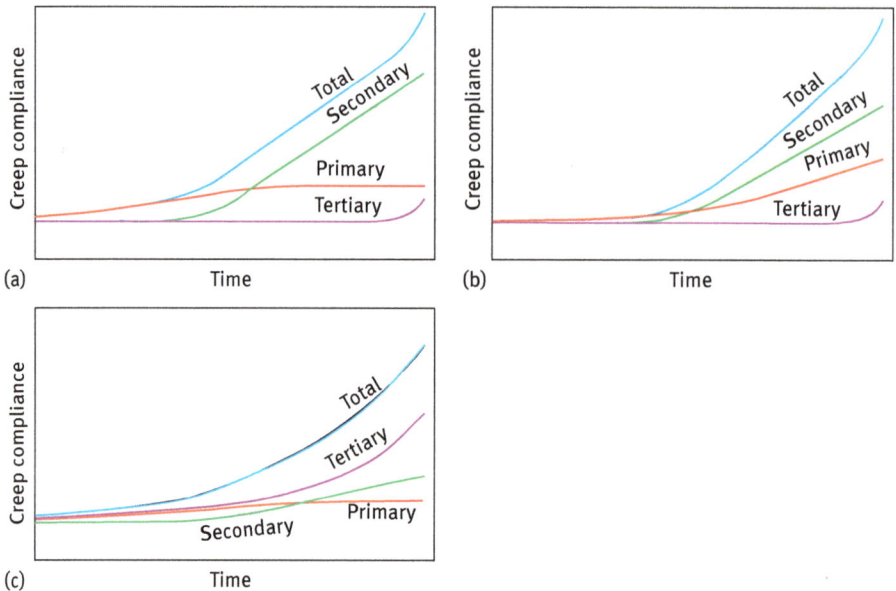

Fig. 7.5: Different possible mechanisms of primary creep, secondary creep, and tertiary creep.

7.4 Creep and relaxation properties of paper

With the overview given in the previous section, we are now ready to discuss the time-dependent nature of paper. The behaviour is quite complex, and to account for all effects would require many parameters and extensive testing. Not only is the global loading response of paper a combination of elastic and inelastic behaviour at various time scales, but also the effects of temperature and moisture content and loading history need to be addressed. Furthermore, the effects of network structure are coupled to the effects of the fibres that constitute the paper. All these effects on the mechanical behaviour are coupled together to give the response measured in the laboratory or in the field.

Although one should develop a model that captures all these effects, it would likely be more cumbersome than instructive. The other extreme in modelling is to separate all effects and look at the general response to each parameter. The problem with the second approach is that one will almost always be able to find contradictory experimental results because of the couplings between so many parameters. Nevertheless, the second approach is more instructive and is adopted here. Naturally, not all papers will behave exactly as described later. The challenge is to place the results in the right context and provide explanations for why the observed behaviour is different than the general view. Almost all the results in this section can be found in the literature, and the reader is referred to Salmén and Hagen (2002), Coffin (2005), and Ketoja (2008) for more specific information.

7.4.1 Creep

The discussion of creep begins with the basic response of paper to a constant uni-axial tensile load. The tensile creep response of paper is similar to that of many polymers. Figure 7.6 provides experimental data obtained by Brezinski (1956) for the creep of handsheets. The response shown in the figure is typical of many papers and illustrates several important aspects of the behaviour. The dimensionless creep compliance is plotted both as a function of logarithmic time (Fig. 7.6a) and linear time (Fig. 7.6b), the latter to more clearly demonstrate the rapid decay of creep rate that occurs.

The important observations are as follows:
1. The rate of creep decays with time (exponential decay in the rate of creep).
2. Increased load results in disproportionately larger creep compliance (non-linearity).
3. At high loads or long times, the rate of change in creep compliance is constant when expressed with respect to logarithmic time.

The second observation indicates that the creep compliance is a function of stress level and that linear visco-elasticity alone is inadequate to capture the total strain versus time response. Observation 3 suggests that a master creep curve can easily be formed from the data, and thus, an equation could be constructed to describe the creep behaviour for a range of times and load levels. In fact, a simple linear shift of the logarithmic time is all that is required for many papers, as shown in Fig. 7.7. In that case, the shift required to form the master curve of creep is given by

$$J(\sigma_2, t) = J(\sigma_1, te^{\lambda(\sigma_1 - \sigma_2)}),$$

(7.21)

where λ is the slope of the shift factor versus stress curve. This equation illustrates how the creep compliance of paper is a strongly non-linear function of load. As the load acts to shift the compliance curve in time, higher load levels are equivalent to allowing longer time for creep. One interpretation is that the creep in paper is a

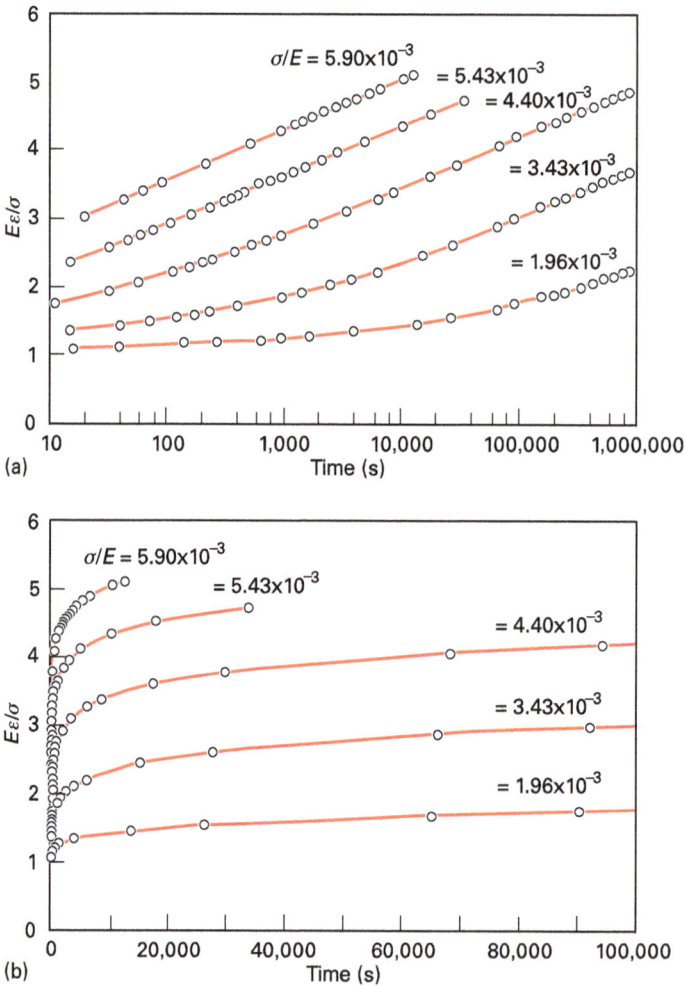

Fig. 7.6: Representation of the data of Brezinski (1956) as the product of the initial modulus and creep compliance for logarithmic time scale (a) and linear time scale, but for a shorter time range (b).

load activated phenomenon. In other words, the characteristic time constants in eqs. (7.19) and (7.20) can be written as

$$\tau = \tau_0 e^{\lambda \sigma}, \tag{7.22}$$

where τ_0 is the characteristic relaxation time at zero load.

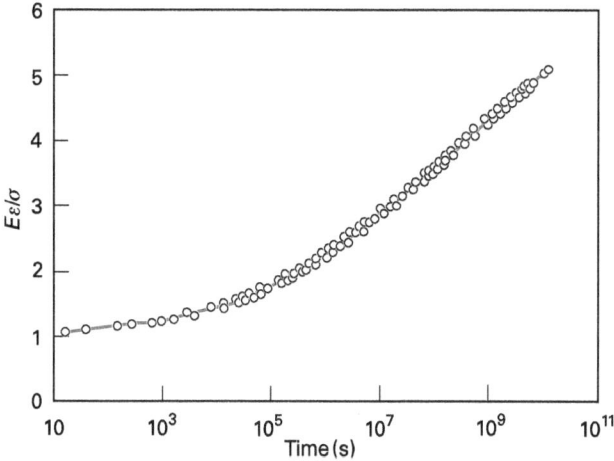

Fig. 7.7: Master curve of creep formed by moving the curves in Fig. 7.6a parallel to the time axis.

An alternative approach to combining the creep curves is to shift the curves shown in Fig. 7.6a vertically. This would work in any regime where the creep is linear with the logarithmic time. For high loads, significant inelastic deformation occurs at times lower than those measured. This deformation can be taken as an instantaneous plastic component. The vertically shifted curves would superimpose, and the remaining creep response could be explained with a linear visco-elastic model. At low loads, the passage of much time would be required to activate the log-linear creep. In other words, once the creep response is activated by load it accumulates rapidly, and the rate of additional creep accumulation is fairly independent of load.

At low load levels, Fig. 7.6a shows two different time regimes of creep. For low loads, Brezinski (1956) found that the initial creep deformation could be fit with a power-law equation. At high loads, the creep was log-linear for all measured time scales. If one uses plasticity to account for the non-linearity of creep, it could be modelled as a strain-activated process. Even though the additional creep has been made linear, the plasticity still leaves the material inherently non-linear to load.

Brezinski (1956) also found that in the primary regime (power–law relationship), creep was largely recoverable and that the secondary creep (log-linear response) was largely non-recoverable. Although it is difficult to completely separate the recoverable and non-recoverable components, it seems reasonable to assume that, in general, the primary power-law creep is recoverable, and the secondary log-linear creep is non-recoverable. At some point in time, a tertiary stage of creep will occur where the creep rate begins to increase until the sample fails. The tertiary creep is typically short in duration and likely occurs because some critical level of damage to the paper has been reached.

A generic compliance curve that captures these observed behaviours at any given load level is

$$J(t) = B_0[1 - \exp(b_0 t^\omega)] + B_1 \log(b_1 t + 1) + B_2 \exp(b_2 t^\zeta), \qquad (7.23)$$

where B_0, B_1, and B_2 are strain magnification factors; b_0, b_1, and b_2 are time magnification factors; and ω and ζ are shape factors. The first two terms represent primary and secondary creep mimicking behaviour observed by Brezinski (1956), and the last term is representative of tertiary creep. The primary creep is written so that it acts as a delayed elastic response. Creep curves obtained from eq. (7.23) were given as the solid lines in Fig. 7.6.

Because we expect at least three types of creep to occur in a given paper, the variation of creep response from paper to paper will differ. Also, the measured response depends on level and duration of load. One should always characterize the paper for creep over several load levels and for sufficient duration so that it can be properly modelled. To separate out recoverable and non-recoverable creep, creep recovery tests should be completed. This requires that one removes load after a specified time and measures the recovery of strain. Recovery may require more time than the initial creep.

7.4.2 Stress relaxation

Johanson and Kubát (1964) investigated the stress relaxation in paper (Fig. 7.8). Similar to creep, when stress is plotted as a function of logarithmic time, very little relaxation initially occurs but at longer times the relaxation is fairly linear with the logarithmic time. The fact that the relaxation modulus is distinct for each level of strain (top graph in Fig. 7.8) again demonstrates the load non-linearity present in paper.

In creep and relaxation tests, the initial loading is assumed to be quick compared to any relaxation processes. For creep tests, accomplished with a dead load weighting, this can be accomplished quite easily. Relaxation testing is typically done on a universal tester, and one has the choice of loading either instantaneously or with a constant strain rate. The choice affects the presentation of results. In the first case, the relaxation modulus (eq. (7.5)) is divided by the elastic modulus. This is equivalent to dividing the stress by the ideal stress that would arise in a linearly elastic material when the strain is applied instantaneously. In the second case, the relaxation stress is divided by the stress achieved when the desired strain was first

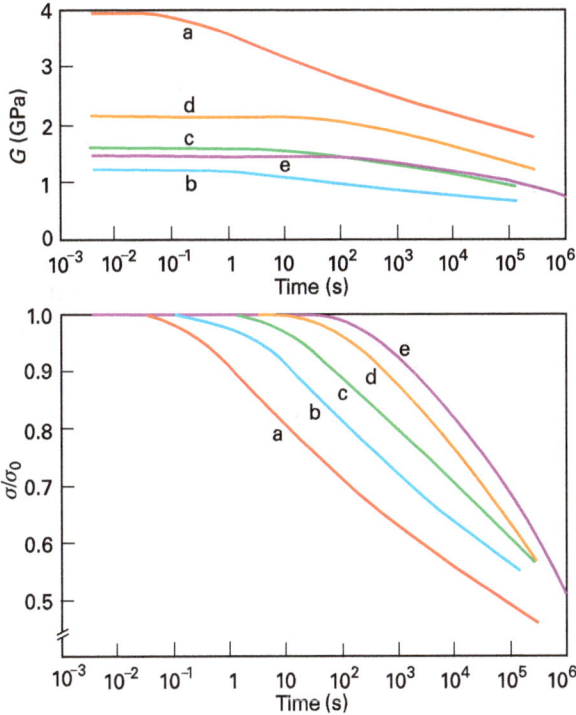

Fig. 7.8: Stress relaxation curves for paper for different strain values, showing the relaxation modulus (top graph) and stress normalized with the initial stress value (b); after Johanson and Kubát (1964).

achieved. This is equivalent to using the linear elastic strain, not the total strain, to calculate the relaxation modulus (eq. (7.5)).

Consider the effect that stress relaxation has on web dynamics (Chapter 6). Under steady-state conditions, the web in an open draw is under a state of constant strain. Because of relaxation, the tension in the web will be reduced compared to when no relaxation occurred. In addition, the magnitude of the stress gradients will be decreased.

7.4.3 Tensile versus compressive creep

Because paper is a fibrous network structure, it is unstable in compression. On various length scales the material will flex, buckle, and fold under compression, in essence making paper inherently weaker in compression than tension. It is also more compliant at larger stress levels and longer times. Figure 7.9 compares tensile and compressive isochrones. At small load levels, the creep responses are similar, but at larger load levels, compression is more complaint. Vorakunpinij (2003) shows that increased paper density prolongs the range of stress levels where tension and compression are similar.

The added deformation from structural rearrangements has the effect that in compression the creep rate does not decrease with time as much as it does in tension. Therefore, instead of the log-linear secondary creep, compressive creep is more like a power-law or even linear with time. Corrugated boxes under compression exhibit a linear secondary creep response (see Fig. 3.26) because of additional panel buckling and concentration of load into the corners of the box, as discussed in Section 3.4. The linearity of the secondary creep simplifies phenomenological models of corrugated board box. On the other hand, applying the techniques covered in Section 4.4 to creep could suggest ways to decrease not only the creep rate of a box under compression but even to get the creep rate to decay with time, in the same way as the constituting papers of the box behave.

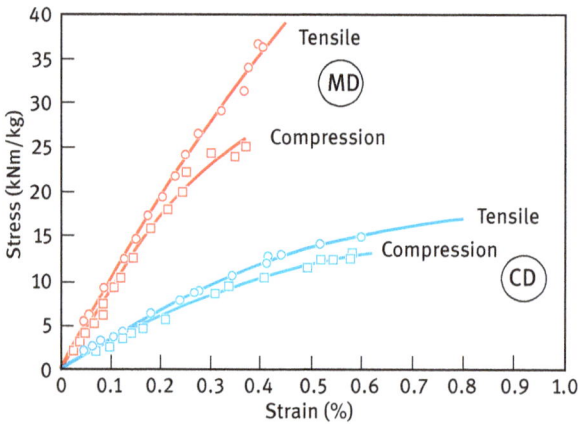

Fig. 7.9: Isochronous curves in tension and compression for both MD and CD creep; after Haraldsson, Fellers, and Kolseth (1993).

7.4.4 Effect of papermaking process and furnish

The basic time-dependent behaviour discussed in the previous three sections is generic, but given the vast variation in paper properties that can be achieved through choice of materials and processing, the time-dependent response of paper can be modified in the manufacturing process. In this section, some effects of processing and furnish on the creep response are introduced.

Brezinski (1956) carried out creep tests on handsheets made at different levels of wet pressing. Increased wet pressing acts to increase paper density and to increase the area of bonding between fibres. Figure 7.10 shows 24 h isochrones for sheets made at different levels of wet pressing. Clearly, higher levels of wet pressing decreased creep compliance. For these sheets, the elastic modulus also increased with increased wet pressing. If the isochrones in Fig. 7.10 are scaled by the initial modulus, they all collapse to the same curve. This suggests that a simple stress magnification can explain these effects. If we define an efficiency factor ϕ as the ratio of elastic modulus to some reference value (Seth and Page 1981), then the normalized creep compliance $J^{norm} = \phi J(\sigma/\phi, t)$ is independent of wet pressing.

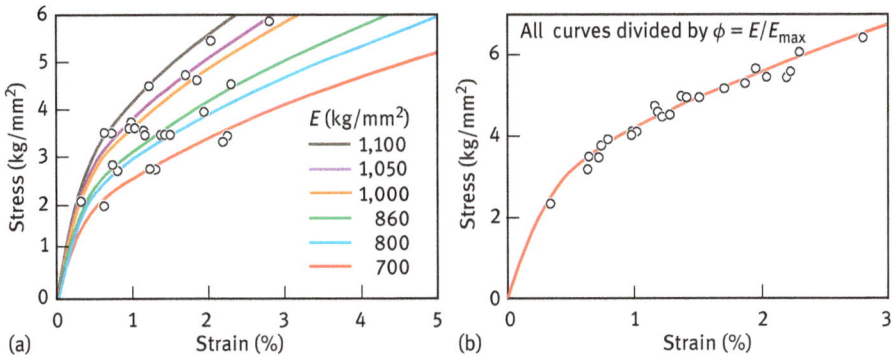

Fig. 7.10: Isochronous creep curves for handsheets made with different levels of wet pressing (a), and the same data divided scaled by the efficiency factor ϕ, based on data from Brezinski (1956).

Moderate wet straining applied during drying has the effect of increasing elastic modulus, decreasing breaking strain, and slightly increasing tensile strength. Shulz (1961) showed that wet straining also decreases creep compliance. If the creep compliance is multiplied with the elastic modulus, a master creep curve can be constructed, as shown in Fig. 7.11 (Coffin, 2005). Here the logarithmic time shift was linearly proportional to the degree of wet straining. Isochronous curves for various wet-strained sheets will likely not superimpose with only an efficiency factor ϕ, but

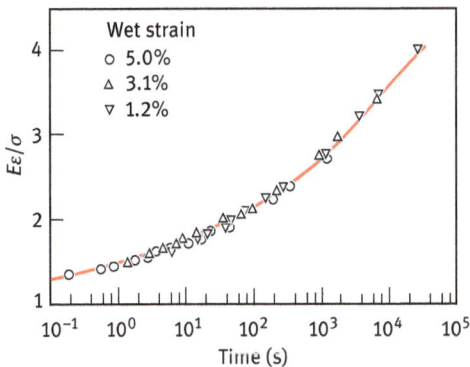

Fig. 7.11: Master curve of creep for sheet of various levels of wet straining, using data from Shulz (1961).

the fact that a master creep curve can be formed implies that the major effect of wet straining is to shift the activation of secondary creep response to longer times.

7.5 Moisture effects

The mechanical response of paper is influenced by the environment. Both changes in temperature and moisture content will change the mechanical properties, can alter the structure, and will likely influence the distribution of stresses in a sheet of paper. The hygrothermal influences on dimensional stability are discussed in Chapter 9. Here we discuss the influence of moisture on the time-dependent properties.

7.5.1 Softening with moisture

As illustrated in Section 2.2, increased moisture will decrease elastic modulus. At moisture contents below 5%, the elastic modulus is fairly constant. For moderate moisture increases above this level, the modulus decreases in approximately a linear fashion. Water is considered to be a plasticizer for paper and, in essence, activates thermal relaxations that normally occur at higher temperatures. Moisture will also affect the time-dependent nature of paper, such as the isochronous curves in Fig. 7.12a. If one scaled stress values with the elastic modulus measured at the same moisture content (Fig. 7.12b), the higher moisture content curves would converge, but it is not clear if they form one curve. At lower moisture contents, creep compliance is lower, but the effect can be removed by exposing the paper to high moisture before the creep test. A sample preconditioned by moisture will show a shift in the low moisture creep behaviour aligning it with creep at higher moisture contents (Fig. 7.13). The main influence of the moisture conditioning is thus to lower the stiffness of paper and release the effects of drying tension applied on the paper machine.

7.5.2 Accelerated creep

Accelerated creep is one of the manifestations of the mechanosorptive effects. The term mechanosorptive is used to describe the coupling between mechanical and moisture effects during changing moisture contents. Accelerated creep occurs when a material is subjected to the combined action of creep load and cyclic moisture content. Figure 7.14 shows the results for two tensile creep tests. For the log-linear curve, moisture was kept constant at a high level corresponding to 80% relative humidity (RH) of the surrounding air. For the second curve, the RH was at first held at

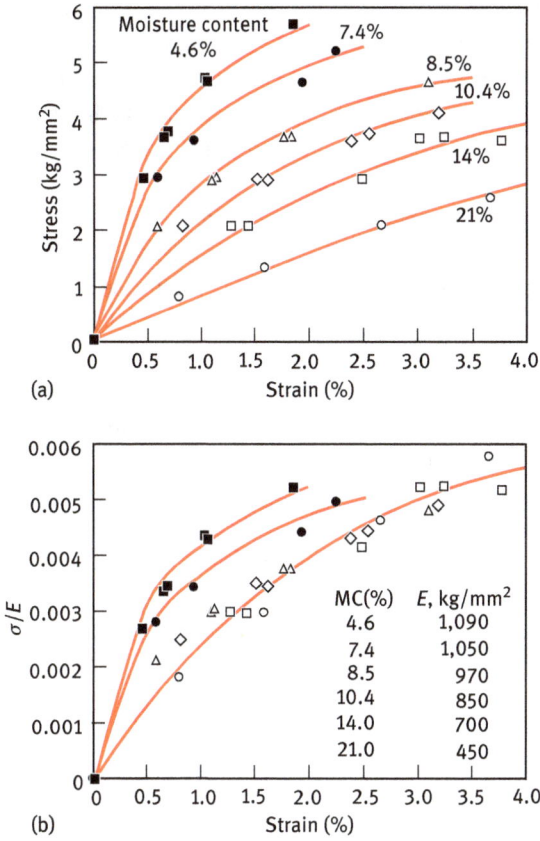

Fig. 7.12: Isochronous creep curves at 24 h for different moisture contents (a) and the same curves with stress divided by the initial elastic modulus, using data from Brezinski (1956).

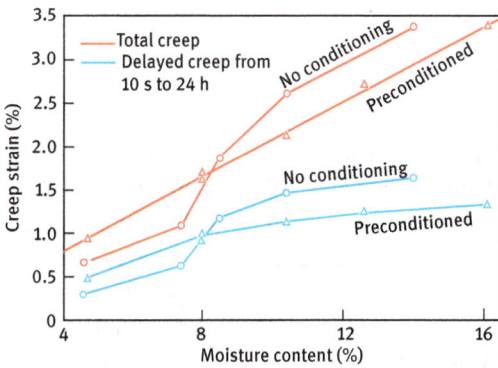

Fig. 7.13: Creep strain against moisture content, with and without preconditioning at high moisture content, using data from Brezinski (1956).

80% and then cycled between 80% and 30% RH for every 2 h. When the moisture cycling begins, two changes in the strain response are noted. First, there is a cyclic strain component due to the change in moisture content of the sample. The cyclic strain results from a combination of the hygroexpansion and the change in elastic modulus. The second effect is an increase in the rate of creep. It is this second effect that is called accelerated creep.

The increased creep rate by moisture cycles leads to shorter lifetime, or the time required for the sample to fail. The box creep curve in Fig. 3.26 was obtained under conditions of cyclic moisture because it is considered a worst-case scenario for box lifetime. Clearly, the increased creep rate is due to a coupling of load and moisture effects. The effect of constant moisture content on creep rate (Fig. 7.14) is not sufficient to explain accelerated creep because in that case, the creep observed under the cyclic humidity should be lower than creep under the constant 80% RH. As shown in Fig. 7.14, almost no creep occurs at the low humidity.

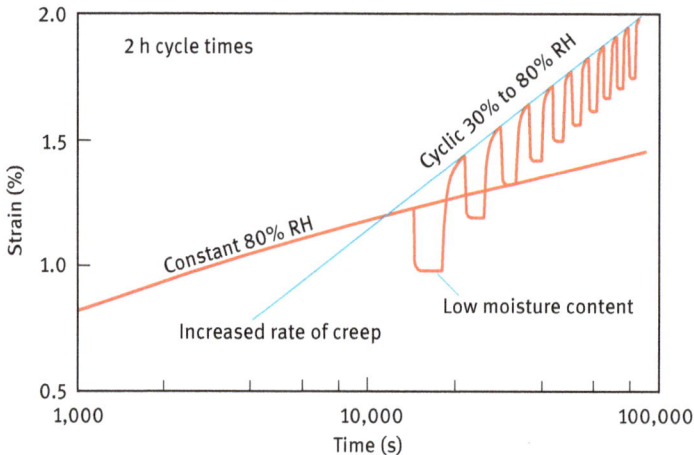

Fig. 7.14: Accelerated creep in paper.

From the viewpoint of mechanics, accelerated creep can be understood fairly easily. When RH is changed, a change of moisture in the paper will happen over some time period. This transport of moisture can be expected to create some moisture gradient across the thickness of the paper. Any change in the local moisture content drives local swelling or shrinkage. Thus, a moisture gradient across the paper creates a gradient in the local hygroscopic strain, the amount of local swelling or shrinkage that the paper would achieve if no restraints were present (see eq. (2.5)). Furthermore, and perhaps more important, paper is heterogeneous, and the hygroexpansivity of the structure will not be uniform even if there is no moisture gradient.

Both moisture gradients and heterogeneity can be expected to create gradients of hygroscopic strains. The change in moisture content combined with the variation of hygroscopic strains will create stress gradients in the paper. Even though the average stress across the sheet (eq. (7.1)) is constant, some local areas will be highly stressed. These stress gradients will persist for some time during and after a moisture change. We have also seen that paper is inherently non-linear with respect to load (Fig. 7.6). This means that highly stressed areas will creep disproportionately more than lightly stressed areas. Therefore, as long as elevated levels of stress exist in the sheet it will produce more creep than expected from the average load.

There are two approaches that could be used to model the mechanism of accelerated creep. The first would be to include a sufficient level of modelling to account for the stress gradients and the non-linear material behaviour. This model would predict transient stress gradients, and the accelerated creep would be a natural outcome. The second approach would be to use only the average stress and moisture values and add a hygroscopic strain component that accounts for the influence of the variations. The first approach is more fundamental, and the second approach is more empirical, but either method could be used to model accelerated creep.

There are two important aspects of accelerated creep that any model should reflect:

1. Although the overall creep rates are lower at lower moisture contents, the acceleration of creep is more severe at lower moisture contents. This is because stress gradients are more severe and persist for a longer time because the material is less compliant.
2. The rate of creep accumulation is affected by the moisture cycle. The acceleration of creep with a given cycle time depends on the time needed for stress gradients to relax. If the cycle time is close to the relaxation time of stress gradients, accelerated creep will be most severe. Longer moisture cycles then reduce accelerated creep.

According to this explanation, there is nothing extraordinary about accelerated creep. In fact, tests (DeMaio and Patterson, 2005) have shown that, also in the case of accelerated creep, the effects of wet pressing can be accounted for with an efficiency factor ϕ. This implies that the same mechanisms are active as in creep at constant moisture and that only the stress distribution is different in the case of cyclic moisture.

In Section 3.3.4, the practical problem of long-term stacking performance of corrugated board boxes was discussed. Accelerated creep is of prime importance to the lifetime of stacked boxes. Box lifetime is discussed in the next section.

7.6 Prediction of box lifetime

The discussion of the stacking strength of corrugated board boxes in Chapter 3 concerned primarily the short-time loading behaviour. Here we consider what happens at long time scales. The useful lifetime of a box is considered to be the time lapse that a box can safely protect its contents in the service environment. In general, during its useful life a box could be subjected to a combination of dynamic loading from transportation, dead loading from storage, and combined hygroscopic and thermal effects from changes in the environment. For this discussion, lifetime is the time of survival for a box subjected to a dead load in a given environment, such as the case with stacking in a warehouse.

7.6.1 Creep response of a box

Consider a box (Fig. 7.15) subjected to a compressive load P that will result in creep. Figure 7.16 shows the creep process and predominant box deformation mechanisms, described by Kellicutt and Landt (1951) and Bronkhorst (1997). When the box is first loaded, the primary deformation will be a flattening of the top flaps until the load distributes fairly uniformly around the perimeter of the box. This deformation can be large, as first noted by Malcolmson (1936). For a well-formed box, the walls will compress uniformly if the load is evenly distributed. In the secondary regime, the deformation is dominated by structural buckling of the box panels and inter-flute buckling of the liners at the corners, similarly to the behaviour under short-time loading (Section 3.4.2). This is the longest time regime of

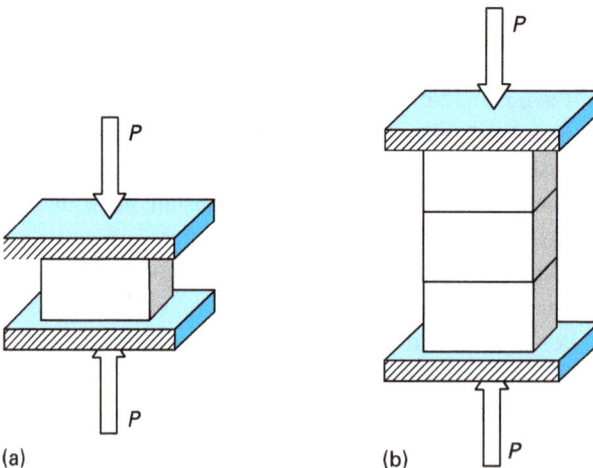

Fig. 7.15: Compressive loading of a single box (a) and a stack (b).

the deformation process. As first pointed out by Little (1943) for the box compression test (BCT), box perimeter and bending stiffness are important factors that would impact this regime. The tertiary phase is the initiation of failure in the box wall due to the formation of a hinge, typically along a corner.

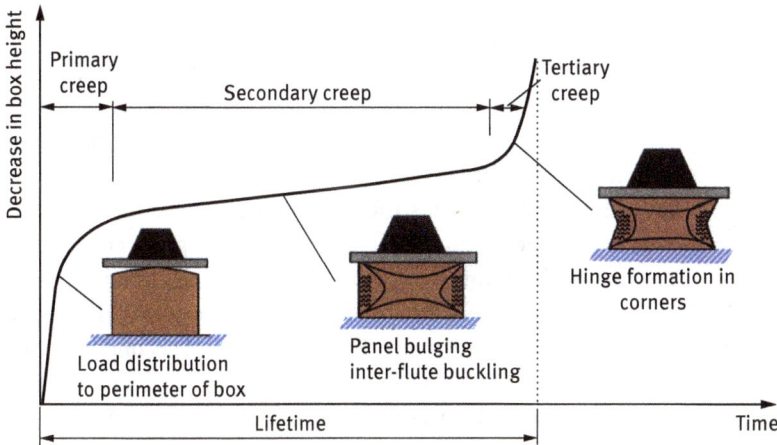

Fig. 7.16: Typical creep response of a box subjected to compression.

7.6.2 Previous equations for box lifetime

Through a series of box lifetime tests, Kellicutt and Landt (1951) found that for loads less than 0.75BCT, the lifetime, t_{BL}, could be expressed as a function of load level according to

$$t_{\mathrm{BL}} = t_{\mathrm{BL}_0} e^{-k_0 P / \mathrm{BCT}}, \tag{7.24}$$

here, t_{BL_0} is the lifetime at zero load and k_0 is a load proportionality factor. Stott (1959) found similar results. In addition, he found that the sensitivity to moisture was greater for lifetime than BCT. Dagel and Brynhildsen (1959) developed a similar equation

$$t_{\mathrm{BL}} = e^{k_1 (\mathrm{BCT}/P - 1)} - 1, \tag{7.25}$$

where k_1 is another proportionality factor. The advantage here is that t_{BL} goes to infinity when $P \rightarrow 0$ and to zero when $P \rightarrow$ BCT. Equation (7.25) was developed from assumptions that the secondary creep response is linear in logarithmic time, that the rate of creep is proportional to the load, and that the failure of a box is triggered at the same displacement regardless of load level.

Observations that the lifetime of a box is dominated by the secondary creep and that the rate of secondary creep is approximately inversely proportional to time led Bronkhorst (1997) to develop a prediction of lifetime as

$$t_{BL} = \frac{\varepsilon^{max,SC}}{\dot{\varepsilon}^{SC}}, \tag{7.26}$$

where $\varepsilon^{max,SC}$ is the amount of secondary creep strain at box failure and $\dot{\varepsilon}^{SC}$ is the average secondary creep rate. Other researchers had developed similar relationships but allowing the secondary creep rate to be raised to some other power than negative one (Bronkhorst, 1997). Equation (7.26) refers to the fact that lifetime correlates very strongly to secondary creep rate.

In fact, testing has shown that whereas lifetime versus P gives large scatter in results, lifetime versus $\dot{\varepsilon}$ reconciles the scatter and provides good correlation (Bronkhorst, 1997). This gives the impression that the key to improving lifetime is to decrease the secondary creep rate, $\dot{\varepsilon}^{SC}$. While this is an important finding, it does not help predict the lifetime of a given box. Both lifetime and secondary creep rates are measured from the same tests and only verify that the behaviour illustrated in Fig. 7.16 and the relationship according to eq. (7.26) holds. Seemingly identical boxes loaded with the same dead load may easily give vastly different lifetimes. While the boxes with the shortest lifetimes tend to have the largest secondary creep rates, we have no means to predict in advance which boxes give the shortest lifetime. Thus, eq. (7.26) by itself is not pertinent for lifetime predictions.

In order to be of predictive value, the equation for lifetime needs to be expressed in terms of a prescribed parameter. For stacking lifetime, the prescribed parameter is load. In essence, this is what box designers do. A box is designed to have certain strength (BCT). In a simple scenario, the stacking load is prescribed by the user, the expected conditions that the box might see are identified, and then the box designer applies a series of safety factors to determine the minimum BCT that the user needs. The term safety factor implies that we need to increase the strength of the box to account for unknown effects. While this is clearly the conservative, safe, and ethical path to take, it also implies unfounded ignorance of the mechanisms of box compression and material behaviour of paper. For a given box, we certainly know why increased loading levels lead to reduced lifetime. Therefore, the idea of safety factors is misleading. These factors are inverses of load reduction factors based on the expected service environment of the box.

Coffin, Wade, and Hussain (2019) have developed an equation that links lifetime to BCT, $\varepsilon^{max,SC}$ and eqs. (7.26) and (7.22). The point of their development was to explore the variability of parameters such as BCT, $\varepsilon^{max,SC}$, and λ on the variability of lifetime. An alternative but similar derivation is given in the next section.

7.6.3 Derivation of a new equation for box lifetime

Equations (7.24) to (7.26) do not satisfactorily capture the effect of load on lifetime. Equation (7.26) relies on two measured quantities and is therefore not valid for design. The approach used by Dagel and Brynhildsen (1959) is reasonable, but eq. (7.25) does not utilize the observation that secondary creep rate is linear in logarithmic time. At failure, the total compressive strain of a box is approximately

$$\varepsilon^{\max} = \varepsilon_0 + \dot{\varepsilon}^{SC} t_{BL}, \tag{7.27}$$

where the primary deformation $\varepsilon_0 = \varepsilon^{\max} - \varepsilon^{\max, SC}$. As discussed in Section 7.4.1, it is reasonable to assume that characteristic time constants have an exponential dependence on load level. The creep compliance would then be expected to be

$$\frac{\dot{\varepsilon}^{SC}}{P} = k_2 e^{\lambda P}, \tag{7.28}$$

where λ is the slope of the shift factor versus stress curve (cf. eq. (7.21)), and k_2 is a constant. One can also assume that ε_0 is proportional to load,

$$\varepsilon_0 = k_3 P. \tag{7.29}$$

Substituting eqs. (7.28) and (7.29) into eq. (7.27) gives

$$\varepsilon^{\max} = Pk_3 \left[1 + \frac{k_2}{k_3} e^{\lambda P} t_{BL} \right]. \tag{7.30}$$

If the load, P, were chosen equal to the BCT that would be measured at the equivalent environmental conditions, then the lifetime would be practically zero, and one gets $\varepsilon^{\max} = k_3 BCT$. Solving eq. (7.30) for the lifetime gives the final expression

$$t_{BL} = \kappa \left[\frac{BCT}{P} - 1 \right] e^{-\omega \frac{P}{BCT}}, \tag{7.31}$$

where $\kappa = k_3/k_2$ and $\omega = \lambda BCT$ are constants to be determined from lifetime tests, P is the applied box load, and BCT is the box compression strength measured at the equivalent environmental conditions.

7.6.4 Accounting for variability

From the literature, it is clear that variability in strength must be considered in lifetime studies. Here we show how eq. (7.31) can be used to assess the effect of variability in BCT on lifetime. Variability in lifetime could arise from all the parameters κ, ω, and BCT, but we consider only the effect of variation in BCT. For a more through treatment, see Coffin, Wade, and Hussain (2019). We used the data of

Challas, Schaepe, and Smith (1994) for 300 BCT tests and fitted a normal distribution function to the normalized BCT values (i.e. the mean set equal to 1). This gave the coefficient of variation of 0.067; we use this as a typical value of variation in BCT. The effect is illustrated in Fig. 7.17. In a stack of three boxes, the chance of having a weak box is about three times greater than the chance with only one box. This shifts the whole distribution to lower values because the strength of the stack is given by the weakest box in the stack. The distribution of BCT values will also affect the variation in stacking lifetime.

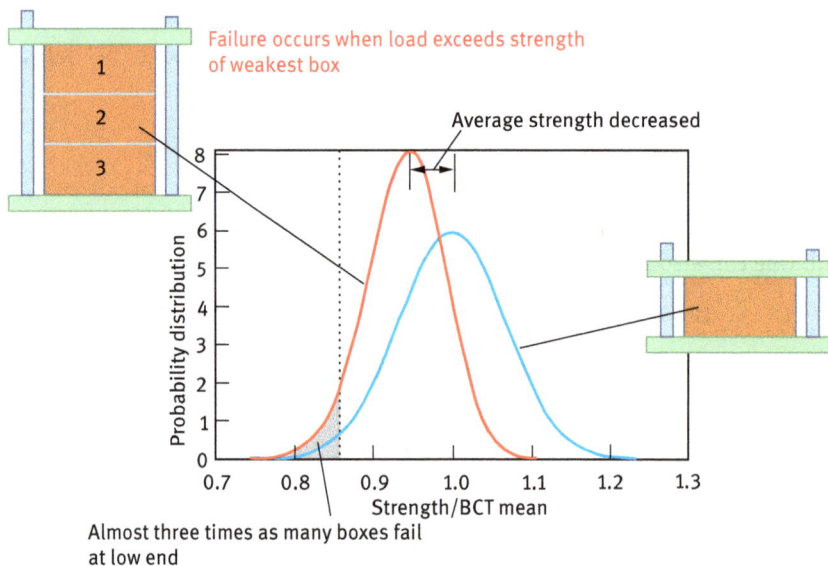

Fig. 7.17: BCT variation affects the strength of stacked boxes.

We then fitted eq. (7.31) to data reported in the literature (Fig. 7.18). We found that $\omega = 20$ is reasonable for all the data, even under conditions of cyclic moisture. In contrast, the value of κ turned out to depend on box construction and moisture conditions, shifting the line in Fig. 7.18 along the vertical axis. Using a similar but different expression for lifetime, Coffin, Wade, and Hussain (2019) found that the value of ω varied between data sets, but for this instructive example, we fix the value at 20. The large variability in lifetime data is apparent in the figure. We then substituted the normal distribution for BCT into eq. (7.30) in order to calculate predictions for the variability in lifetime. Figure 7.19 illustrates the result. For the load of BCT/2, the average lifetime is 45 days, but 2.5% of the boxes should fail in less than 7 days. This drastic reduction in the box lifetime arises entirely from the distribution of BCT values, with the coefficient of variation equal to 0.067. In a stack, the frequency of premature failures would be even bigger.

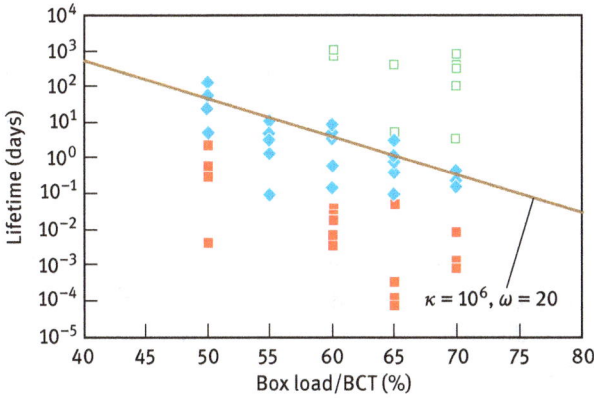

Fig. 7.18: Box lifetime versus box load from Koning and Stern (1977), each type of symbols representing one box type. The straight line is the fit of all data to eq. (7.31).

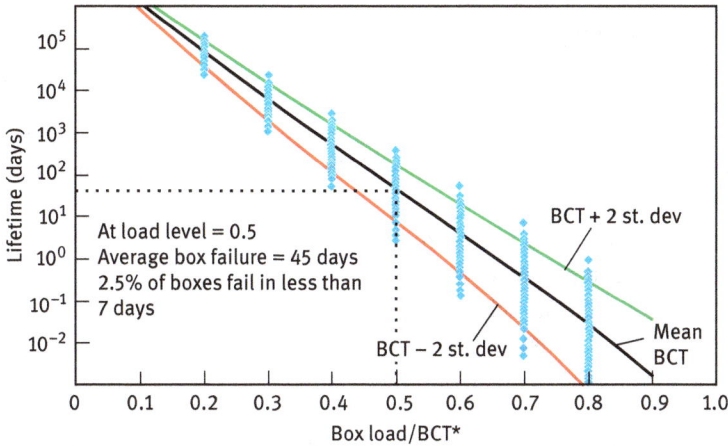

Fig. 7.19: Predicted variability in box lifetime calculated from eq. (7.31) with parameters obtained from Fig. 7.18.

Through the constant ω, box lifetime is inherently tied to the creep response of the box, and the average response appears to be predictable. Damage, flaws, poor load distribution, and variable moisture content will affect lower BCT and, in turn, have a dramatic effect on lifetime. Thus, the actual stacking failures cannot be predicted from the average properties of boxes. The effect of load level to lifetime appears to directly come from the large non-linear effect of load on creep rate, which magnifies the effect of BCT variation on box lifetimes. This explains some of the high variability in experimental data of lifetime versus load. It also suggests that the majority of stacking failures may better be attributed to the failure of a few weak boxes. Coffin, Wade, and Hussain (2019) surmised that the coefficient of variability in lifetime (COV_{LT}) could be approximated from the coefficient of variability of both maximum strain $COV_{\varepsilon max,SC}$ and BCT (COV_{BCT}) as

$$COV_{LT} = \sqrt{(COV_{\varepsilon max,SC})^2 + (COV_{BCT})^2 \left(1 + \omega \frac{P}{BCT}\right)^2}. \quad (7.32)$$

Equation (7.32) shows how ω amplifies the variability in lifetime depending on the load level. This subject of variability in strength and creep is more thoroughly discussed in Chapter 8 with specific discussion of stacking performance and lifetime distribution in Sections 8.2.2 and 8.4.5.

7.7 Concluding remarks

Accounting for the time-dependent nature of the mechanical response of paper proves to be useful for many practical problems. After studying this book, the reader should better appreciate that stress relaxation and creep should be accounted for in web mechanics and box lifetime. In addition, the reader should envision that problems in fracture, bending, and even micromechanics could account for the time-dependent nature of paper. Overall, if you account for time dependence in material behaviour, you must do the following:

- define the time scales involved in the problem
- separate recoverable and non-recoverable deformation
- consider linear and non-linear behaviour
- account for differences in tension and compression
- know that time-dependent stress distributions impact responses (e.g. accelerated creep)
- consider that multiple factors may influence observed performance (e.g. variability in lifetime)

This chapter provided an introduction to the basics of time-dependent behaviour. If one approaches a problem from a fundamental point, then appropriate constitutive equations must be introduced and robust testing must be completed to determine the values of model parameters. For example, the creasing and folding problem of Chapter 4, the crack tip models of Chapter 5, or the fluting problem presented in Chapter 9 could all include time-dependent behaviour by extending the constitutive equation.

Additionally, this chapter demonstrated that observations of creep could be used to develop empirical equations that could prove useful, such as the prediction for box lifetime. The advantage of this approach is that a simple equation could be used to help guide box designers, similar to the McKee equation for BCT discussed in Chapter 3.

Finally, no phenomena can be considered solely on its own when it comes to understanding practical problems. The example given in Section 7.5 shows that although a reasonable prediction for lifetime can be developed, without consideration of the large variability in BCT strength, such an equation would prove to be of little use. Those who are able to integrate information on time dependence with information on fracture, dimensional stability, web mechanics, variability, and micromechanics will be able to solve practical problems facing the paper and related industries.

There is still much fundamental research on creep and relaxation that is necessary. Future testing should include stages of recovery so that recoverable and non-recoverable deformations can be separated. It would be helpful to study the time dependence of recoverable deformation separately from the time dependence of non-recoverable deformation. Once this is established one could more effectively study the effects of furnish and processing on the creep response.

Very little work on the time-dependent response of paper under multi-axial states of stress has been published. In order to develop appropriate models, studies and data are required. In addition, the effect of previous loading histories on the future response of paper requires more study. It would be possible to develop an entire body of work on developing the appropriate test methods to fully characterize time-dependent stress–strain behaviour.

Studies of the effect of moisture and cyclic moisture on time-dependent behaviour of paper are still needed. The coupling of moisture and mechanical loads is a complex problem, and appropriate tests are needed from which results can better elucidate mechanisms and behaviours. The reader who wishes to gain true insight into the behaviour of paper cannot ignore the time-dependent nature of paper.

Literature references

Brezinski, J.P. (1956). The creep properties of paper. Tappi J. 39(2), 116–128.

Bronkhorst, C.A. (1997). Towards a more mechanistic understanding of corrugated container creep deformation behavior. J. Pulp Pap. Sci. 23(4), J174–J181.

Challas, J., Schaepe, M. and Smith, C.N. (1994). Predicting package compression strength geometry effect. Final Report Project 3806 to the Containerboard and Kraft Paper Group of the American Forest and Paper Association.

Coffin, D.W. (2005). The creep response of paper. In Proceedings of the Advances in Paper Science and Technology, Trans. of the 13th Fundamental Research Symposium, I'Anson, S.J., ed., (Cambridge), II, 651–747.

Coffin, D.W., Wade, K. and Hussain, S. (2019) Links between BCT and Lifetime for Corrugated Boxes, Proceedings of the 2019 International Paper Physics Conference (Atlanta, GA, USA) 160–174.

Dagel, A.V. and Brynhildsen, H.O. (1959). Tidens inverkan på papplådors belastingsförmåga. Sv. Papperstidn 62(3), 77–82.

DeMaio, A. and Patterson, T. (2005). Influence of fiber–fiber bonding on the tensile creep of paper. In Proceedings of the Advances in Paper Science and Technology, Trans. of the 13th Fundamental Research Symposium, I'Anson, S.J., ed., (Cambridge), II, 649–775.

Johanson, F. and Kubát, L.J. (1964). Measurements of stress relaxation in paper. Sv. Papperstidn. 67(20), 822–832.

Haraldsson, T., Fellers, C. and Kolseth, P. (1993). The edgewise compression creep of paperboard: New principles of evaluation. In Proceedings of the Products of Papermaking, Transactions of the Tenth Fundamental Research Symposium, Baker, C.F., ed., Oxford, 601–637, FRC.

Kellicutt, K.Q. and Landt, E.F. (1951). Safe stacking life of corrugated boxes. Fibre Contain 36(9), 28–38.

Ketoja, J. (2008). Rheology. Chapter 8. In: Paper physics, 2nd Edition Niskanen, K., ed. Helsinki: Finland: Paperi ja Puu Oy.

Koning, J.W. and Stern, R.K. (1977). Long-term creep in corrugated fiberboard containers. Tappi J. 60(12), 128–131.

Little, J.R. (1943). A theory of box compressive resistance in relation to the structural properties of corrugated paperboard. Pap. Trade J. 116(24), 275–278.

Malcolmson, J.D. (1936). The value of the compression test for corrugated boxes. Fibre Contain 21, 14–15.

Salmén, L. and Hagen, R. (2002). Viscoelastic properties. Chapter 2. In: Handbook of physical testing of paper, 1, 2nd Edition Mark, R.E., Habeger, C.C., Borch, J. and Lyne, M.B., eds. New York, NY, USA: Marcel Dekker.

Shulz, J.M. (1961). The effect of straining during drying on the mechanical and viscoelastic behavior of paper. Tappi J. 44(10), 736–744.

Seth, R.S. and Page, D.H. (1981). The stress-strain curve of paper. In Proceedings of The Role of Fundamental Research in Paper Making, Transactions of the Seventh Fundamental Research Symposium, Brander, J., ed., (Cambridge), 421–452, FRC.

Stott, R.A. (1959). Compression and stacking strength of corrugated fibreboard Containers. APPITA 13(2), 84–88.

Vorakunpinij, A. (2003). The effect of paper structure on the deviation between tensile and compressive creep responses. Doctoral Dissertation. Institute of Paper Science and Technology, Atlanta, GA.

Ward, I.M. and Sweeney, J. (2004). An introduction to the mechanical properties of solid polymers, 2nd Edition. Chichester, West Sussex, UK: John Wiley and Sons.

Tetsu Uesaka

8 Statistical aspects of failure of paper and board

8.1 Introduction

Failures of paper and board happen both in manufacturing and in end-use. The phenomena are typically manifested, on the one hand, as web breaks on paper machine and in printing houses, and on the other hand, collapses of container box in warehouses and logistic chains. Although these problems do not occur frequently, once they occur, they have serious economic consequences. Therefore, current strength specifications of paper and board all concern these less-frequently occurring problems.

An important aspect of these problems is that the *systems*, to which the paper and board are subjected, are of prime importance, that is, paper machines, printing presses, and their associated systems. Paper or board is only part of the system. Therefore, web breaks and box collapses are considered to be the performance of *systems*, rather than simple paper strength or board strength, which will be elaborated later in this chapter.

Another important point is that these problems are *rare* phenomena, and normally exhibit enormous variability in their occurrences, as we will see in the following sections. This statistical nature of the problems makes a direct "cause–effect"-type investigation, such as pilot-scale or real trials, difficult and expensive.

In this chapter, we first discuss such systems and statistical nature of failure of paper and board by taking the two examples: (1) web breaks on paper machine or in printing presses and (2) collapses of container boxes. These practical examples may help illuminate the problem and define scientific problems. Based on this preparation, we then discuss basic approaches for treating statistical and systems failures.

8.2 Practical examples

8.2.1 Web breaks in printing press and on paper machine

In printing houses in North America and Europe, web breaks typically happen with a frequency of a few breaks per 100 rolls. This means that one may find breaks at every 350 km of the printed paper, assuming that the press runs 40 in (1.02 m) diameter rolls and the break frequency is three breaks per 100 rolls. This number, however, varies considerably from one printing house to another. For example, some printing houses

Tetsu Uesaka, Mid Sweden University, Sundsvall, Sweden

https://doi.org/10.1515/9783110619386-008

occasionally encounter more than 10% of web breaks (10 breaks per 100 rolls), whereas others are beginning to show the performance below 1%. In an extreme case, such as Japanese printing houses, they can run the presses at a few breaks per 10,000 rolls, or less (Uesaka, 2005). These observations suggest the large influence of how the printing press is operated. It has been well recognized in early days of runnability research (Page and Bruce, 1985) that printing houses create much more difference in runnability than paper suppliers, and that web breaks, as seen as a time series, are essentially Poisson's process, that is, a randomly occurring process.

Similar phenomena are seen even in web breaks on paper machine. In European and North American paper mills, web breaks occur typically a few breaks per day. However, on a "bad" day, this number climbs up to more than 10 breaks per day. In another extreme, for Japanese paper mills, it is not unusual to find the break frequency of a few breaks *per month* or even less (Miyanishi and Shimada, 1998).

These anecdotes suggest the extreme importance of the *systems*, such as the way in which the production equipment is inspected and maintained, and the way the paper web is handled. Such protocol has been developed over the years based on experiences and understandings of the web transport system, the nip mechanics, and continuous improvement efforts. (The mechanics and dynamics of web transport systems, as related to tension control and system stability, are discussed in Chapter 6, and stresses produced in printing nips, which also affect breaks, are discussed in Chapter 10.)

Then, a natural question is, at a given system (specific press or paper machine conditions), what controls break frequency. Does paper matter?

One common notion is "The stronger the paper, the better the performance." In this case, we imply "strength" as standard strength parameters (tensile, tear, tensile energy absorption, burst, and stretch) as well as fracture toughness parameters. However, the typical experience is that there are no significant differences in "strength" between failed rolls and survived rolls, or between "good" months and "bad" months. This has been observed in the earlier work of runnability study (Page and Bruce, 1985) and also in a number of investigations by paper mills and printing houses. Correlations between the strength and break frequencies were found only when an extremely large number of rolls are collected and analysed, for example, in the order of 10,000 to 50,000 rolls (Deng et al., 2007a; Ferahi and Uesaka, 2001; Page and Bruce, 1985; Page and Seth, 1982). Similar observations have been reported also for paper machine runnability. For example, Miyanishi and Shimada (1998) found that strength parameters did not appear as principal factors that control web breaks on paper machine. These results suggest, again, that system conditions, such as how printing presses and paper machines are operated and maintained, have a *dominant* effect on web breaks.

Another often-cited cause for web breaks is "defects". Here we refer to those that are visible, such as edge cuts, calendar cracks, slime holes, wrinkles, bursts, thin spots, and stickies. The most notable example is reeling defects, such as crepe wrinkles and bursts, which are frequently found in the edge rolls taken from the last set of a jumbo reel. Since these are sometimes found in the outbreaks of runnability problems, major efforts have been made in improving winding–reeling operation. These incidents also prompted a series of investigations to directly observe the relation between the presence of defects and web breaks. Moilanen and Linqdvist (1996) installed a web inspection system in the in-feed section of a commercial gravure printing press to relate defects going into the press and web breaks. They observed 76 breaks in total, but 73 breaks had no relation to the defects detected by the inspection system, and noted that only three breaks *might* have been caused by defects. Ferahi and Uesaka (2001) took 50 broken samples randomly from a commercial printing press, and inspected with a special optics to detect any type of defects. Again, the result showed no indication of any specific defects, except one break (among 50 breaks) associated with crepe wrinkle.

These results may seem somewhat contradictory to the common belief that web breaks are caused by defects. However, the recent data obtained by printing houses are supporting these investigations. Figure 8.1 is break statistics data obtained by a well-maintained commercial printing house in North America (Deng et al., 2007b). The largest contributor to web breaks is, as discussed earlier, press-related causes, and the second largest is "unknown", in which the printing house did not see any obvious defects in paper webs or obvious problem in press operation. The third cause is creasing, which is another issue of a non-uniform web tension. It should be noted that the most familiar defects are all tailing in this break cause statistics. Another US printing house also reported a similar result (Uesaka, 2005).

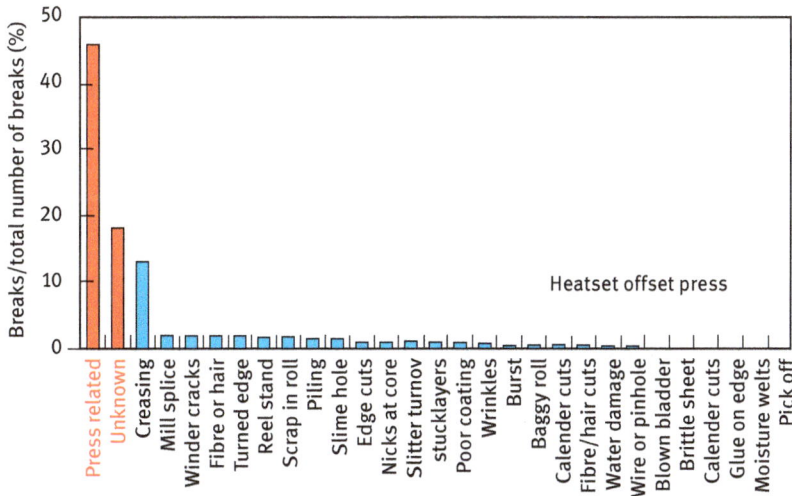

Fig. 8.1: Break cause statistics for heatset offset presses (Deng et al., 2007). Reproduced with permission from Pulp and Paper Technical Association of Canada (PAPTAC).

Therefore, in the overall statistics, such defect-driven breaks are actually minority. More importantly, those defect-related runnability problems are most effectively solved by implementing best practices rigorously. From the point of investigating the problem, the key question may be what constitutes the unknown breaks. Invisible defects are often cited to explain unknown breaks. However, they are rarely defined and are difficult to distinguish from the natural structures that the paper possesses, such as mass distributions and fibre orientations in various length scales, fibre/pigment agglomerates and their spatial distributions, and fibre damages. It may, therefore, be more appropriate to consider paper as a disordered material, rather than a uniform continuum but with defects.

From this perspective, it is natural to expect that strength of paper varies statistically. Figure 8.2 shows cumulative distributions of tensile strength for two different size specimens. The cumulative distribution function (CDF) $Q(T)$ is the probability that the paper fails at a tension equal to or less than T. A typical range of average tension in printing presses is also shown in the figure. As the specimen size increases, the distribution shifts to lower tension (i.e. the paper becomes weaker), and the shape of the distribution also changes. (We will discuss the significance of the cumulative distribution in later sections.) It should also be noted that average pressroom tensions are far below the typical strength that the paper exhibits.

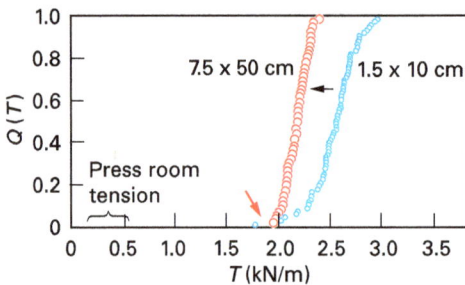

Fig. 8.2: Cumulative distribution functions of tensile strength of a newsprint for two different size specimens (1.5 cm × 10 cm and 7.5 cm × 50 cm). Tensile strength is expressed as tensile force/width.

Therefore, in order to predict web breaks, it is necessary to know (1) the strength distributions of the paper web with the sizes equivalent to the paper roll or the paper machine width, and (2) their distribution shapes, particularly in lower tails. In Section 8.3, we will discuss approaches taken to answer these questions.

8.2.2 Stacking performance of container box

One of the end-use performance parameters of container boxes is lifetime: it is the time period during which the container box can sustain its structure without any collapse under various mechanical loading conditions and also under varying environmental (temperature and humidity) conditions. Like web breaks, lifetime also exhibits *systems* and *statistical* nature. A typical example is shown in Fig. 8.3 (Nyman, 2004). Here, Box A, designated by blue, has a box compression test (BCT) value of 2.44 kN, whereas Box B, designated as red, has a value of 2.92 kN. Box B has a higher strength than Box A. However, when they are subjected to creep tests under different loads, Box A (weaker box) showed *longer* lifetime than Box B (stronger box) at the same applied loads (left figure). Another important point is the variability. Each data point shown in the left figure is an average of 10 box tests. When all individual data points are plotted, the right figure is obtained. At the lowest applied load, the shortest lifetime was about 24 h, whereas the longest one was more than 300 h.

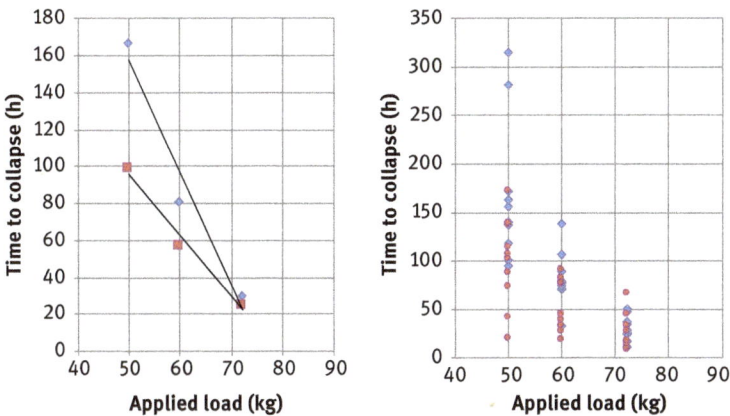

Fig. 8.3: Time to collapse of container boxes (B-flute). Box A is designated by blue, and Box B by red. (The original data is the courtesy of Ulf Newman based on his PhD work (Nyman, 2004)).

Such phenomena are not unusual for box lifetime. Kellicutt and Landt have determined creep lifetime of boxes by applying dead load under different atmospheric conditions (Kellicutt and Landt, 1951), and reported that increasing dead load shortens the lifetime, and in the range where the dead load is less than 75% of the static compression strength, each decrease of 8% dead load extended lifetime, on average, by about "eight times". Besides, the lifetime data showed huge scatters. Koning and Stern (1977) also performed similar creep tests at a 26.7 °C temperature and a 90% relative humidity, and reported the lifetime varying from 125 to 2,817 h at a given 50% dead load of the maximum compressive strength. This large scatter of lifetime data prompted researchers to take the secondary creep rate, instead of lifetime, to organize the data (Bronkhorst, 1997; Koning and Stern, 1977) based on the empirical relation found by Monkman and Grant (1956). (See also Chapters 3 and 7.) However, the plots still showed large scatters (Leake and Wojcik, 1989; Morgan, 2005; Popil and Schaepe, 2008), for example, 40–50% coefficient of variations (COVs) (Popil and Schaepe, 2008). In the end-use, it is these few "bad" boxes that create catastrophic collapse of an entire stack of boxes.

It should be noted that these data in the above references were collected under *nominally* constant loading, temperature, and humidity conditions, and for the same type of boxes. In practical situations, however, the loading histories vary widely, for example, impact loadings, such as dropping, falling, and handling abuses, as well as high-/low-frequency vibrations. Therefore, the mechanical conditions given to container boxes should be regarded as *dynamic* and *random* loadings, which includes spikes as well as very slow changes (Brandenburg and Lee, 1988). Besides, temperature and humidity vary in logistic chains, and these give another (environmental) loading to boxes through thermal and hygrostrains. (See also Chapters 7 and 9.) Therefore, the statistical nature of end-use conditions is of extreme importance to lifetime of container boxes.

To summarize, the problems of web breaks and box failure have the following common features:

– Systems failure (paper or board is part of the system): particularly the end-use conditions have large, sometimes dominant, impacts on the performances;
– Accordingly, large statistical distributions of the performance parameters (web break frequency and lifetime);
– The applied load is generally much lower than ultimate strengths that are routinely measured.

8.3 Statistical approaches for materials/systems failure

In this section, we introduce statistical approaches to analyse the distributions of strength parameters, and their dependence on the system size (scaling law). We first discuss time-independent statistical failure models that may be relevant to the problems of web breaks, and then we proceed to more general time-dependent statistical failure models that are intended for creep failure-type phenomena for container boxes.

8.3.1 The chain model

The simplest model, but most informative model, may be a chain model where elements are linked in a series, as shown in Fig. 8.4.

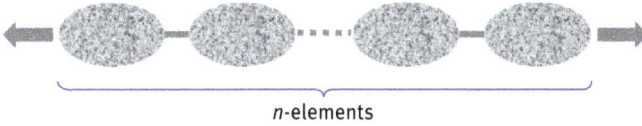

n-elements

Fig. 8.4: The chain model.

The "element" can be part of the material or of a big system. Suppose we apply tension or tensile stress T, the failure of the entire chain occurs when its weakest element fails. This is often called "weakest link" hypothesis, which appears in the early work of extreme value statistics (Fisher and Tippett, 1928a; Weibull, 1939). We denote here a CDF (Cumulative Distribution Function) of the element $Q_0(T)$, and the corresponding CDF for the entire chain of the n elements $Q_n(T)$. We assume each element is statistically independent and subjected to the same tension T. Then the survival probability of the chain, that is, $1 - Q_n(T)$, is the probability of a combined event of the survival of all elements in the chain. In other words,

$$1 - Q_n(T) = [1 - Q_0(T)]^n. \tag{8.1}$$

It is immediately clear that, as n increases (as the length of the chain increases), the survival probability of the entire chain decreases, since $1 - Q_0(T) < 1$. In other words, for a large number of elements n, the distribution of the entire chain strength is mainly determined by the low tail of the element distribution $Q_0(T) \ll 1$. For large n, eq. (8.1) can be approximated as

$$Q_n(T) \approx 1 - \exp\{-nQ_0(T)\}. \tag{8.2}$$

Equation (8.1) or (8.2) represents an important scaling law, called weakest link scaling (WLS). For example, eq. (8.2) is transformed to

$$\ln(-\ln(1 - Q_n(T))) = \ln(n) + \ln(Q_0(T)). \tag{8.3}$$

By plotting the left-hand side of the above equation as a function of $\ln(T)$ (or T), the size effects appear as a vertical shift $\ln(n)$ of the curves, as shown in Fig. 8.5. This type of plotting is called the Weibull plot.

From eq. (8.2), we can derive various distribution functions by choosing an appropriate functional form for $Q_0(T)$. However, there is no way of a priori knowing the element distribution $Q_0(T)$. One of the functional forms that may describe the low tail of

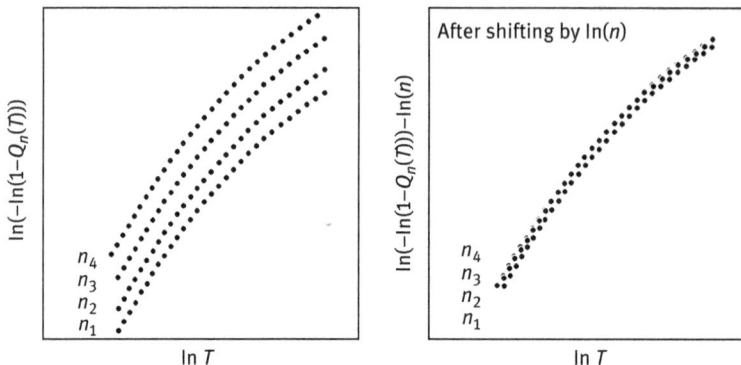

Fig. 8.5: Weakest link scaling plot (the Weibull plot).

$Q_0(T)$ is a power-law form $(T/T_0)^m$ (a m-time-differentiable, monotonically increasing function). Applying this form, we obtain the well-known Weibull distribution,

$$Q_n(T) \approx 1 - \exp\left\{ -n\left(\frac{T}{T_0}\right)^m \right\}, \tag{8.4}$$

where m is called the Weibull shape parameter (or simply the Weibull exponent). Although the Weibull distribution is probably the most frequently used form in engineering mechanics, it is only one of the possible extreme-value distributions, as discussed by Fisher and Tippet (1928a).

The most important point of the above model is the WLS (Weakest-link Scaling) law (eq. (8.1)). Although the WLS is often, mistakenly, called the "Weibull" theory, it is independent of whether the distribution is the Weibull distribution or not. Furthermore, the WLS is not limited to the one-dimensional case, such as the chain model above, but appears even in a general three-dimensional system. We will discuss this more in later sections.

8.3.2 Bundle model

In real materials, the failure of one weakest spot does not necessarily lead to the failure of the entire system. It normally requires an accumulation of damages of a certain size. Bundle model, where elements are placed in parallel, as shown in Fig. 8.6, can capture such more realistic failure.

Here the failure of each element is assumed to be statistically independent. This model was first analysed by Daniels (1945) to obtain strength statistics of fibre bundles, and, later, has been investigated extensively by a number of researchers as a model of fibre-reinforced composites with various extensions (Curtin, 1999a, 2000;

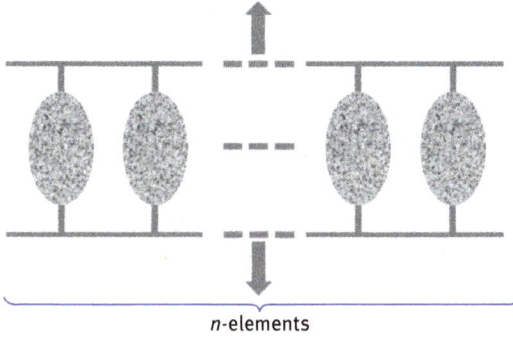

n-elements

Fig. 8.6: Bundle model.

Curtin and Scher, 1991, 1992; Harlow and Phoenix, 1978, 1979, 1982; Lu et al., 2002; Phoenix et al., 1997; Phoenix and Raj, 1992; Wu and Robinson, 1998). In spite of its simple structure, the model shows rich characteristics of statistical failure that are equally found in other more complex models.

In this model, we can consider two extreme load-sharing mechanisms: global load sharing (GLS) and local load sharing (LLS). The GLS assumes that the failure of one element results in equal sharing of the released load by all other elements, like a bundle of fibres separately loaded. The LLS, on the other hand, assumes that the load sharing happens only at the neighbouring element(s). For example, if one element, neighboured by other fibres on both sides, fails, then the other two neighbouring fibres carry the additional load that the failed element released. The LLS mimics the typical stress enhancement effect by cracks.

In the case of GLS, we can obtain analytical solutions (Daniels, 1945). The most important result is that the COVs (Coefficient of Variations) of the strength is scaled with $1/\sqrt{n}$. Therefore, as the size of the bundle n increases, the distribution of the bundle becomes narrower and narrower, that is, the strength distribution of a larger system becomes sharper. The average of the strength is, however, independent of the size of the bundle (under certain restrictions), unlike the chain model above. In other words, the larger the system does not mean the lower the strength. It is indicative that, by diffusing damage concentrations, the system becomes more uniform, in terms of strength distribution, and can avoid strength loss even if its size becomes larger. Interestingly, this phenomenon was actually observed for wet-web strength measured at a pilot-scale straining machine (Ora, 2012). Such sheets tend to be much larger than standard tensile specimens, and highly dissipative because of the wetness. The wet-web strength in this case follows the Gaussian distribution, instead of the typical extreme-value distributions, such as Weibull and Gumbell distributions. The measured strength distributions were extremely narrow: 2–3% in terms of the COV or 20–40 in terms of the Weibull modulus (Ora, 2012).

In the case of LLS, which may be more interesting in relation to realistic structures, Wu and Leath (1999a) obtained exact recursive formulas for strength distributions under different boundary conditions of a general bundle model. As the system
size (bundle size) n increases, the CDF first moves to higher tension and then moves
back to the lower tension. In other words, unlike the chain model, when the size n
is relatively small, the strength actually *increases* with increasing the system size.
After the critical size, say n_c, this trend is reversed. What is most important is that
for $n > n_c$, the WLS, eq. (8.1), emerges and the average strength starts decreasing
with the size. We find that the WLS is a rather *general, asymptotic property* of
strength when the system size becomes large. The emergence of WLS does not necessarily mean that the failure is controlled by the pre-existing weakest spot. Figure 8.7
shows the Weibull plots of the strength distributions of a model fibre-reinforced
composite with different system sizes (Ibnabdeljalil and Curtin, 1997). The number
of elements (called "fibres") in the bundle was varied from 100 to 900, and the
distribution curves were shifted according to the size by following eq. (8.3). As
shown in the figure, the WLS indeed appeared when the system size reached 400.
It should also be noted that the plots are slightly curved with heavier tails in both
ends, suggesting a non-Weibull distribution, as will be discussed later.

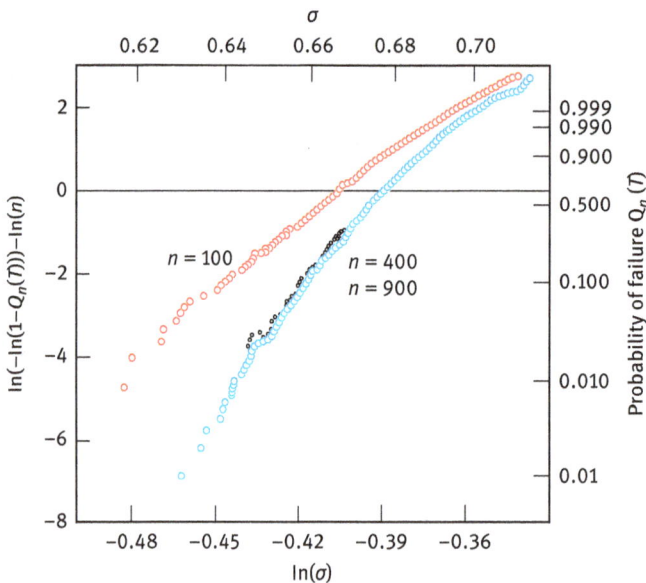

Fig. 8.7: The Weibull plots of strength distributions of bundles with different sizes. Reprinted from
(Ibnabdeljalil and Curtin, 1997) with permission from Elsevier.

8.3.3 Network models

The extension of fibre bundle models to two dimension or three dimension is conceptually straightforward, by using lattices or other networks (Fig. 8.8).

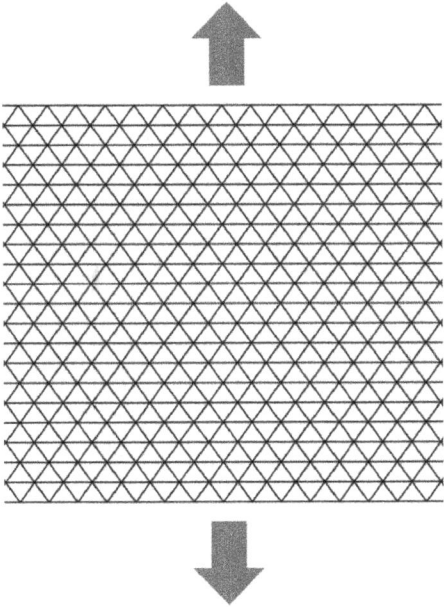

Fig. 8.8: Lattice network model (triangular lattice).

However, the investigation of the asymptotic size scaling and the final distribution, called characteristic distribution, for $n \to \infty$ is far more difficult, as it gains more degrees of freedom, more calculations of combinatory, and requires a large number of realizations. Therefore, early developments started from a relatively small network of triangular lattice (Curtin and Scher, 1990, 1991), or a simple network of electric fuses, called "random fuse model", a network of electric fuses subjected to increasing voltage where each fuse has a finite threshold value of current drawn from a uniform distribution (De Arcangelis and Herrmann, 1989; De Arcangelis et al., 1985). Later, researchers started introducing larger system sizes, more number of realizations, and more realistic failure characteristics, such as stochastic failure characteristics, into network elements (Curtin, 1999b; Curtin et al., 1997; Curtin and Scher, 1997; Mattsson and Uesaka, 2015, 2017; Newman and Phoenix, 2001).

Although these models considerably vary in terms of structures and failure characteristics of the element, they share universal features observed in those fibre bundle models developed earlier. First, the emergence of WLS as the system size increases. In other words, eq. (8.1) becomes valid once n surpasses a certain value. Of course, it *does not* necessarily mean that the weakest spot present in the initial configuration triggers breaks. The weakest spot may be first damaged, but because of stress/structure inhomogeneities, the damages are created in other spots as well. With increasing load, these damages grow, coalesce (or may even stop growing), depending on the local inhomogeneous structures. For a large enough system where WLS appears, these damages form somewhat *independent* clusters of various sizes. The system failure occurs when the largest cluster among these damages starts failing, like an avalanche. This is a qualitative explanation for why the WLS appears, although the initial weakest spot does not necessarily trigger a break. As shown later, paper seems to belong to this general class of quasi-brittle, disordered materials.

Another important universal feature is that the characteristic distribution, the distribution obtained as $n \rightarrow \infty$ when applied to the WLS ($Q_0(T)$ in eq. (8.1)), is not a Weibull distribution. It has two heavier tails in both ends in the Weibull plot, and is expressed by the double exponential form (later called, DBL distribution) (Duxbury et al., 1987). This was found in the random fuse model for electric breakdown (Duxbury et al., 1987), the fibre bundle model for mechanical breakdown (Wu and Leath, 1999b), and recently the time-dependent failure model (lifetime) (Mattsson and Uesaka, 2015, 2017). It is interesting to note that this distribution approaches, but very slowly, the Gumbel distribution as $n \rightarrow \infty$ (Shekhawat, 2013; Wu and Leath, 1999b), one of the extreme-value distributions derived by Fisher and Tippett (1928b). However, practically, it is difficult to distinguish, experimentally, this distribution from the more popular Weibull distribution when only a limited number of distribution data is available (e.g. the range of probability from 0.01 to 0.99). Therefore, the Weibull distribution is considered to be an engineering approximation of real distributions. It should also be noted that the Weibull distribution gives a conservative (higher) estimate of failure probability in the lower tail of the distribution, which may be a favourable engineering property.

8.3.4 Time-dependent, statistical failure

The above discussion is based on the notion that the failure happens when the stress (or force) reaches a certain threshold value ("strength"). However, it is well known that, even if the stress (or force) is well below the strength, failure can still happen with time, as typically seen in fatigue failure and creep failure. It is also known that such time-dependent failures are highly variable. Generally, failure phenomena are not only dependent on stress (force) histories, but also are probabilistic. In this section, we introduce a phenomenological formulation of time-dependent statistical failure by Coleman (1957, 1958), and derive some relevant formula to box performance. This theory has been extensively used later in theoretical and numerical studies of time-dependent statistical failures (Mahesh and Phoenix, 2004;

Newman and Phoenix, 2001; Phoenix et al., 1988; Phoenix and Tierney, 1983), and, unlike empirical formulae of creep and fatigue failures, it is useful to organize data from different loading histories (constant loading rate, random loading, creep, cyclic loading, etc.) in a unified way.

Suppose that the material (or the system) is subjected to a loading history $f(s)$ $(0 < s \le t_B)$, and fails at time t_B. The CDF of lifetime t_B (the probability that the material fails in less than time t_B), $F(t_B)$ is given by (Coleman, 1958; Mattsson and Uesaka, 2018),

$$F(t_B) = 1 - \exp\left\{ -\left[\int_0^{t_B} \left(\frac{f(t)}{S_c} \right)^\rho dt \right]^\beta \right\}, \tag{8.5}$$

where S_c, ρ, and β are material constants, which will be described later. Although Coleman's formulation is sometimes misunderstood as a creep formula, it deals with more general loading histories and allows to calculate important parameters of statistical failures under different loading histories. For example, for a creep test $f(t) = f_0 = \text{constant}$, the lifetime distribution is given by

$$F(t_B) = 1 - \exp\left\{ -\left(\frac{f_0}{S_c} \right)^{\rho\beta} t_B^\beta \right\}. \tag{8.6}$$

It is immediately apparent that the lifetime distribution follows the Weibull distribution with its exponent β. The exponent is a measure of uniformity: the higher the β, the more uniform (or less scatter) the lifetime. For $\beta \gg 1$, the distribution becomes more symmetric and is close to a bell shape, whereas, for $\beta \le 1$, the distribution becomes highly skewed towards zero lifetime. From eq. (8.6), the median value of lifetime, $F(t_{B,\text{median}} = 0.5)$, is calculated as

$$t_{B,\text{median}} = (\ln 2)^{1/\beta} \left(\frac{f_0}{S_c} \right)^{-\rho}. \tag{8.7}$$

In experiments, the median is more easily calculated than the mean, because typical lifetime distributions have a long tail in the longer lifetime range, which is difficult to obtain in creep tests. Equation (8.7) indicates that plotting the logarithm of a median lifetime as a function of the logarithm of creep load gives a linear relationship with the slope of $-\rho$. Therefore, at a given creep load, ρ is a measure of durability: the higher the ρ, the more durable the material. The parameter S_c is a measure of short-term strength. As shown in eq. (8.6), S_c is the creep load at which the material fails less than the unit time, with the probability of 0.63, that is, $1 - 1/e$.

A more common strength test than creep is a constant loading rate test or a constant displacement rate test. If the material shows a linear response, both are identical. In the case of the constant loading rate test, $f(t) = \alpha t$, where α is the loading rate, the distribution of the strength $f_B(= \alpha t_B)$ is given by

$$G(f_B) = 1 - \exp\left\{ - \left(\frac{1}{(\rho+1)S_c^\rho \alpha} \right)^\beta f_B^{\beta(\rho+1)} \right\}. \tag{8.8}$$

In this case, the mean of the strength, $f_{B,\text{mean}}$ is

$$f_{B,\text{mean}} = \left\{ (\rho+1)S_c^\rho \alpha \right\}^{\frac{1}{\rho+1}} \Gamma\left(1 + \frac{1}{\beta(\rho+1)} \right), \tag{8.9}$$

where $\Gamma(\cdot)$ represents the gamma function. These equations show important implications of typical strength distributions. First, eq. (8.8) shows that the strength distribution follows the Weibull distribution. Second, its exponent is $\beta(\rho+1)$, instead of β for creep. Since the typical range of ρ is 20 – 50, the exponent of creep lifetime distribution is considerably lower than that of strength. This explains why creep lifetime always shows enormous scatters as compared with strength. Equation (8.9) indicates that plotting the logarithm of mean strength against the logarithm of loading rate α yields a linear relationship with the slope $1/(\rho+1)$. In other words, the time dependence of strength is related to the parameter ρ. This is consistent with the molecular interpretation of ρ given by Phoenix and Tierney (1983),

$$\rho = \frac{U_0}{kT}, \tag{8.10}$$

where U_0 is a potential barrier to thermal fluctuations of atoms, and kT is the thermal energy with Boltzmann constant, k, and absolute temperature, T. In other words, higher temperature or more thermal fluctuation gives lower ρ, and thus more time dependence.

It is interesting to note that, although the timescales of creep failures and typical strength tests are very different, both are controlled by the same set of material parameters ρ, β, and S_c. Therefore, once these three parameters are determined, one can predict failure probabilities under different loading histories. Recently, an experimental procedure using the constant loading rate tests and numerical protocols have been developed to determine these three parameters (Mattsson and Uesaka, 2018). This method can characterize time-dependent statistical failures without performing more time-consuming and difficult creep tests.

Lastly, it may be questioned whether Coleman's theory is applicable to wider classes of materials, other than single fibre (one-dimensional material). Coleman's formulation is based on three postulates: (1) weakest link hypothesis, (2) a breakdown rule that describes the rate of breakdown (damage) as a function of force, and (3) a probabilistic failure criterion, which assume a power-law form with a damage variable. In the case of fibre network systems, it was shown that the WLS did appear with increasing the system size (Mattsson and Uesaka, 2015), and that the breakdown rule indeed follows the power law (Mattsson and Uesaka, 2013), which is one of the cases that Coleman proposed (Coleman, 1958). For the third postulate, the assumption leads to a Weibull distribution for both strength and creep lifetime. However, as discussed earlier, for two- and three-dimensional systems, both strength and creep lifetime distributions deviates from the Weibull distribution. They asymptotically approach the so-called DBL distribution when increasing the system size. Although the Weibull distribution is still a good approximation within the typical, experimentally accessible range of probability (0.01–0.99), any parameter obtained by fitting with the Weibull distribution tends to be dependent on the system size. Therefore, it should be noted that the material parameters, particularly β and S_c, are inherently system-size dependent (Mattsson and Uesaka, 2017).

8.4 Statistical failure of paper and board

In paper mills or box plants, the basic quality control parameter is currently the mean of the strength properties of choice, and strength distributions are rarely measured. Accordingly, specifications to the strength properties to ensure performance, such as web runnability and lifetime, are currently based on past experiences or on an empirical safety factor (the ratio of the strength to the operating load). Therefore, the next step would be to develop a rational control strategy by taking into account the statistical aspects of the performance.

In this section, we take an example of the paper web failure, and introduce the applications of some of the concepts of statistical strength models discussed above.

8.4.1 Distributions of paper strength

Strength distributions have been measured by using standard laboratory testing devices or pilot-scale web strainers, and have been compared with various distribution forms, such as the Weibull distribution, double-exponential distribution, the Gumbel distribution, and log normal distribution (Gregersen and Helle, 1998; Korteoja et al., 1998; Sears et al., 1965; Uesaka and Ferahi, 1999; Wathen and Niskanen, 2006).

Figure 8.9 shows a typical Weibull plot for newsprint sample of the standard size test specimen. Generally, the Weibull distribution (either two-parameter or three-parameter distribution) tends to fit to the experimental data well, but not necessarily gives the best fit among the different distributions (Korteoja et al., 1998). However, it was noted that distinguishing the fits of the different distributions is generally difficult, because the difference is subtle in the normal range of probability where the experimental data are available. Usually the lowest tail of the CDF obtained by experiments or trials is in the range of 10^{-2} to 10^{-3}, which is still much far higher than

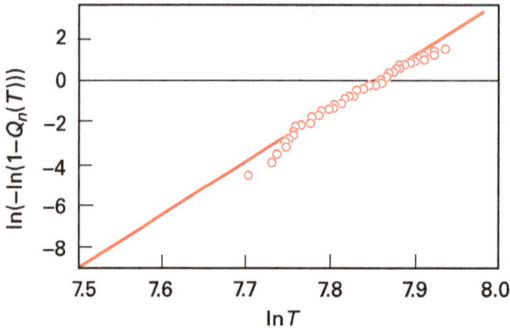

Fig. 8.9: Strength distribution of a newsprint in the Weibull plot, measured using the standard test specimens.

the probability range for realistic web breaks, for example, the probability of 10^{-9} (Uesaka and Ferahi, 1999). Figure 8.10 shows the same Weibull plot, but obtained from a larger width specimen (25 cm) using an automatic, continuous web strainer. Again, the majority of the points fits well to the two-parameter Weibull distribution, except some data scatter in the low tail (Wathen and Niskanen, 2006).

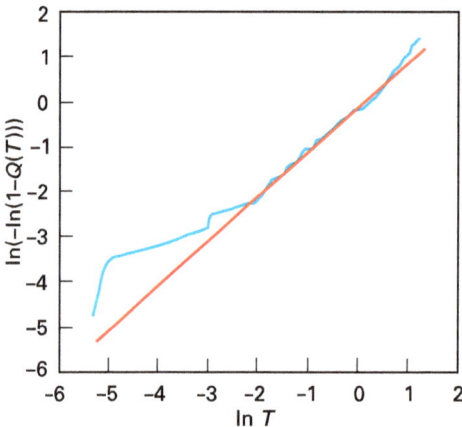

Fig. 8.10: Strength distribution of large-size specimens. Redrawn from Wathen and Niskanen (2006). Reproduced with permission from Pulp and Paper Technical Association of Canada (PAPTAC).

In the experimental determination of strength distributions, there has been increasing awareness on the importance of experimental conditions: (1) the size of the specimen, which we will discuss in more details; (2) sampling frequency and total sampling length; and (3) uncertainty of equipment. For the second factor, strength variations generally originate from various sources with a long spectrum of time/

length scales in papermaking. Therefore, the auto correlation of strength properties can exist, to a significant extent, at least within 1 hr of production. Therefore, random sampling over more than 1 hr period of production is important. Secondly, the break frequency in printing presses is in the order of 350 km, as discussed earlier. Therefore, in order to capture the variations in this length scale, it is impossible to utilize normal sampling procedures used for standard strength testing or even for tests using web strainers. The current best procedure for determining the strength distributions is to analyse data collected by an automatic testing device for every jumbo reel for an extended time period (e.g. months or a year). These data are often measured at several points across the paper machine width, and, thus, one can obtain cross-machine direction profiles of the strength distributions. The small specimen size, however, poses the scaling issue, as will be discussed later, but the sampling frequency and sampling length are ideal, and the data are reliable as long as the equipment is properly adjusted and maintained. For the third factor (uncertainty of equipment), it was observed that the determination of the strength distribution parameters, such as the Weibull exponent, is much more demanding than that of average strength, because slight misalignments of loading device, the choice of load cell of different capacities, and other equipment defects sensitively affect the distribution.

8.4.2 Size effects of strength distribution (strength scaling)

The size effect of strength distribution is critical in predicting web breaks in practical situations, because a typical size of the paper web is much larger than the standard tensile specimen size (e.g. 1.5 cm width). As discussed earlier, many material systems exhibit the WLS with increasing the system size. The question is at which size the material starts showing the WLS (eq. (8.1)).

Figures 8.11 and 8.12 show the Weibull plots of the strength distributions for the samples with varying lengths (Fig. 8.11) and varying widths (Fig. 8.12) for newsprint samples (Hristopulos and Uesaka, 2004).

In these figures, the curves are already shifted according to eq. (8.3) to evaluate the applicability of WLS. In the length direction, the WLS appears to hold even at 25 mm length, as all the curves tend to collapse to a single curve after vertical shifting, such as shown in Fig. 8.11. This means that as the length increases, the strength decreases according to the WLS. On the other hand, in the width direction (Fig. 8.12), this collapse of the curves happened only after 20 mm width, suggesting that in order to ensure the WLS, one requires a specimen of at least 20 mm width. Therefore, in order to estimate strength distributions of a realistic size roll using the WLS, the standard size specimen of 15 mm width is not sufficient, requiring, at least, 20 mm width and length in the case of newsprint.

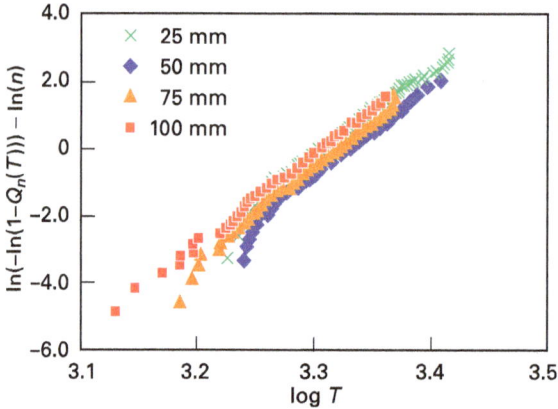

Fig. 8.11: The Weibull plots for different length specimens (Hristopulos and Uesaka, 2004). Reproduced with permission from American Physical Society (APS).

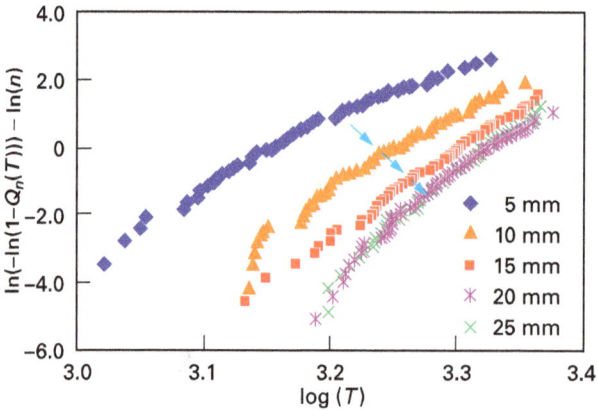

Fig. 8.12: The Weibull plots for different width specimens. (Hristopulos and Uesaka, 2004). Reproduced with permission from American Physical Society (APS).

8.4.3 Factors controlling strength distributions

The question of what constitutes the lower tail of the strength distribution has been the subject of active discussions. In the early days of research, shives in mechanical printing papers have been regarded as the culprit of web breaks, and thus constituting the low tail of the distribution. Sears et al. (1965) and Adams and Westlund (1982) observed that shive-driven breaks did dominate the cause of breaks in trials conducted by a pilot-scale web strainer. Even in small sample size testing, such as in Fig. 8.2, shives were indeed observed for the samples failed at a tension in the lower portion of the distribution curves, as indicated by the red arrow. However, in a real printing press environment, it is surprisingly rare to find shives in the failed samples. For example, the direct inspection of the 50 break samples randomly collected from a printing house showed no indication of shive-driven breaks (Uesaka, 2005), as noted earlier. Second, over the last 30 years there has been a drastic reduction of shives in the mechanical pulps, but there have been no drastic changes in the break rate. This discrepancy between the earlier literature and the direct investigations in printing houses has been a puzzle, and it may be explained by comparing the difference in the tension level used in the literature (Adams et al., 1982) and in printing presses (Fig. 8.2). The tension used in the literature is much higher than the tension applied in printing presses, in order to accelerate breaks. The tension range happened to be in the lower portion of the distribution curves (the red arrow in Fig. 8.2). In this specific range, the examination of failed samples showed the presence of shives or shives-like defects. However, at a much lower tension, such as the tension in the printing systems, there is no guarantee that shives still trigger the breaks in the same manner. Instead, it is rather natural to expect a different mechanism. Recall that the initial weakest spot does not necessarily determine the final break for disordered systems. In addition, the probability that the occasional tension spikes hit this specific tension level of the shives-related failure is also very low. Therefore, the phenomena observed in an elevated stress level, so-called accelerated break tests, may not have direct relevance to those that may happen at a much lower stress level in printing presses.

Formation (Section 2.5.2) has been suspected as another factor controlling the distribution of the strength, specifically the Weibull shape parameter. The effect of formation on *mean* strength has been extensively studied and reported in the early literature (Norman and Wahren, 1976). The speculation is that lower basis weight areas form *weak* spots that will be the cause of low stress failure, and thus decrease mean strength and increase the strength distribution. However, micro-mechanistic studies did not support this notion (Korteoja et al., 1996, 1998; Wong et al., 1995, 1996). The low basis weight areas are sometimes shielded and reinforced by high basis weight areas, so that they do not necessarily become weak spots. Besides, the complex interactions among different length scales of disorders make formation effects even more obscured. Hristopoulos and Uesaka (2003b) also estimated the effect of formation in a *macro*-length scale (i.e. visual formation) on strength distribution, and concluded that the effect is insignificant. Korteoja et al. (1998) and Wathén and Niskanen (2006) again reported very weak relations with formation. It was shown experimentally that the Weibull shape parameter, a measure of strength uniformity, does not vary systematically with visual formation of newsprints (Fig. 8.13).

However, when handsheets are prepared from the fibre suspensions of different consistencies, the Weibull shape parameter of the resulted sheets sharply decreased with deteriorating formation (Hristopulos, Uesaka, 2003a), such as shown in Fig. 8.14. In this case, formation varied in the size of 1 mm × 1 mm, much less than the characteristic length scale of the strength distribution (i.e. the typical length scale for the onset of WLS). Currently, the factors controlling the strength variations, particularly in the low tail of strength distribution, are still an open question.

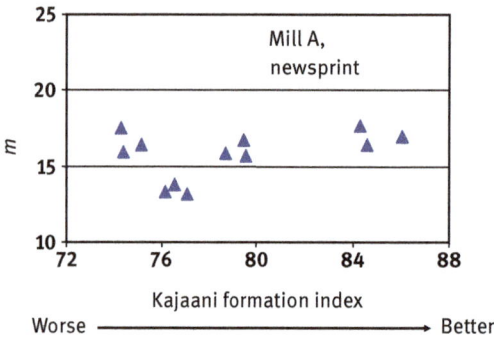

Fig. 8.13: The Weibull shape parameter *m* for newsprints of different visual formation values measured by commercial equipment.

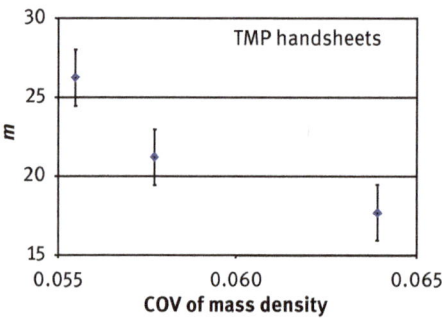

Fig. 8.14: The Weibull shape parameter *m* as a function of the coefficient of variation (COV) of mass density for handsheets prepared from fibre suspensions of different consistencies. The COV of mass density was measured by commercial equipment with the 1 mm × 1 mm size.

8.4.4 Web break prediction

With the above preparation, we will attempt to predict web breaks in a printing press. We assume the paper web as a chain of elements, and each element size satisfies the WLS requirement, as discussed above. This assumption is valid when the size of the reference specimen exceeds the critical length at which the WLS emerges. Another assumption is that the paper web has no macroscopic artificial defects. That is, we deal with the cases of "unknown" breaks and "press-related" breaks in Fig. 8.1.

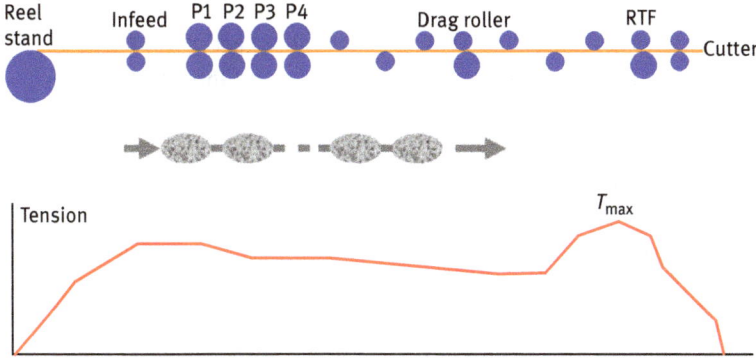

Fig. 8.15: Paper web running through a printing press system with tension variations.

As illustrated in Fig. 8.15, while the paper web runs through the printing system, it is subjected to tension variations in space-wise and time-wise. (The tension variation data have been collected at different sections of various printing systems (Deng et al., 2008).) The probability of the ith element survives while running through this process is

$$1 - Q_0(T_{\max}(i)),\qquad(8.11)$$

where $T_{\max}(i)$ is the maximum tension that the ith element would experience by going through the entire process. Then the probability that all elements contained in the chain (say, one roll) is

$$\prod_i(1 - Q_0(T_{\max}(i)))\qquad(8.12)$$

Therefore, the probability that the roll fails, BR, is

$$\mathrm{BR} = 1 - \prod_i(1 - Q_0(T_{\max}(i))).\qquad(8.13)$$

For

$$Q_0(T_{\max}) \ll 1,\qquad(8.14)$$

$$\mathrm{BR} \approx N_{\mathrm{roll}} \cdot \mathrm{Ex}\{Q_0(T_{\max}(i))\},\qquad(8.15)$$

where N_{roll} is

$$N_{\mathrm{roll}} \equiv \frac{A_{\mathrm{roll}}}{A_{\mathrm{ref}}}\qquad(8.16)$$

and A_{roll} is the total area of the paper in the roll and A_{ref} is the area of specimen used for obtaining strength distributions. In eq. (8.15), Ex{.} denotes the operation of taking the expectation for the variations of $Q_0(T_{max}(i))$. From the data of strength distributions and tension variations, we can evaluate the expectation in eq. (8.15), and thus estimate break frequency, BR (Uesaka, 2005; Uesaka and Ferahi, 1999).

Figure 8.16 is a sample calculation of the number of break frequency per 100 rolls for the papers with two different average strengths. It should be noted that in this calculation the printing press tension was assumed to be constant over time and in space, which is, of course, not the case in reality.

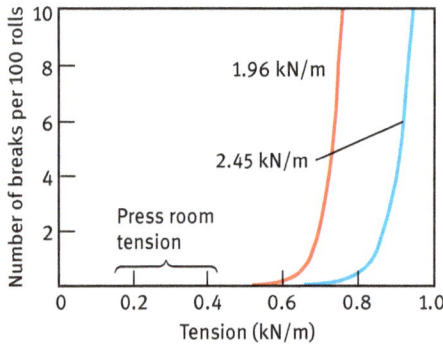

Fig. 8.16: Prediction of web breaks when tension variations are absent for the Weibull modulus $m = 19$, and average tensile strength $T_{avg} = 1.96$ kN/m and 2.45 kN/m.

As shown in Fig. 8.16, without tension variations, there is virtually no break even for the low-strength paper, in the normal range of pressroom tension. In Fig. 8.17, break rate is calculated as a function of strain variations (time-/space-wise) from its mean. Break rate is a very sharp function of the strain variation. These sample calculations illustrate that the printing press tension variations indeed have a great impact on web breaks. At the same time, these curves are sensitively affected by the strength uniformity of the paper web, that is, the Weibull shape parameter. The more uniform the strength distribution, the less the web break frequency, as expected. Field data showed that performance differences among different papers often appear also in the statistics of *press-related* breaks (Uesaka, 2005). Paper still matters!

8.4.5 Lifetime distribution of containerboard and container box

The current design criteria of container boxes or boards are essentially based on compression strength from BCT or edge-wise compression test. However, as shown in Fig. 8.3, the actual creep performance is controlled by additional two factors: one is durability ρ, which influences the relationship between lifetime and load, and

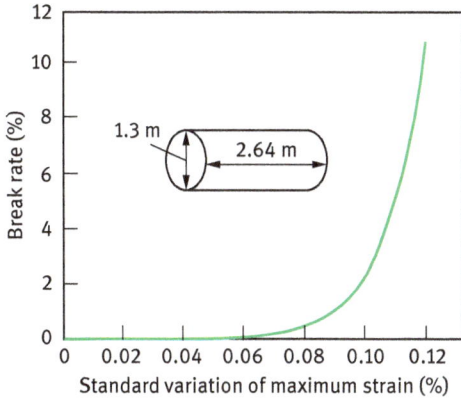

Fig. 8.17: Prediction of web breaks as a function of strain variations in a printing press. The average tension applied in the printing press is 0.2 kN/m, average tensile strength: 2.5 kN/m, strain to break: 1.1%, the Weibull modulus: 15.

the other is reliability β, which is the Weibull shape parameter of lifetime distribution, and is associated with the scatter of lifetime in Fig. 8.3 (right figure). Stronger boxes do not necessarily guarantee better performance in stacking situations. Then, the basic question is what controls these two parameters ρ and β.

First, converting defects created in corrugating processes may be mentioned as a possible source of variations. However, in the case of creep lifetime, containerboard itself also plays a very significant role. Figure 8.18 shows frequency histograms of both strength and lifetime, which are normalized by the means of each property (Mattsson and Uesaka, 2015). The strength distribution is typically more symmetric with the COV, 8–10%. However, the lifetime distribution is highly asymmetric and the COV can easily reach 100% or even 200% (i.e. $\beta \leq 1$). Clearly, the COV of containerboard far exceeds the one for boxes. This very high variation of creep lifetime, as compared with the one for strength, is not surprising from the point of the aforementioned relationship between the Weibull shape parameter $m(=\beta(\rho+1))$ for strength and the same parameter β for creep lifetime. This discrepancy is not special to containerboard or paper, but is widely seen in many materials. Figure 8.19 shows the comparison of different materials, including containerboards, carbon/glass/Kevlar fibres, and their corresponding composites, in terms of the two parameters β and ρ (Mattsson and Uesaka, 2018). Carbon fibres and their composites are characterized by their high durability ρ but low reliability $\beta(\beta < 1)$, whereas Kevlar and cellulose fibre materials are less durable, but possess better reliability.

At this stage, there has been no systematic accumulation of experimental data regarding structure–property relationships for these parameters, but there is some numerical data available for lattice network systems (Mattsson, Uesaka 2017). At a

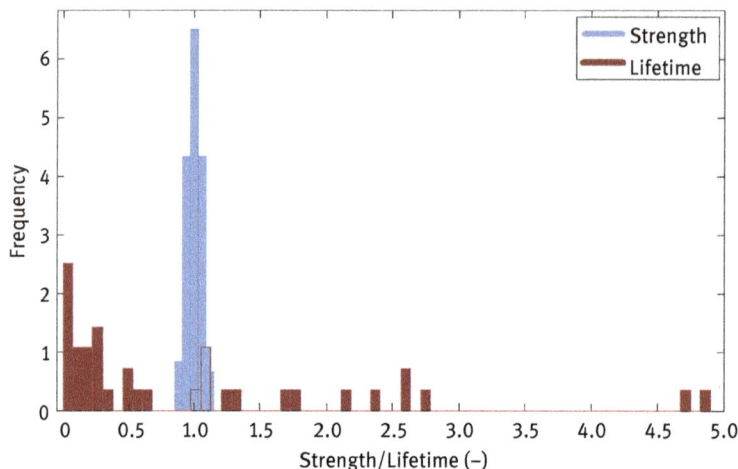

Fig. 8.18: Frequency histograms of strength distribution and lifetime distribution of linerboard. Strength and lifetime are normalized by their mean values.

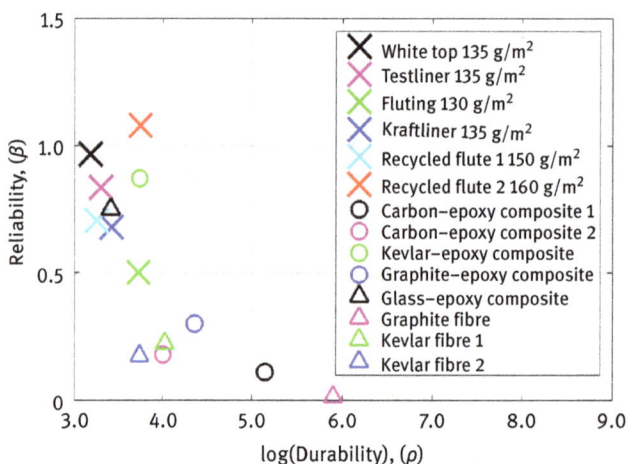

Fig. 8.19: Reliability parameter β and durability parameter ρ for different materials.

given structure, it was shown that increasing the durability/brittleness ρ of the fibre indeed decreases the reliability β of the network, as shown in Fig. 8.19. The reliability of the network is influenced by both fibres and structures. Among them, structural disorders and variations of elastic stiffness of fibres do contribute significantly to the deterioration of the reliability β. In order to design new performance with these parameters, more data are obviously necessary.

8.5 Concluding remarks

The problems described in this chapter, web breaks and box collapses, are probably two of the most recognized applications of paper mechanics. Understandably, researchers started tackling the problems by investigating the strength characteristics of paper (or board/box). However, it is immediately obvious that these problems are rarely the problem of paper, but the result of complex interactions between paper/board and end-use environments (printing presses, paper machines, stacking configurations, transport conditions, etc.), and thus the investigation of the material alone did not lead to any conclusion. In other words, it requires *systems' understandings*. Another important aspect of the problems is the *statistical nature*: web breaks and box collapses are rare phenomena, which occur with very low probabilities. Therefore, researchers are required to look at distributions, particularly tails of the distribution, rather than mean values. These systems and statistical aspects of the problems make direct cause–effect experiments/trials very difficult, if not impossible, demanding new approaches. One of such approaches, presented in this chapter, is a combination of theoretical, numerical, and experimental approaches, focusing on statistical mechanics, instead of empirical statistics. The result of this approach has been encouraging, clearing some of the outstanding problems one by one, and providing new insights for solving problems in a cost-effective way. Challenges would be the developments of high-throughput experimental machines to determine strength distributions, and a fibre network model that will allow access to the low probability tail of the distributions. These goals are now becoming within the reach with the advents of artificial intelligence-inspired robotics and ever-increasing high-performance computing.

Literature references

Adams, R.J. and Westlund, K.B. (1982). Off-line testing for newsprint runnability, 1982 Int. Print. Graph. Arts Conf., CPPA Technical Section, Quebec City, 13–18.
Brandenburg, R.K. and Lee, J.-L. (1988). Fundamentals of packaging dynamics, MTS systems corporation, Minneapolis, MN, USA.
Bronkhorst, C.A. (1997). Towards a more mechanistic understanding of corrugated container creep deformation behaviour. J. Pulp Pap. Sci. 23, J174–J180.
Coleman, B.D. (1957). Time dependence of mechanical breakdown in bundles of fibers. I. Constant total load. J. Appl. Phys. 28(9), 1058–1064.
Coleman, B.D. (1958). Statistics and time dependence of mechanical breakdown in fibers. J. Appl. Phys. 29(6), 968–983.
Curtin, W.A. (1999). Stochastic damage evolution and failure in fiber-reinforced composites. Adv. Appl. Mech. 36, 163–253.
Curtin, W.A. (2000). Dimensionality and size effects on the strength of fiber-reinforced composites. Compos. Sci. Technol. 60, 543–551.
Curtin, W.A., Pamel, M. and Scher, H. (1997). Time-dependent damage evolution and failure in materials. II. Simulations. Phys. Rev. B 55(18), 12051.

Curtin, W.A. and Scher, H. (1990). Brittle fracture in disordered materials: A spring network model. J. Mater. Res. 5(03), 535–553.

Curtin, W.A. and Scher, H. (1991). Analytic model for scaling of breakdown. Phys. Rev. Lett. 67(18), 2457–2460.

Curtin, W.A. and Scher, H. (1992). Algebraic scaling of material strength. Phys. Rev. B 45, 2620–2627.

Curtin, W.A. and Scher, H. (1997). Time-dependent damage evolution and failure in materials. Theory, Phys. Rev. B, 55(18), 12038–12050.

Daniels, H.E. (1945). The statistical theory of the strength of bundles of threads. Proc. Royal Soc. London A 183, 404–435.

De Arcangelis, L. and Herrmann, H.J. (1989). Scaling and multiscaling laws in random fuse networks. Phys. Rev. B 39(4), 2678.

De Arcangelis, L., Redner, S. and Herrmann, H.J. (1985). A random fuse model for breaking processes. J. Phys. Lettres, Les Editions de Physique 46(13), 585–590.

Deng, X., Ferahi, M. and Uesaka, T. (2007). Press room runnability. A comprehensive analysis of press room and mill database. Pulp Pap. Canada, Pulp and Paper Technical Association of Canada, Montreal, Canada 108(2), T39–T48.

Deng, X., Parent, F., Manfred, T., Aspler, J.S. and Hamel, J. (2008). Pressroom draw variations and its impacts on web breaks, Part 1. Newspaper press trials, Proc. 35th Int. Res. Conf. Iarigai, Iarigai, Valencia, Spain.

Duxbury, P.M., Leath, P.L. and Beale, P.D. (1987). Breakdown properties of quenched random systems: The random-fuse network. Phys. Rev. B 36(1), 367.

Ferahi, M. and Uesaka, T. (2001). Pressroom runnability. Part 2: Pressroom data analysis, PAPTAC Annu. Conf., Montreal, Quebec, Canada.

Fisher, R.A. and Tippett, L.H.C. (1928). Limiting forms of the frequency distribution of the largest or smallest member of a sample. Proc. Cambridge Philos. Soc. 14, 180–191.

Gregersen, O.W. and Helle, T. (1998). On the assessment of effective paper web strength, Chem. Eng., Norwegian University of Science and Technology, Trondheim.

Harlow, D.G. and Phoenix, S.L. (1978). The chain-of-bundles probability model for the strength of fibrous materails I: Analysis and conjectures.. J. Compos. Mater. 12, 195–214.

Harlow, D.G. and Phoenix, S.L. (1979). Bounds on the probability of failure of composite materials. Int. J. Fract. 15, 321–336.

Harlow, D.G. and Phoenix, S.L. (1982). Probability distributions for the strength of fibrous materials under local load sharing I: Two-level failure and edge effects.. Adv. Appl. Prob. 14, 68–94.

Hristopulos, D. and Uesaka, T. (2003). Factors that control the tensile strength distributions in paper, 2003 Int. Pap. Phys. Conf., PAPTAC, Vctoria, British Columbia, Canada, 5–17.

Hristopulos, D. and Uesaka, T. (2004). Structural disorder effects on the tensile strength distribution of heterogeneous brittle materials with emphasis on fiber networks. Phys. Rev. B 70(064108), 1–18.

Ibnabdeljalil, M. and Curtin, W.A. (1997). Strength and reliability of fiber-reinforced composites: Localised load-sharing and associated size effects. Int. J. Solids Struct. 34(21), 2549–2668.

Kellicutt, K.Q. and Landt, E.F. (1951). Safe stacking life of corrugated boxes. Fibre Contain. 36, 28–38.

Koning, J.W. and Stern, R.K. (1977). Long-term creep in corrugated fiberboard containers. Tappi 60 (12), 128–131.

Korteoja, M., Lukkarinen, K.K., Kaski, K., Gunderson, D.E., Dahlke, J.L. and Niskanen, K. (1996). Local strain fields in paper. Tappi 79(4), 217–223.

Korteoja, M., Salminen, L.I., Niskanen, K.J. and Alava, M.J. (1998). Statistical variation of paper strength. J. Pulp Pap. Sci. 24(1), 1–7.

Leake, G.H. and Wojcik, R. (1989). Influence of the combining adhesive on box perfprmance. Tappi 8, 61–65.

Lu, C., Danzer, R. and Fisher, F.D. (2002). Fracture statistics of brittle materials: Weibull or normal distributions. Phys. Rev. E 65, 671021–671024.

Mahesh, S. and Phoenix, S.L. (2004). Lifetime distributions for unidirectional fibrous composites under creep-rupture loading. Int. J. Fract. 127(4), 303–360.

Mattsson, A. and Uesaka, T. (2013). Time-dependent, statistical failure of paperboard in compression, In: I'Anson, S.J. (ed.), Adv. Pulp Pap. Res. Trans. 15th Fundam. Res. Symp., The Pulp and Paper Fundamental Research Society, Cambridge, UK, 711–734.

Mattsson, A. and Uesaka, T. (2015). Time-dependent statistical failure of fiber networks. Phys. Rev. E 92(4), 42158.

Mattsson, A. and Uesaka, T. (2017). Time-dependent breakdown of fiber networks: Uncertainty of lifetime, Phys. Rev. E 95(5), 53005.

Mattsson, A. and Uesaka, T. (2018). Characterisation of time-dependent, statistical failure of cellulose fibre networks, Cellulose, 25(5), 2817–2828.

Miyanishi, T. and Shimada, H. (1998). Using neural networks to diagnose web breaks on a newsprint paper machine. Tappi J., TAPPI 81(9), 163–170.

Moilanen, P. and Linqdvist, U. (1996). Web defects inspection in the printing press – Research under the Finnish Technology Program. Tappi J. 79(9), 88–94.

Monkman, F.C. and Grant, N.J. (1956). An empirical relationship between rupture life and minimum creep rate in creep rupture tests. Proc. ASTM 56, 593–620.

Morgan, D.G. (2005). Analysis of creep failure of corrugated board in changing or cyclic humidity regimes, 59 Appita Annu. Conf. Exhib., Appita, Auckland, New Zealand, 393–400.

Newman, W.I. and Phoenix, S.L. (2001). Time-dependent fiber bundles with local load sharing. Phys. Rev. E 63, 21507–21520.

Norman, B. and Wahren, D. (1976). Mass distribution and sheet properties of paper, Fundam. Prop. Pap. to its end uses, Trans. Symp. London BPBIF.

Nyman, U. (2004). Continuum Mechanics Modelling of Corrugated Board, Report TVSM, Ulf Nyman, Structural Mechanics, Box 118, SE-221 00 Lund: Sweden.

Ora, M. (2012). The effect of web structure on wet web runnability, Doctoral thesis, Dep. For. Prod. Technol., School of Chemical Technology, Aalto University, Finland.

Page, D.H. and Bruce, A. (1985). Pressroom Runnability, Newspr. Press. Conf. (Chicago, IL, USA CPPA ANPA).

Page, D.H. and Seth, R.S. (1982). The problem of pressroom runnability. Tappi J. 65(8), 92–95.

Phoenix, S.L., Ibnabdeljalil, M. and Hui, C.-Y. (1997). Size effects in the distribution for strength of brittle matrix fibrous composites. Int. J. Solids Struct. 34, 545–568.

Phoenix, S.L. and Raj, R. (1992). Scalings in fracture probabilities for a brittle matrix fiber composite. Acta Met. Mater. 40(11), 2813–2828.

Phoenix, S.L., Schwartz, P. and Robinson, H.H. (1988). Statistics for the strength and lifetime in creep-rupture of model carbon/epoxy composites. Compos. Sci. Technol., Elsevier 32(2), 81–120.

Phoenix, S.L. and Tierney, L.J. (1983). A statistical model for the time dependent failure of unidirectional composite materials under local elastic load-sharing among fibers. Eng. Fract. Mech., Elsevier 18(1), 193–215.

Popil, R.E. and Schaepe, M.K. (2008). Dependence of corrugated container lifetime on component properties, Prog. Pap. Phys. Semin., TKK (Helsinki University of Technology), Espoo, Finland.

Sears, G.R., Tyler, R.F. and Denzer, C.W. (1965). Shives in newsprint: The role of shives in paper web breaks. Pulp Pap. Mag. Canada 66(7), T351–T360.

Shekhawat, A. (2013). Fracture in Disordered Brittle Media, Cornell University.

Uesaka, T. (2005). Principal factors controlling web breaks in pressrooms – Quantitative evaluation. Appita J. 58(6), 425–432.

Uesaka, T. and Ferahi, M. (1999). Principal factors controlling pressroom breaks, TAPPI Int. Pap. Phys. Conf., TAPPI Press, San Diego, CA, 229–245.

Wathen, R. and Niskanen, K. (2006). Strength distributions of running webs. J. Pulp Pap. Sci. 32(3), 1–8.

Weibull, W. (1939). The phenomenon of rupture in solids, The Royal Swedish Institute for Engineering Research, Stockholm.

Wong, L., Kortschot, M.T. and Dodson, C.T.J. (1995). Finite element analyses and experimental measuremnt of strain fields and failure in paper, 1995 Int. Pap. Phys. Conf., TAPPI, Niagara-on-the-lake, Ontario, Canada, 131–135.

Wong, L., Kortschot, M.T. and Dodson, C.T.J. (1996). Effect of formation on local strain fields and fracture of paper. J. Pulp Pap. Sci. 22(6), J213–J219.

Wu, B.Q. and Leath, P.L. (1999). Failure probabilities and tough-brittle crossover of heterogeneous materials with continuous disorder. Physical Review B, Phys. Rev. B 59, 4002–4010.

Wu, E.M. and Robinson, C.S. (1998). Computational micro-mechanics for probabilistic failure of fiber composites in tension. Compos. Sci. Technol. 58, 1421–1432.

Part III: **Reactions to moisture and water**

Artem Kulachenko
9 Moisture-induced deformations

9.1 Introduction

In this chapter, we discuss the impact of moisture by considering cases in which moisture induces out-of-plane deformations. Such undesirable deformations are a nagging quality problem for the papermakers and, in most cases, are caused by changes in moisture. We will consider three phenomena, namely, curl, fluting, and cockling.

9.2 Curl

Instead of presenting a wordy definition of what the curl is, let us start the discussion by making a simple experiment demonstrating this phenomenon. Take an A4 copy paper. Make sure it is flat. Cut two 10 × 2 cm strips into two orthogonal directions (Fig. 9.1). Then take a slightly wet paper towel and rub it once against one of the sides of each strip.

Fig. 9.1: Demonstration of paper curl: wetting the paper strips (left), reaction to the wetting (centre), and final dry state (right).

Observe carefully what happens. Both strips bulge out, but in a different manner. Strip number one has a bulge along the longest dimension and strip number two across it. Watch carefully what happens next. After half a minute or so, the curvature in both strips is reduced and they nearly restore the original flat shape, but only for a short while. They continue to bend and now they bulge out in the opposite direction. Remember, there is no external force applied at this point. As previously, the orientation of the bulge is different in the two strips. Two minutes after the start of the experiment, the strips are permanently deformed. You leave these strips untouched for a day or a week and nothing will change – they will remain curled. This phenomenon is called paper curl.

Artem Kulachenko, KTH Royal Institute of Technology, Stockholm, Sweden

https://doi.org/10.1515/9783110619386-009

There are a number of questions that we can formulate based on these experiments. Let us state some of them:

1. Why does the paper deform upon moistening?
2. Why does the paper react to applied moisture by bending?
3. Why does it bend first away from the moisture application?
4. Why does it deform differently in the strips taken in two orthogonal directions?
5. Why does the strip end up having a permanent curl?
6. Why does the curl change the direction of bending during the deformation process?

We will explain the physics behind curl by answering these questions one by one.

9.2.1 Moisture-induced deformations

Our first question is: *why does the paper deform once it is exposed to moisture?* The main constitutive component of paper is the wood fibre that is hydrophilic. The wood fibres readily absorb and desorb moisture. It can be demonstrated by weighing the fibres under changing relative humidity (RH). RH is defined as the ratio of the current partial pressure to the saturation vapour pressure of water at a given temperature. Figure 9.2 shows how the fibre moisture content (the per cent of moisture in relation to the total dry weight of the fibre) changes with RH at 20 °C. The first line is the desorption isotherm from the "never-dried" state. The second line corresponds to subsequent absorption followed by desorption (third line).

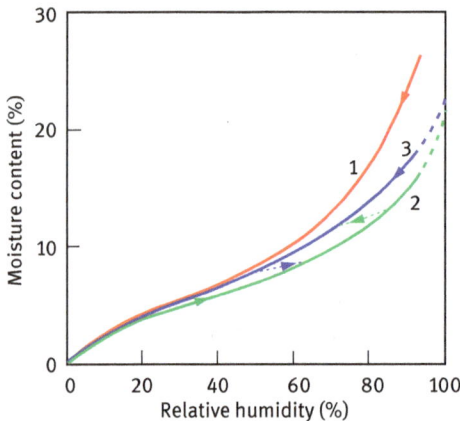

Fig. 9.2: Moisture content versus relative humidity in a sulphite pulp at 20 °C, after Christensen and Giertz (1965). Reproduced with permission from The Pulp and Paper Fundamental Research Society (www.ppfrs.org).

There are a few interesting things to note here. First, there are three distinct stages in the sorption process shown in Fig. 9.2. In the initial stage, corresponding to low partial pressure in the region from 0% to 20% RH, there is a steep change in moisture content. In the second stage, from 20% to 70% RH, the dependence is almost linear. Finally, in the third stage of high partial pressure in the region from 70% to 100% RH, there is again a sharp change in moisture content. Various theories have been used to explain S-shape of the sorption isotherm (e.g. a multilayer water sorption theory (Brunauer et al., 1938) or the solution theory (Hailwood and Horrobin, 1946) for two hydrates). The solution theory presents the so-called Hailwood–Horrobin sorption equation, which relates moisture content, temperature, and RH, and it is often used to describe wood drying.

The second interesting aspect to notice is the presence of hysteresis in the sorption isotherms. At equal RH, the fibre may have lower moisture content during absorption than during desorption. Finally, the third interesting fact is that the initial desorption curve (1) lies above the subsequent desorption curve (3). Let us try to understand why.

The fibre is a composite structure itself. The water can access only the hydroxyl (functional -OH) groups of the fibre. Hydroxyl groups arranged in a coherent manner have a tendency to form strong hydrogen bonds with each other. The water cannot penetrate mutually bonded regions. In the fibre, however, there are plenty of non-crystalline, amorphous molecular regions, which are not arranged in an ordered fashion and thus remain accessible. Furthermore, the fibre has larger discontinuities often referred to as pores. The water can penetrate these regions and form hydrogen bonds with accessible hydroxyl groups (Walker, 2006).

By absorbing water, fibres swell and increase in dimension. When losing moisture, the fibres shrink. During shrinkage, the constituents of the fibre are drawn close together and are able to form hydrogen bonds. During this process, some pores in the fibre structure are closed and the water below may not escape through them. On the other hand, not all of these newly formed hydrogen bonds can be broken again during the subsequent rehydration and fewer are available for bonding with water. This implies that some pores on the fibre surface are permanently closed during the first desorption. Consequently, the initial desorption curve lies above all the subsequent desorption curves.

The adsorption curve always lies below the desorption curve because the adsorbed water molecules have to penetrate and push apart the consolidated fibre wall constituents. The desorbed water molecules, on the other hand, cannot escape easily because some channels and pores are being closed during desorption. The work dissipated during sorption is seen through hysteresis.

What happens when a fibre swells in a sheet of paper? In the network of fibres, swelling of an individual fibre affects neighbouring fibres to which it is connected. As a result, the paper expands upon applied moisture and shrinks when the moisture is removed. This effect is called hygroexpansion and is very similar to thermal expansion in other materials, with moisture being a counterpart of temperature.

Figure 9.3 shows hygroexpansion strain in a freely dried paper sheet presented versus cycling RH and moisture content.

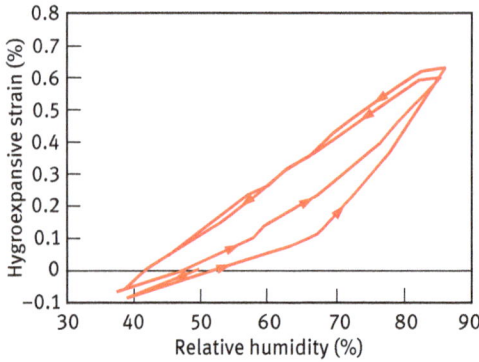

Fig. 9.3: Hygroexpansion strain in a freely dried sheet versus: a) cyclic relative humidity and b) cyclic moisture content, data of Uesaka et al. (1992). Reproduced with permission from Pulp and Paper Technical Association of Canada (PAPTAC).

There is an interesting observation to be made: the hysteresis is only seen in the plot with respect to RH and not with respect to the moisture content. The hysteresis originates solely from the absorption/desorption of moisture by the constitutive fibres (Fig. 9.3).

The hygroexpansion strain ε^{hyg} enters constitutive relations identically to the thermal expansion strain ε^{th} and other inelastic strains ε^{in} (such as plastic or creep strain). In the one-dimensional case, it can be expressed as

$$\sigma = E\varepsilon^{el} = E\left(\varepsilon^{tot} - \varepsilon^{hyg} - \varepsilon^{th} - \varepsilon^{in}\right), \tag{9.1}$$

where σ is the stress, E is the elastic modulus, and ε^{el} is the elastic strain. See eqs. (9.5) and (9.6) for the three-dimensional case.

So, we know now that paper reacts to moisture by changing dimensions. The next question is now: *why does the paper react to applied moisture by bending?* Bending is always associated with a strain gradient across the thickness. For example, in the case of pure bending with a small deflection and a linear elastic response of the material, the through-thickness distributions of the axial strain and stress will take the shape depicted in Fig. 9.4.

There are three important things to note here:
1. The distribution of strain and stress is antisymmetric with respect to the middle plane (the neutral plane passing through the middle of the sheet).
2. In the absence of other loads, the total strain is equal to the elastic strain.
3. The stress is proportional to elastic strain.

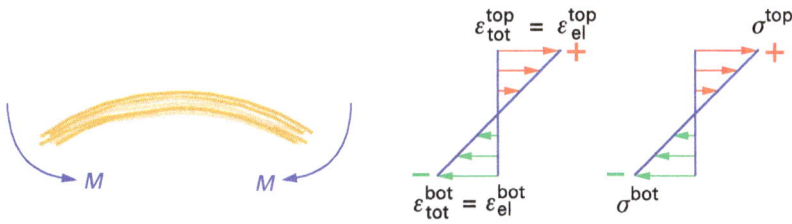

Fig. 9.4: Strain and stress distribution in case of pure bending.

Now we have to understand where the through-thickness strain gradient comes from when the moisture is applied. One of the classic examples of such a gradient from the mechanical textbooks is a bimetallic beam under thermal load (Roark and Young, 1975). In this example, the beam is made of two or more layers having different thermal expansion properties. When the temperature is increased, one layer expands more than the other, creating a strain gradient and the beam bends. Making an analogy with this case, we can speculate about the variation of the elastic properties of paper in the thickness direction. Paper does constitute a layered structure and such variations naturally exist. However, if this is the only reason for bending, then the paper should always bend towards the same side regardless of where the moisture is applied. A quick test will reveal, however, that this is not the case, and the paper consistently bends towards the side of applied moisture. Let us try to find out *why the paper initially always bends away from the direction of applied moisture* (Fig. 9.1)?

We apply moisture from the upper side of the paper strip. This means that the moisture will be absorbed by that side first. Paper is a porous structure and water transport inside the paper is controlled by diffusion, which does not happen instantly. Because of the delay in moisture diffusion, the upper side is affected by the moisture before the bottom side. These effects include changes to the material properties as well as hygroscopic swelling. Most importantly, for us, the differential swelling through the sheet causes the part of the sheet further away from the moisture to act as a constraint, prohibiting deformation. As a result, a through-thickness strain gradient appears, and the bending takes place. Had the moisture been equally absorbed across the thickness of paper, this would not have happened. Unfortunately, ensuring the equilibrium in moisture absorption in paper has proven to be very challenging (Ketoja et al., 2001).

Will it be possible to reproduce the same phenomenon by heating a sheet of metal from one side? If the temperature–moisture analogy is correct, then we should be seeing exactly the same effects with the metal exposed to one-sided temperature change. It appears to be difficult since the heat transfer in metals happens faster than the moisture transfer in paper. It is practically impossible to create a sufficient temperature gradient through the thickness to demonstrate this phenomenon unless extreme conditions, such as quenching, are used.

Let us consider closer how the through-thickness strain and stress profile look like in case of non-uniform moisture distribution. Imagine that we have a linear elastic paper. The paper is initially dry, which means the initial moisture content is zero. Now, a one-sided application of moisture resulted in a linear through-thickness moisture distribution. The paper is assumed to be unconstrained. Figure 9.5 shows the total and elastic strain distribution in this case.

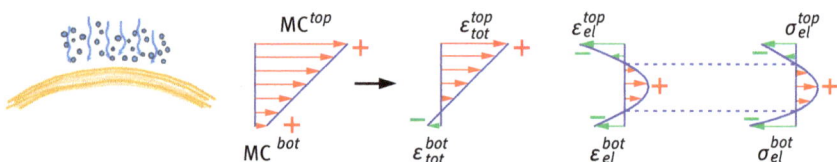

Fig. 9.5: Strain and stress distribution created by a linear distribution of moisture content (MC).

Although the deformed paper may look similar to that from the case of pure bending, there are a number of important differences:

1. The total strain is asymmetric. In fact, the total strain can be positive throughout the thickness.
2. The elastic strain is no longer the only contribution to the total strain. The total strain is now the sum of elastic strain and hygroexpansion strain (eq. (9.1)) in which the hygroexpansion part is greater than the elastic one.
3. There are negative elastic strains and stresses on the expanding top side. Negative means that this side of paper is under compression. These compressive strains develop purely due to internal constraints. The top side expands but the expansion is restrained by the layer of paper below it.

    It is similar to the situation in which the material expands between two walls. Despite the expansion, the stress in the material will be compressive. In terms of eq. (9.1), it means that the total strain, ε^{tot}, is positive but the elastic strain, ε^{el}, is negative because $\varepsilon^{el} = \varepsilon^{tot} - \varepsilon^{hyg} < 0$.

    The material in the bottom layer does not expand as much as the upper layer since the relative moisture change is lower in the bottom. The bottom layer experiences compression due to the global bending. Only the material in the middle of the paper has positive stresses.
4. Although the stress and the elastic strain are related by a constant factor (the elastic modulus), this factor varies at different points of the sheet because the elastic modulus itself is dependent on the moisture content. The elastic modulus of paper decreases with moisture and can be reduced by nearly a factor of two upon increasing moisture content from 5% to 10% (Fig. 9.6). This means

that at an equal elastic strain, regions having higher moisture content will be less stressed.

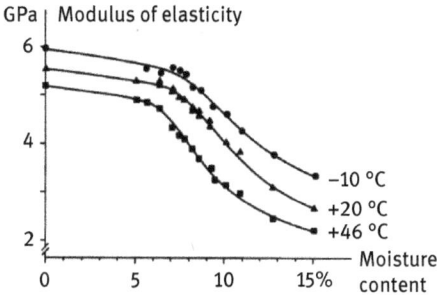

Fig. 9.6: Dependence of the elastic modulus of paper on the moisture content at different temperatures (Nissan et al., 1977).

So, now we know that the moisture causes a dimensional change. We also know that the paper bends due to the moisture gradient developed because of the delay in absorption across the thickness. Now the next question is: why does the paper bend differently in the strips taken in the two orthogonal directions? The first thing that we can recognize is that the commercial paper is not isotropic. Paper properties become anisotropic during paper manufacturing. The directions of papermaking are usually not marked on the office paper, so how can we distinguish between the different directions by doing a curl experiment? There are more fibres oriented in the machine direction (MD) compared to the cross-direction (CD), which is reflected in a higher elastic modulus along the MD. So, it is natural to assume that the paper will also expand more in the MD. It turns out to be the opposite.

The largest contribution to the hygroexpansivity has long been attributed to changes in the fibre dimensions in the transverse direction during drying. This is because the fibres exhibit a significantly larger relative dimensional change in the transverse direction in response to moisture content alternation. One of the pioneering works with direct observation of the transverse hygroexpansivity of softwood pulp fibres with the help of microradiography was presented by Tydeman et al. (1966). Despite the essentially two-dimensional observation allowing the visual detection of the fibre width change, the authors used the optical density of the image on the radiograph to correlate it with the fibre thickness change. By doing this, they could observe both the width and thickness change. The reported values for the fibre width shrinkage averaged around 30% for the beaten pulp and 20% for unbeaten from fully soaked to dry state were used widely since then. It was also observed that the fibres contracted even more in their thickness during drying, which was explained by the collapse of the fibre structure. The changes in the fibre wall thickness for fibres with open lumen could not be detected at that time. The fibre wall thickness has later been reported to be an important parameter with thinner wall thickness resulting in greater hygroexpansion (Uesaka and Moss, 1997, Pulkkinen et al., 2009).

Knowing that the dimensional changes on the fibre level happen mostly in the direction of fibre thickness, in other words, across the fibre, it is easier to accept that the hygroexpansion is higher in the CD. The experimental result in Fig. 9.7 confirms that. It shows the dependence of hygroexpansion coefficients in the MD and CD on the ratio of paper's elastic modulus in these directions for the sheets dried without constraints.

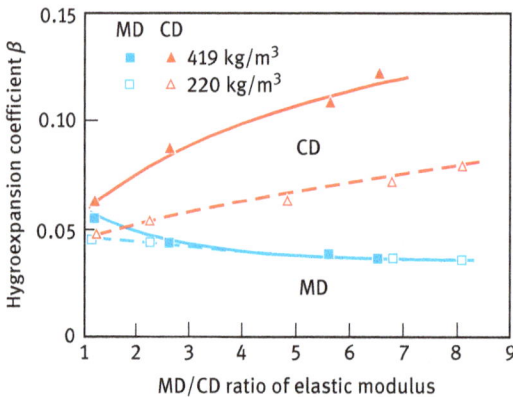

Fig. 9.7: MD (squares) and CD (triangles) hygroexpansion coefficient β (eq. (2.5)) against fibre orientation anisotropy, measured as a function of the MD/CD ratio of elastic modulus in laboratory sheets of two densities, 419 kg/m^3 (black symbols) and 220 kg/m^3 (white symbols). The strain is assumed to be given in per cent. Data of Uesaka (1994). Reproduced with kind permission from Springer New York LLC.

Indeed, in a typical commercial paper, the hygroexpansion coefficient in the CD is approximately three times higher than in the MD. Does this prove that the transverse shrinkage of the fibres dominates the overall hygroexpansivity? With a theoretical framework based on composite theory, Uesaka (1994) showed that a large fraction of the hygroexpansion should come from the longitudinal deformation of the fibres. This was later confirmed by the three-dimensional fibre network simulations of Motamedian and Kulachenko (2019), who estimated that the role of the longitudinal hygroexpansion of the fibres is comparable to the transverse one. The significance of the longitudinal deformation of the fibres is indirectly confirmed by the fact that the hygroexpansion coefficient of fibre networks is nowhere near the transverse hygroexpansion of the fibres even in the case of dense sheets.

Apart from the anisotropy in fibre orientation, there is yet another contribution to the differences between MD and CD. Paper dried freely has a significantly greater hygroexpansion coefficient then the one dried under constraints. In the commercial paper machine, paper is dried under constraints in the MD and has limited constraints in the CD. The higher hygroexpansion coefficient for the freely dried sheets has been explained by two factors: increased bonding and increased stress transfer. The latter is believed to be associated with the changed configuration of the bonded area. Using a micromechanical approach (Uesaka and Qi, 1994), combined with an earlier developed theory (Uesaka, 1994), Uesaka showed that if the fibres develop out-of-plane orientation in the bonded region, it can enhance the stress transfer and, therefore, results in large hygroexpansivity of the entire network. On the contrary, fibres stretched during drying will have a greater probability of forming connections without a significant out-of-plane orientation. Another explanation of the higher hygroexpansion coefficient and yet lower elastic modulus of freely dried paper is offered by Salmén et al. (1987), who attributed the difference to the orientations of the hemicellulose polymer chains affected by the drying stresses. They prove it by bringing the hygroexpansivity of constrained dried sheets to the level of freely dried sheets through exposing them to the humidity cycling (Fig. 9.8). Such "reconditioning", according to their explanation, releases dried-in stresses and affects the hemicellulose structure by making it more reactive to moisture changes.

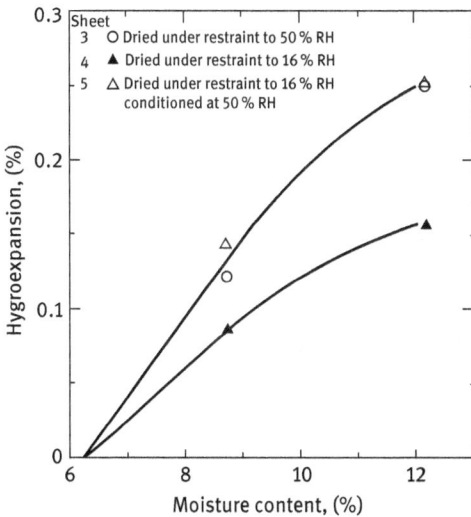

Fig. 9.8: Hygroexpansion versus moisture content for the sheets dried freely, under restraints with and without reconditioning (Salmén et al., 1987).

Figure 9.8 also shows that the dependence of the hygroexpansion strain on moisture is generally non-linear and thus requires moisture-dependent hygroexpansion coefficients. In the range of moisture relevant to practical curl problems (0–15%), the hygroexpansion coefficient can essentially be assumed constant.

What about thermal expansion? The effect of thermal expansion is rarely considered in paper. The thermal expansion coefficients of paper range between 4 and $15 \cdot 10^{-6}/°C$ (de Ruvo et al., 1973), which is comparable to that of stainless steel ($11–13 \cdot 10^{-6}/°C$). This means that changing the paper temperature by 1,000 °C will cause a strain change of 0.1% (with a thermal expansion coefficient of $10 \cdot 10^{-6}/°C$). The equivalent strain change can be achieved by changing the moisture content with just 1% (with hygroexpansion coefficient 0.1). Thus, hygroexpansion has a much greater influence on dimensional changes in paper than thermal expansion.

9.2.2 Effect of moisture history

Now we know that the differences in curl appearance between the MD and CD strips are caused by inherited anisotropy of paper properties as well as drying anisotropy. Both, having the fibres aligned mostly along the MD and having the constraint drying conditions in the MD, contribute to the decreased hygroexpansivity in this direction. We can move on to the next question, namely, *why does the strip end up having the permanent curl?* After we apply water, some part of it is absorbed in the direction of thickness causing hygroexpansion, but at the same time water begins to evaporate from the paper surface. In fact, there is a very small probability that at any point during this process, the moisture in the sheet is equally distributed throughout the thickness.

Apart from causing expansion, the added moisture reduces elastic properties of paper, so the actual stress state is very difficult to predict accurately. However, even under these circumstances, if the paper was 100% elastic, it would recover its flat shape once the original moisture content is restored. The moisture should indeed restore over time and become uniform, but we know already that the curl does not disappear. The fact that there are some permanent deformations suggests that the material underwent some irreversible inelastic changes.

Typically, inelastic deformations are described with plasticity or creep. Both can be used interchangeably or simultaneously for practical problems (Mäkelä, 2007). Creep formulations are more popular for describing paper, because they offer more flexibility – for example, creep may happen at any stress, while plastic deformation typically requires a definition for yield stress at which plasticity sets on. Creep allows for stress relaxation and delayed elastic recovery (Gates, 1963, Algar, 1965), which is not permitted by classical plasticity. Plasticity, on the other hand, has an explicit measure for residual strain and it can be combined with creep in cases when the time dependency needs to be incorporated. We will use plasticity to explain the residual curl in this case. Alternative creep formulations can be found in literature (Uesaka, 1991, Lu and Carlsson, 2001b, Uesaka, 2001).

How large are the moisture-induced irreversible deformation? To get some reference, you can first create a residual curl by subjecting a paper sheet to pure bending. You may try to roll a paper into a tube and let it go. You will see some residual curvature developed. The smaller the radius of the tube, the more severe the residual curl. However, the curl always develops in the direction of rolling – unlike our case when the paper switches the direction of curvature during the test. Also, you need to bend it a lot to get the residual curl similar to the one from moisture. This tells that the moisture-induced strains in our experiment are very large.

Even knowing that there have been some irreversible inelastic deformations, we still lack the answer to the question: *why does the curl change its direction during the test*? This is a difficult question that is occasionally asked during dissertation seminars on topics related to moisture–paper interaction. Indeed, the side that absorbed the moisture first had to expand more than the opposite side, so it is counterintuitive – why did it end up being shorter in the end? We found out already that there are negative stresses on the expanded upper part (Fig. 9.5). The compressive stresses may actually be substantial enough to promote negative plastic deformations. Plastic and hygroexpansion strains are both inelastic in nature (eq. (9.1)), but in contrast to hygroexpansion strain, the plastic strain is unrecoverable, so it will stay permanent even when the load is removed.

Having a permanent negative strain on the expanding side means that this side will be shorter. According to the stress diagram in Fig. 9.5, the bottom side is also under negative stress (which may actually be higher than the one on the top side because the elastic modulus on the bottom is higher due to lower moisture). So, why does the top side have a larger negative unrecoverable strain in the end? It brings us to an important material characteristic of paper, that is, under equally large stresses, the plastic strain appears earlier in a region with higher moisture. Figure 9.9, for example, shows the tension–strain curves at different RH in tension (positive strain) and compression (negative strain). The first interesting detail from Fig. 9.9 is the difference in the tensile and compressive behaviour. Having the same elastic modulus in tension and compression, paper reaches the yield stress at a lower strain when compressed.

However, the detail we will focus on is that a higher RH decreases the elastic modulus and the yield stress. For a quantitative demonstration of the effect of plastic strain, we make use of the stress–strain curves in Fig. 9.9a and show the zoom-up view of the CD compressive stress–strain curves at 10% and 90% RH in Fig. 9.10.

Assuming that unloading is completely elastic, one can see that after exposure to a tensile loading, T_u, plastic strain is created at 90% RH, but not at 10% RH. Alternatively, unloading from a strain ε_u, the plastic strain is larger at 90% RH than at 10% RH. Applying these observations to the stress distributions in Fig. 9.5, one

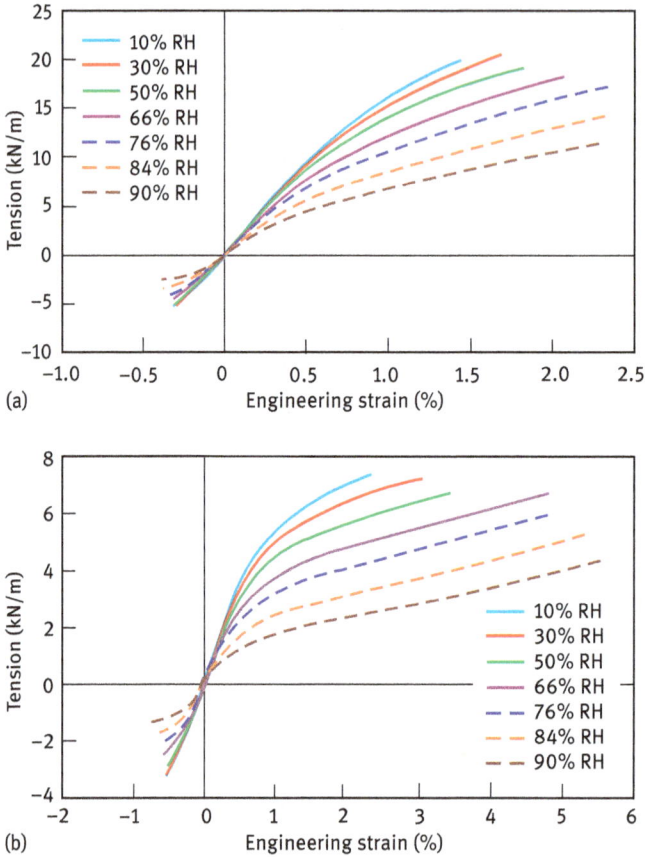

Fig. 9.9: (a) Tension–strain curves in MD and (b) CD at different relative humidities (RH) for a paper made of chemical pulp (data courtesy of Petri Mäkelä).

can understand how the difference in moisture content can create a clearly larger permanent compressive strain on the top side than on the bottom side. The same happens in our experiment with the paper strips.

The effect of moisture on inelastic deformation is similar to the effect of temperature in metals or plastics, where the increase in temperature reduces the yield stress. In our case, both upper and bottom sides of the paper are under negative stress (Fig. 9.5), but the upper side will have larger irreversible strains developed during this process because it has higher moisture content throughout the entire process. How can we predict the moisture distribution in the thickness direction?

The moisture distribution as a function of position and time can be calculated by solving a diffusion equation which is the easiest but not the only one way of

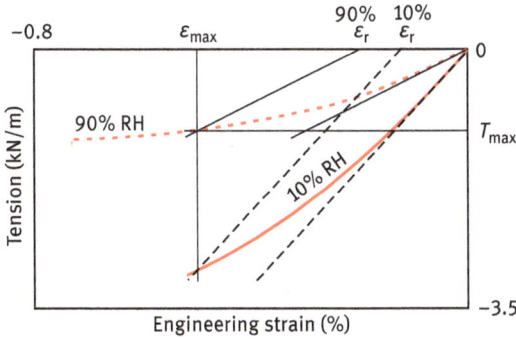

Fig. 9.10: Compressive tension–strain curves in CD of the paper in Fig. 9.7 at 10% RH and 90% RH.

computing it. For a one-dimensional case, the diffusion equation can be written as (Youngs, 1973)

$$\frac{\partial c}{\partial t} = \frac{\partial}{\partial z}\left(D\frac{\partial c}{\partial z}\right), \tag{9.2}$$

where c is the moisture content, z is the thickness coordinate, and D is the diffusion coefficient. The diffusion coefficient is strongly dependent on moisture and temperature, both of which increase it. The diffusion coefficient is usually estimated by fitting the solution to experimental data (Crank, 1956). Figure 9.11 shows the dependence of the diffusion coefficient on moisture. When large temperature changes are involved, the moisture transport is connected to heat transport and, therefore, more equations are involved in the solution.

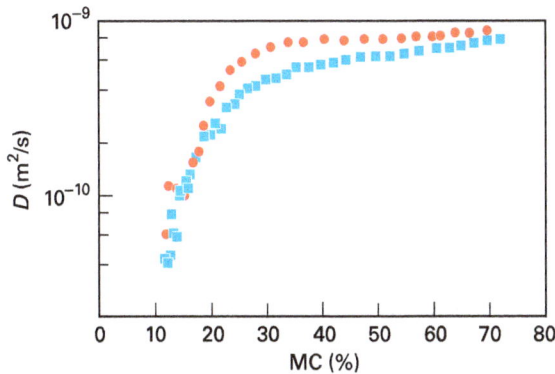

Fig. 9.11: Diffusion coefficient against moisture content at room temperature for paper made of an unrefined (circles) and refined chemical pulp (squares) (Topgaard and Söderman, 2002). Reproduced with kind permission from Springer Science + Business Media B.V.

Continuing the moisture–temperature analogy, eq. (9.2) is similar to the heat equation. It is suitable for relatively slow absorption and desorption. The boundary conditions can be described through the moisture content which is calculated from the equilibrium relations between ambient RH and moisture content similar to that depicted in Fig. 9.2. Figure 9.12 shows a solution of the diffusion equation for the case of one-sided moisture application and drying with assumed initial and boundary conditions (Uesaka, 1989).

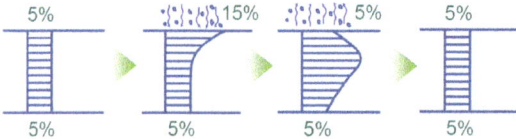

Fig. 9.12: Moisture distribution during one-side moisture application and drying calculated with the diffusion equation (Uesaka 1989).

As the moisture penetrates deeper in the sheet and towards the bottom layer, causing expansion, the negative stresses follow as we learned already. When the sheet starts to dry, the moisture moves back towards the side where it was applied and evaporates there. Since moisture transport inside the paper is, in this case, controlled by diffusion and the diffusion rate increases with increased moisture content, the moisture escapes through the upper, wetter part. Thus, at the end of the drying, when the moisture content is restored, there is an asymmetric profile of residual strains inside the sheet with larger negative residual strains on the top side. Larger negative strain means that the top side is shorter, which is manifested in the observed permanent curl towards the upper side.

Usually, the moisture transport equations and equations governing the deformations of the sheet are solved independently. The moisture history is determined and then applied to the sheet as an external load. There is, however, a coupling between the stresses and moisture content. Stretching the paper increases the moisture content and this is attributed to the increase of the accessible hydroxyl groups exposed upon stretching of fibres (Olsson and Salmén, 2001). Coupling stresses and moisture complicates the solution considerably, but due to the relatively weak nature, such a coupling is never considered.

Thus, the short answer to the question about the reversed direction of curl is that it is due to *specific* moisture and stress history. Accounting for the moisture history is very important when you need to interpret the impact of moisture on dimensional changes in paper. Indeed, even the preceding moisture and stress history of the paper we used (Fig. 9.1) may influence the magnitude of curl observed, but it cannot explain the direction that is always towards the side of moisture application.

The stress history in paper is created in the manufacturing process (Section 2.5). In particular, the paper web is under MD tension during the drying process. Using simple terminology, this creates internal stresses in the paper. These stresses can be released when paper is moistened the first time because moisture softens paper. The release of dried-in strains is manifested through the irreversible shrinkage after the first moistening–drying cycle of a paper that was initially dried under restraint (Fig. 9.3). In a freely dried sheet, the irreversible shrinkage is insignificant (Fig. 9.3). Subsequent moisture changes happen without considerable irreversible changes in hygroexpansion. The release of the eventual dried-in strains will influence curl if the dried-in strains are non-uniform in the thickness direction of paper. Indeed, one sometimes adjusts the temperatures of the last drying cylinders on the paper machine so that the paper comes out curl free.

Let us summarize the considered curl case demonstrated in Fig. 9.1 by going through the deformation process once again:

1. The water is applied, and it is absorbed by the top side of the sheet.
2. The fibres in the paper react to water by swelling and the paper expands. The expansion is non-uniform through the thickness because the moisture is unequally distributed.
3. Non-uniform expansion creates a strain gradient in the thickness direction making the top side longer. As a result, the paper bends downwards.
4. Non-uniform expansion also causes considerable negative stresses on the top side of the sheet which leads to negative plastic strains.
5. The water evaporates from the sheet escaping from the wetter top side because the diffusion is faster in the regions of elevated moisture.
6. After the water escapes, the top side has larger negative residual strains meaning that this side is shorter. Consequently, the paper bends upwards and remains curled.

9.2.3 Interesting case: competing curls

Typically, because of the anisotropic nature of paper, there is a principal direction of the curl.[1] Paper usually curls along the CD. What will happen if a paper has isotropic in-plane properties? Isotropic sheets can be produced in a handsheet mould. When the fibre suspension is poured onto the mould, the fibres orient randomly in-plane, and the paper receives more or less isotropic in-plane properties.

Let us consider a simple case: an initially flat square sheet of paper with isotropic in-plane properties. The moisture is increased uniformly, and we assume that the curl appears due to the variation of hygroexpansion coefficients through the thickness. Let us assume that the upper layer has a lower hygroexpansion coefficient than the bottom

---

[1] The case considered here was inspired by the work of Nordström et al. (1998) and Lu and Carlsson (2001a).

layer. This type of curl is called the structural curl. We do not consider any type of inelasticity in this case. Moreover, because of symmetry, we will only consider a quarter of the sheet (Fig. 9.13). All the analyses described further were performed with the finite-element method.

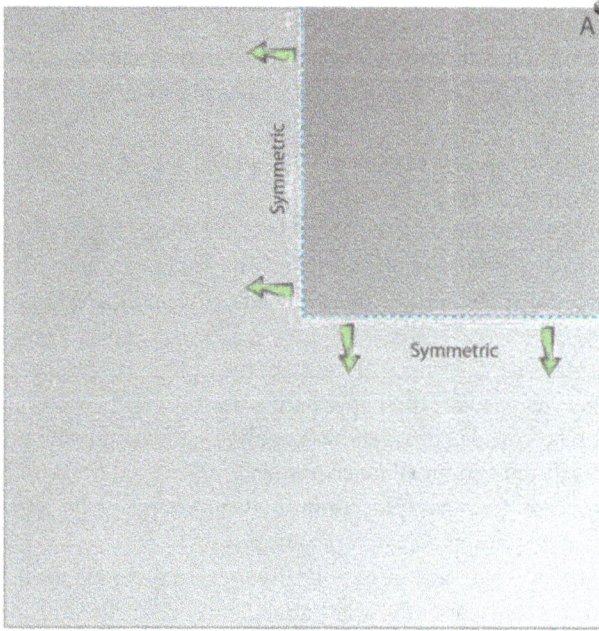

Fig. 9.13: Square paper sample in a "competing curl" experiment, and only the upper right part will be considered in the analysis.

Let us follow the development of deformation upon increased moisture. Figure 9.14a shows the deformed state after the moisture was slightly changed (the dotted frame shows the initial plane state). It is possible to see that both edges lift up, which can be interpreted as a presence of curl in two directions of the sample. At some point, however, one of the directions "prevails" and rolls over the other (Fig. 9.14b). The transformation was sudden, which is typical to a buckling-type bifurcation.

It is rather difficult to anticipate buckling in this situation because buckling requires an in-plane compressive force. Where does it come from, in this case, when the structure expands in all the directions? Another puzzling thing is that the sheet bends along the Y-axis and not along the X-axis. Everything is identical in these two directions, but the paper prefers one of them, for some reason. Before these questions are answered, it is good to ensure that we do deal with buckling bifurcation. Normally, a linear buckling analysis can be helpful to get an estimate of a

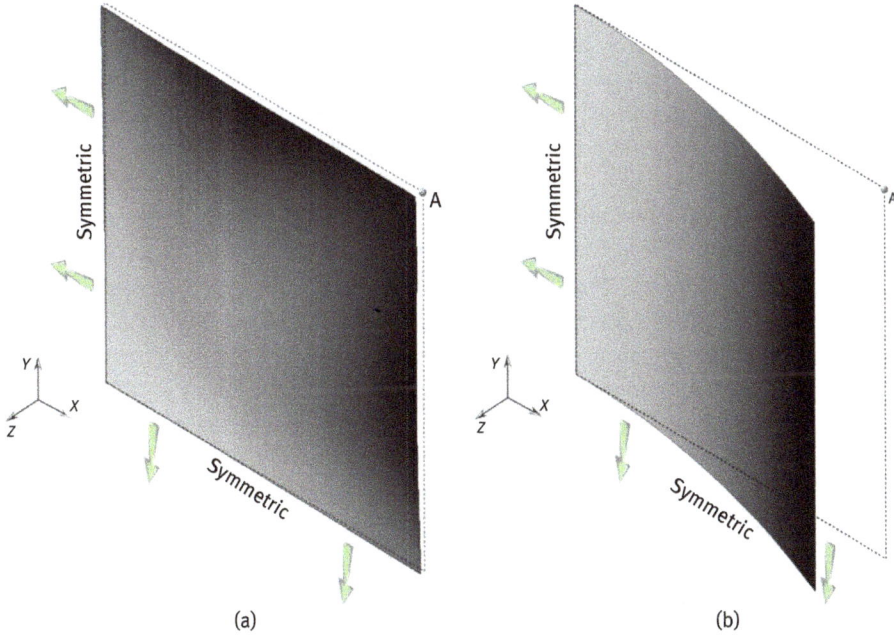

Fig. 9.14: Deformation of square paper sample computed by non-linear analysis prior to buckling bifurcation (a) and post-buckling form (b).

buckling load. In a finite-element method, for example, such analysis is typically done in two steps. The first step is a linear static analysis with a unit force. The static analysis returns to the so-called stress–stiffness matrix, which represents the addition to the stiffness of the system depending on the applied load. The linear buckling load can then be calculated in the second step by adding an unknown multiplier of the stress stiffness to the regular stiffness and solving the eigenvalue problem

$$([\mathbf{K}] - \lambda_i[\mathbf{K}_{ss}])\{\mathbf{u}\}_i = 0 \tag{9.3}$$

where $[\mathbf{K}]$ is the stiffness matrix of the system, $[\mathbf{K}_{ss}]$ is the stress–stiffness matrix, λ_i are the load multipliers (buckling load if a unit force was used in the static analysis), $\{\mathbf{u}\}_i$ is the eigenvector of displacements representing the buckling shape, and, finally, i is the index of the eigenmode (the number of eigenmodes is equal to the rank of the system matrices). The solution of eq. (9.3) is the buckling load and the buckling shape. The post-buckling amplitudes cannot be determined by the linear buckling solution. Furthermore, the linear solution usually overestimates the buckling load because it does not account for large deflections associated with buckling. Neither can it account for material non-linearities. In order to estimate the post-buckling amplitudes, a non-linear solution procedure should be used. In the non-linear procedure, the changes in the system are introduced in a number of steps, for example, the moisture is gradually changed from the original to the final value. At each step, the non-linear equilibrium equations are solved. Sometimes, ideal modelling systems may be more stable than the real systems, which are full of imperfections. A perturbation in a form of geometrical imperfection can be introduced in the model prior to the solution in order to be able to capture instability. The load step chosen in the non-linear solution should be adequately lower than the first buckling load determined by the linear buckling analysis. If the first buckling load is unreasonably high, it may indicate that the system would rather fail first than buckle. Nevertheless, this can also be misleading because the conditions for buckling bifurcation can sometimes develop during the deformation process when the deformed structure is different from the original. This is exactly what happens in our case. Let us now consider the procedure for the non-linear solution.

Usually, buckling manifests itself through sudden changes in geometry upon a small increment of force. We have moisture as a body force in our case. It seems natural then to observe the out-of-plane displacement of the corner point A (Fig. 9.13) as a function of moisture change. Figure 9.15 shows how the displacement in the Z-direction of point A changes with the increased moisture (both scales are normalized with respect to the maximum values at the end of the loading phase). A sudden change in a slope indicates the onset of buckling. It should also be reflected in the sudden change of elastic energy of the system. Observing the energy change has an advantage in case the point to monitor cannot be defined. The disadvantage, however, is that it is difficult to measure elastic energy, and in case of local buckling, for example, the changes in energy corresponding to local buckling are too small to be captured.

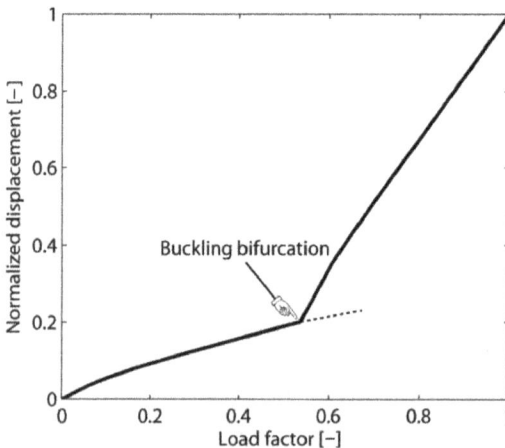

Fig. 9.15: Normalized displacement of the corner of the paper sheet (point A in Fig. 9.13) as a function of moisture change.

Let us first discuss the origin of compressive forces. The compressed areas in the system can be shown with the third principal stress. Figure 9.16 visualizes the distribution of the third principal stress in the middle plane of the sheet just prior to buckling instability. The dark areas correspond to the maximum absolute value of the third principal stress. The arrows show the direction of the third principal stress.

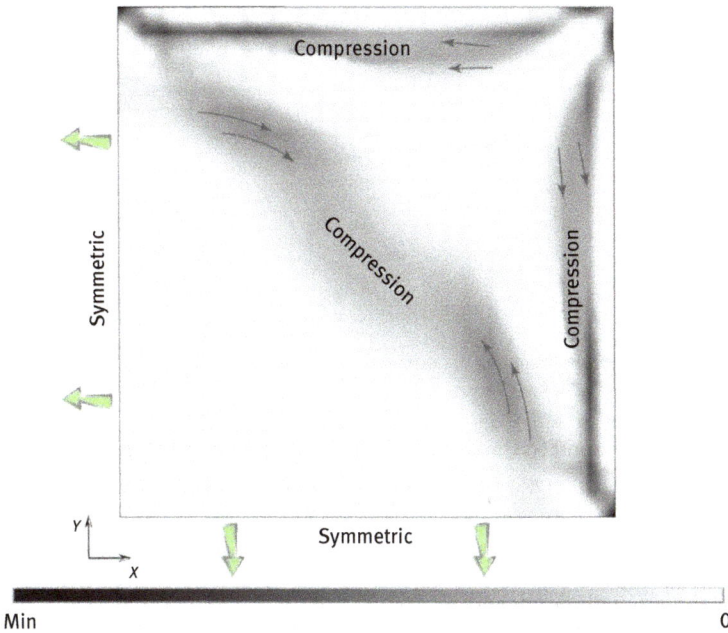

Fig. 9.16: Third principal (compressive) stress in the middle surface of the paper.

The presence of negative stresses can be explained by the fact that paper wants to expand equally large in all directions, but the geometry of the sample does not allow for this, creating internal in-plane constraints. Paper escapes these constrains through buckling. If the stresses are plotted in a cylindrical coordinate system with the origin in the centre of the entire sample (or left bottom corner in Fig. 9.13), then it becomes clear that the hoop stresses in this coordinate system are the dominating component in creating the observed compression.

In Fig. 9.16, there are actually three compression fronts in the sheet: one in the middle and two near the edges. The edge fronts are equally large, and both have absolutely the same chances to "prevail". The fact that the sample prefers one of them is a matter of small perturbations. Think about a thin ruler that you compress with a finger in the length direction until it buckles (Fig. 9.17).

Fig. 9.17: Ruler buckling experiment.

You can make it easily bend to the left or to the right by very small adjustments of how you apply the force. The post-buckling form is not unique. It means that just a small change in the system can change the direction of the curl in our case.

In the above-mentioned analyses, we assumed that the sheet is perfectly flat. What happens if it is not so? How sensitive is the post-buckling shape for initial non-flatness? If we assign a non-zero Z-coordinate to the right upper corner of the sheet, then at a certain magnitude of non-flatness, the post-buckling shape takes a different form (Fig. 9.18) in which the upper corner leans towards the centre.

It clearly tells us that the initial non-flatness may affect the post-buckling shape and eventually the buckling load. This is an important conclusion for packaging applications, where the walls of the packages are rarely flat after converting. What will happen if the paper is longer in one direction or the sheet is circular? Will we still see buckling? If so, how will the post-buckling shape look like? We leave this intriguing question for the readers' consideration.

In the earlier considered example, both curl components, K_x and K_y, have the same sign. What happens if K_x and K_y have opposite signs? Imagining a situation when the paper has a tendency to bend upwards along the X-direction and downwards along the Y-direction can be difficult. Thus, a reasonable question can be *how is that possible in the first place?* One scenario of this is having different fibre orientations in the two layers of the sheet as illustrated in Fig. 9.19.

When this sheet is subjected to a moisture change, the first layer will have higher hygroexpansion in the X-direction while the second layer will expand more in the

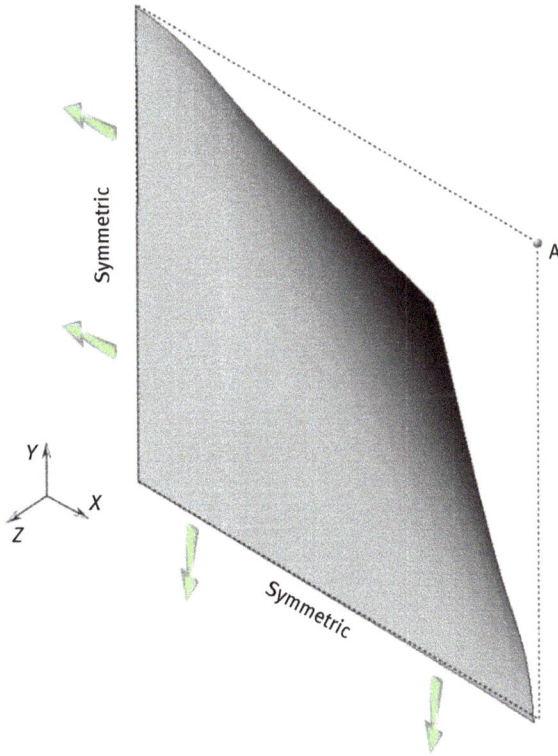

Fig. 9.18: Post-buckling shape in the presence of initial non-flatness.

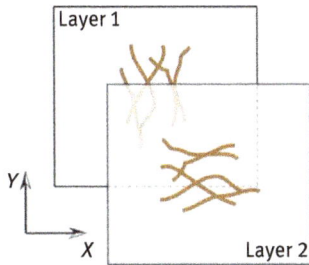

Fig. 9.19: Two-sidedness in the form of different fibre orientation in two layers of a sheet.

Y-direction. As a result, the sheet will bend downwards with $K_y < 0$ and upwards with $K_x > 0$. Similar to the previous case, both curls are equally strong and only one is to remain in the end. This time, however, the turning point comes almost immediately at only a tiny fraction of the moisture change applied in the previous case (Fig. 9.20). Obviously, the compressive stresses could not have developed enough to cause buckling. So, what is the source of this instability? The answer is not that simple. It is interesting that this instability can only be predicted by a non-linear solution. The

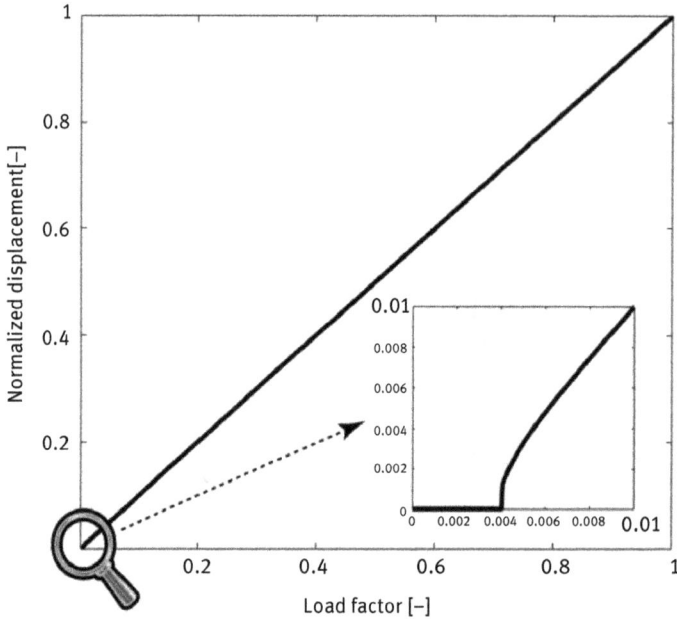

Fig. 9.20: Normalized displacement of the corner of the square paper sheet (point A in Fig. 9.13) as a function of moisture change. The total moisture change was identical to the previous case of "competing curl" with the curl components having the same sign.

linear solution will actually never predict one of the curl components prevailing. Figure 9.21 shows (a) linear and (b) non-linear solutions of the system at equally large moisture change. The corner point A in the linear solution has not deflected at all.

We observed that even small initial imperfection can change the solution. Are there multiple solutions to the same problem? The solution of a linear elastic problem is unique according to Kirchhoff's theorem (Fung and Tong, 2001). The uniqueness can be lost in a number of ways. For example, when the material becomes unstable (case of a plastic hinge) or when the equilibrium equations are changed during the deformation process. The latter can happen when non-linear members are included in the equilibrium equations. Once we account for large deflections in the analysis, the second-order terms can be significant enough to introduce the basic changes into equilibrium at some point. What we deal with then is a case of non-linear bifurcation (Finot and Suresh, 1996, Dunn et al., 2002). An illustrative example of this situation is a ball on a sharp hill (Fig. 9.22). Under gravity, the ball is in equilibrium in Position 1, but it is easy to imagine this equilibrium is unstable and in reality, the ball will end up being in either Position 2 or 3. Such a situation is called bi-stability.

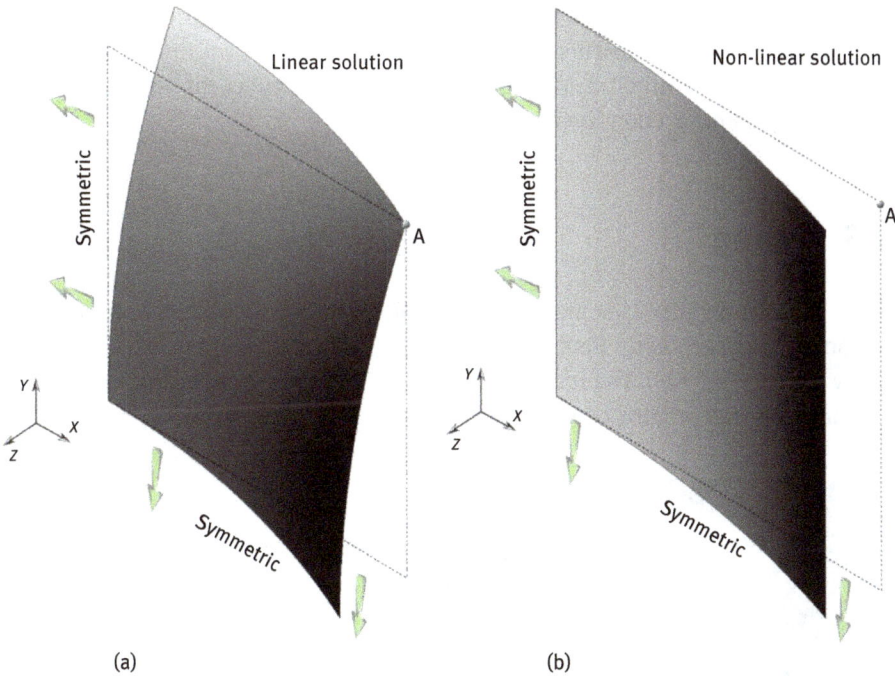

Fig. 9.21: Deformation of square paper sample using linear (a) and non-linear (b) solutions of the problem of competing curls.

Fig. 9.22: Graphical representation of bi-stability.

Making an analogy with our case, the linear solution predicts equilibrium Position 1, but the non-linear solution captures either 2 or 3, which is more realistic as one might guess. The most important points that we can draw from this example are:

1. For thin structures undergoing large deformation, the linear analysis is often unable to predict the deformation pattern and amplitudes.
2. The conditions for bifurcations can develop during the deformation process. It is sometimes impossible to predict instability based on the initial undeformed configuration.

3. The solution is not unique after the bifurcation point. It depends on small perturbations and the presence of imperfections.
4. The presence of initial deformations or stresses can affect the solution and instability point in a non-linear analysis.

9.3 Fluting

Let us consider another example of dimensional instability caused by paper–moisture interaction, namely, *fluting*. Fluting is a waviness you may find on the inked areas in a glossy journal (Fig. 9.23). The waviness is oriented in the MD. It often appears on paper in connection with printing operations involving two-sided, heavy ink coverage.

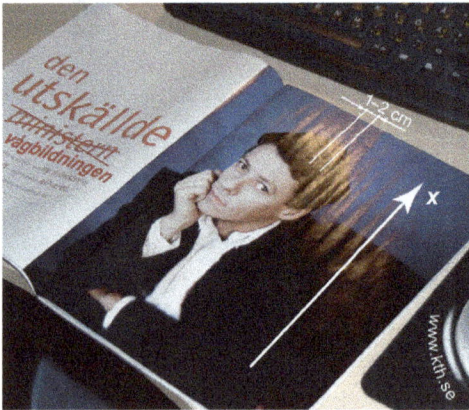

Fig. 9.23: Typical appearance of fluting in a magazine. Reprinted from Kulachenko et al. (2007) with permission from Elsevier.

Before we can investigate this phenomenon, let us list some facts about fluting. Fluting occurs only in web-fed, heatset offset printing. This type of printing covers most of the higher quality printing market. Fluting usually appears with a wavelength of 1–2 cm.

Fluting has been puzzling paper physicists for years, and as a result, there are a number of publications discussing it and yet new investigations on fluting can still be found in many journals. The most difficult part has probably been recreating the conditions the paper experiences in the heatset machine. As we know already, the deformation of the web is history dependent and having the wrong loading history may lead to misleading conclusions.

Despite various views on this phenomenon, there are some observations, which are shared by everyone who has dealt with fluting. The most important observation is that tension and heat are required to cause fluting. In the web-fed heatset offset printing process, the paper web is subjected to tension. Ink and sometimes water are applied onto the paper web in the printing stations and then dried intensively by applying heat in the drying section. This time, we will introduce questions one by one and the first question is: *what is the role of the tension in creating fluting?*

9.3.1 Effect of tension on fluting

Paper and other thin materials may wrinkle due to tension alone (Fig. 9.24). It can easily be demonstrated by pulling a piece of cloth, for example. It seems first like a simple case of buckling. It is buckling indeed but it is not so simple. Buckling occurs in the presence of compressive forces. In the case of tension, we do not explicitly create compression.

Fig. 9.24: Tension wrinkles. Reprinted from Kulachenko et al. (2007) with permission from Elsevier.

Surely, paper can contract due to the Poisson effect, just like many other materials. However, there are no forces acting in the width direction associated explicitly with this contraction alone. So, why does the paper buckle?

There are basically two reasons for tension wrinkling. The first reason is the boundary conditions. Most often you cannot pull the paper without restraining it in the width direction. Such restraints alone can cause negative normal stresses in paper. The second reason is the non-uniformity of the strain across the width. The tension profile of the web across the width is often non-uniform and this can also cause negative stresses (Habeger, 1993). In other words, the boundary conditions and non-uniformities prevent a free contraction of paper and as usual, paper escapes constraints through buckling.

The tension can create waviness. The wavelength, however, depends on the dimensions of the sample and the distance between the constrained boundaries. The larger the length-to-width ratio, the longer the wavelength of the waviness. With distances between boundaries typical to printing applications, it is practically impossible to create a 2 cm fluting wavelength by applying tension. It was noted, however, that applying heat, in addition to tension, creates a fluting pattern with smaller wavelength than tension alone (Hung, 1984, 1986). On the other hand, upon applying heat only, the out-of-plane distortions of the paper do not exhibit the wavy fluting appearance (Strong, 1984). The question is now what has changed with applied heat? Why does it create sharper waves?

Heat removes the water from the web. As a result, the web shrinks. The rate of water removal is extreme – the web should be dried in less than 1 s in the drying section. At a fixed span between the boundaries, the shrinkage will increase the tension. It has been theoretically demonstrated that the higher the tension, the lower the wavelength (Habeger, 1993, Coffin, 2003). The web tension in the dryer usually varies from 250 to 500 N/m (Falter and Schmitt, 1987). Hung (1984, 1986) show, however, that a thin web may experience fluting upon high-intensity drying even at low tension (100 N/m and less). Falter and Schmitt (1987) reported that the web tension could not practically be decreased to the level where the fluting is acceptable. In other words, increased tension upon heating and consequent moisture loss cannot solely explain the decreased wavelength. If it is not the tension only *what is the role of moisture in creating fluting*?

9.3.2 Effect of moisture on fluting

Paper may take up and lose moisture during the printing process. The moisture is applied prior and during the printing. The moisture is removed during drying. As we noted previously, moisture changes may cause considerable expansion or shrinkage of paper, particularly in the CD direction. The areas of the printed product that are covered with ink experience larger moisture changes during printing as, naturally, more water is being applied to the printed areas.

During drying, on the other hand, the ink layer prevents the evaporation of moisture (Simmons et al., 2001). This means that there is differential expansion or shrinkage between image and non-image areas during printing. As a result, shear strains develop at the boundaries between image and non-image areas. These shear strains give rise to compressive forces, which lead to buckling (Fig. 9.25).

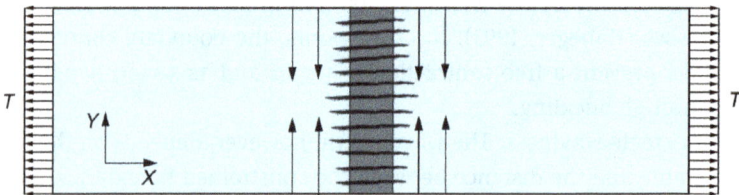

Fig. 9.25: Tension wrinkles caused by differential shrinkage. Reprinted from Kulachenko et al. (2007) with permission from Elsevier.

A number of authors used differential shrinkage as an explanation of fluting. The fluting wavelength was reported to be dependent on the dimensions of the inked area. Fluting decreased with printing image length (Hirabayashi et al., 2005). It was, however, inconsistent with some experimental observation. For example, Mochizuki and Aoyama (1981, 1982) observed an increase in fluting with the length of the printed area. Furthermore, the printing trials, which were inspired by the explanation based on differential shrinkage, showed that fluting occurs even in the case when the entire area of the web was covered by ink. Moreover, in cases where there was no ink applied, the fluting could also be seen right after the drying section, but it relaxed soon after. Figure 9.26 shows, for example, the web right after the drying section. There is no ink applied.

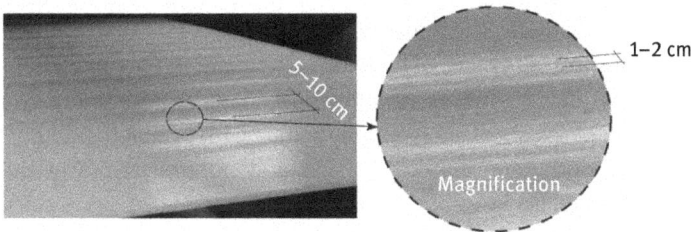

Fig. 9.26: Printing trial showing both fluting (1–2 cm wavelength) and tension wrinkles (5–10 cm wavelength) coexisting on the same web. Reprinted from Kulachenko et al. (2007) with permission from Elsevier.

There is waviness on the web with a wavelength of 5–10 cm. These are typical tension wrinkles. At a closer look, however, there is another waviness superposed on the tension waves. This waviness has a significantly lower wavelength (1–2 cm), close to that reported as fluting. The question is now *what causes the small waves typical to fluting?* There is another question arising from this observation, namely, if the waviness appears everywhere, *why is it visible only on in printed areas in the final product?* At this point, we need to bring in the effect of heterogeneity intrinsic to paper.

9.3.3 Effect of small-scale strain variations on fluting

As we have demonstrated, the theories of fluting based on tension wrinkles and differential hygroexpansion fail to explain the field observations. Tension alone cannot create such a small wavelength, and fluting can be seen even in the case where there is nothing printed on the web.

There is one important thing that was not accounted for in these theories – the inhomogeneous nature of paper. Due to the stochastic structure paper is non-uniform in the plane. Paper density, thickness, porosity, fibre orientation varies in paper on the microscale level, and it is reflected in the physical properties of the paper as a material.

Paper naturally has strain variations even under a uniformly applied load. The moisture changes associated with printing do not happen uniformly either. Due to porosity variations, the paper absorbs and loses moisture non-uniformly in the plane. At the same time, due to fibre orientation variation, the changes in dimensions do not occur evenly from point to point. The implication of uneven dimensional change is that certain parts of the paper are compressed locally. If the compressive forces exceed the critical level, local buckling occurs. The local buckling cannot happen without disturbing the surroundings. Let us demonstrate this with the help of numerical analysis.

We will consider a sheet of paper, which will be represented as a continuum with orthotropic material properties. The sheet will be subjected to a uniform tension without applying any constraints in the direction orthogonal to the direction of tension to avoid tension wrinkles. This means the sheet may contract freely. We will consider only a quarter of the sheet and replace the rest with symmetric boundary conditions (Fig. 9.27). Under the applied tension and no constraints in the width direction, the sheet with perfectly uniform properties may not buckle since there is no source of compressive forces. Now we will introduce spatial strain variations. We will decrease the moisture uniformly apart from the randomly distributed circular spots each having a dimeter of 6 mm, where the moisture will not be changed (Fig. 9.27). The fraction of the total area covered by the spots will be 20%. The effect of the moisture is accounted for with hygroexpansion coefficients. The decrease of moisture with this set-up will create a non-uniform strain field under which compressive forces will act on the spots. We are interested in capturing the buckling instability and post-buckling shapes and amplitudes.

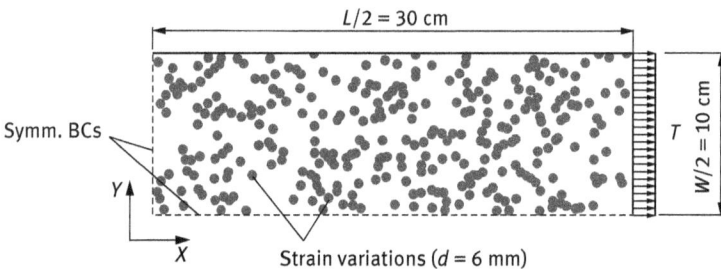

Fig. 9.27: A quarter of the web with spatial strain variations in the form of random pattern of "wet" spots and symmetric boundary conditions. The side with applied tension is restrained in the out-of-plane direction. Reprinted from Kulachenko et al. (2007) with permission from Elsevier.

Figure 9.28 shows the results of simulations after the moisture was decreased by 1%. Despite the random distribution of the spots, the analyses show that the local strain variations can cause rather regular, wavy out-of-plane distortions oriented in the direction of the tensile load. The observed out-of-plane deformations can be, in fact, characterized as fluting because of the characteristic 1–2 cm wavelength. Its waviness was created by a combination of non-homogeneous in-plane strain field and tension.

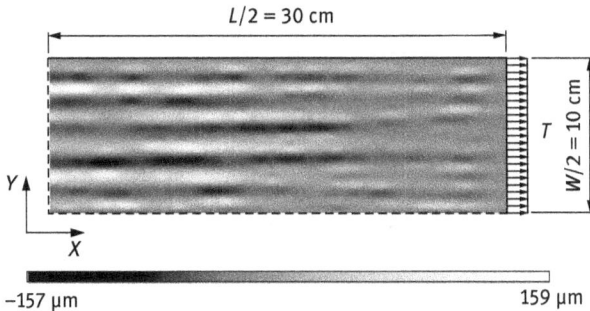

Fig. 9.28: Out-of-plane deformations caused by the "wet" spots local strain variations in Fig. 9.27. Reprinted from Kulachenko et al. (2007) with permission from Elsevier.

The magnitude of the deformation in Fig. 9.28 corresponds to the 1% moisture difference between the spots and the rest of the paper. The instability itself was already triggered when the difference was about 0.5%. With total moisture loss reaching 3–6% during convection drying (air drying) one may wonder how realistic the 0.5% (and ultimately 1% difference in moisture) is. In fact, the convection drying is reported as one of the major players in creating fluting. For instance, infrared (IR) drying causes a less intensive fluting pattern (Hung, 1984, 1986). What is so special with convection drying?

We will use a simple experiment to demonstrate the effect of the drying technique. A paper sheet 70 × 22 cm was cut from the roll, subjected to the high humidity of 90% for 1 h and clamped by two parallel jaws in a tensile tester. Being under the same tension and boundary conditions, the sheets were dried with two different methods: powerful IR lamps and an air-blower. IR drying generally provides more uniform moisture removal, since there is no convection and only little interaction with the paper structure in contrast to convection drying.

In both cases, the papers wrinkled; however, the appearance of the wrinkles was different. With fast IR drying, the wrinkled pattern was irregular. The dominant, most visible wavelength was 3–4 cm (Fig. 9.29a). The air dryer altered the wrinkle appearance drastically (Fig. 9.29b). A regular waviness with the wavelengths of 1–2 cm appeared across the whole width. It appeared almost immediately as the heated air was applied.

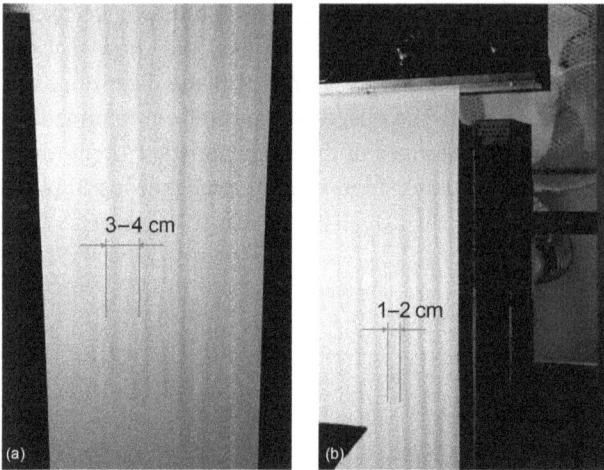

Fig. 9.29: Tension wrinkles created by fast IR drying (a) and convection drying (b). The 70 × 22 cm² paper sheets were subjected to 90% RH for 1 hr before tension and drying were applied. Reprinted from Kulachenko et al. (2007) with permission from Elsevier.

At the end of the drying, the paper sheet in both cases had more or less the same uniform moisture levels, but very different out-of-plane deformation patterns. As mentioned earlier, the convection drying promotes more local drying non-uniformities through the interaction with the paper structure. Having this in mind, the observed results collaborates very well with connection between fluting appearance and small-scale drying non-uniformities described with strain variation spots in the performed simulation.

So, now we know that small-scale drying non-uniformities promote fluting. The next question we need to consider is: *what is the mechanism of fluting retention in the printed areas*? It is impossible to capture the residual deformations in the simulation when the paper is modelled as a linear elastic material – once the uniformity of moisture is restored, the out-of-plane deformation will disappear in the model. As in the case of curl, it is important to capture the entire loading history, because paper undergoes irreversible inelastic changes, which depend on the loading history.

9.3.4 Fluting retention

It has been shown that fluting appears in both image and non-image areas during printing. Why is the waviness visible only in the printed areas of the final product? Perhaps, it is only an optical effect? We can perform a simple test to ensure that the perception of the waviness on the white and coloured areas is different. The colour clearly "highlights" the out-of-plane deflection. Let us measure the out-of-plane

deflections. Figure 9.30 shows the measurements of out-of-plane deflection performed by chromatic aberration. Despite the obvious fluting pattern (seen by the
naked eye), the measurements show the presence of other wavelengths in the deflection pattern. These wavelengths could also be picked up by the naked eye. After fil-
tering out the wavelength longer than 2 cm, the fluting pattern becomes more
apparent. It shows that despite the presence of residual waviness in the unprinted
areas, the amplitudes of the waviness there are much lower than those in the printed
areas. So, more pronounced fluting in the printing areas is not just an optical effect.

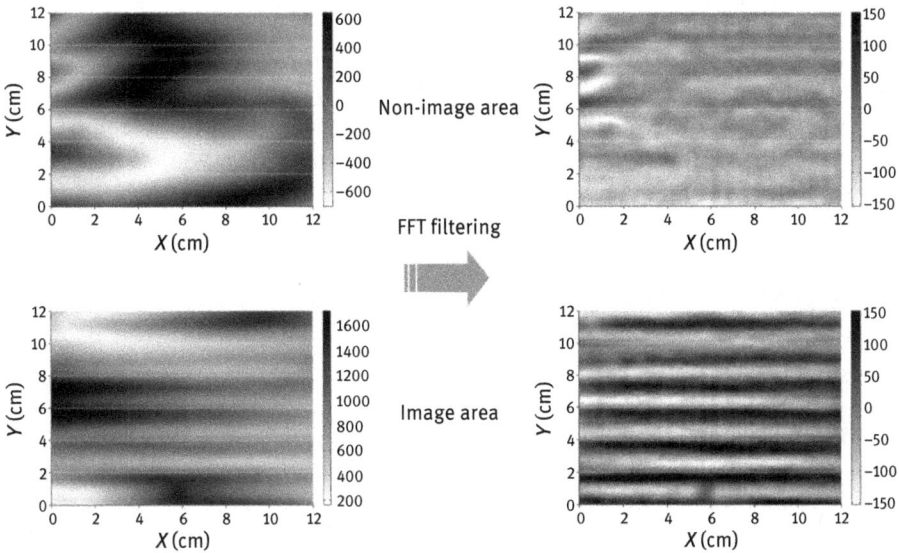

Fig. 9.30: Optical measurements of fluting amplitudes.

The measurements carried out in a printing press show also that the fluting amplitudes
are nearly the same in printed and non-printed areas right after the drying section
(MacPhee et al., 2000). *Why does the paper accumulate more irreversible deformations
in the printed areas?* A plausible explanation of this may have to do with the ink layer
itself. The ink is applied when the paper is deformed, and it solidifies on the printed
areas forming a layer of a finite stiffness. This layer initially conforms to the deformation configuration, and it may prevent the paper from returning to its original flat state.
The stiffness of the ink layer can be easily estimated by comparing the bending stiffness of printed and unprinted samples. The thickness of the ink layer needed for such
estimation can be measured from cross-sectional cuts by microscopy. The estimation
shows, however, that the ink stiffness is relatively low compared to paper stiffness
and even in the worst-case scenario in which the ink solidifies immediately on the
deformed paper, the ink stiffness is not sufficiently large to retain the fluting shape

(Kulachenko et al., 2007). So, if it is not the ink layer that preserves the fluting by fixing the waviness by added bending stiffness on top of the deformed sheet, what is it?

9.3.5 Effect of the drying temperature on fluting

One of the important clues to the discrepancy between printed and non-printed areas was given by Hung (1984, 1986) and later by MacPhee et al. (2000), who demonstrated that the printed areas attract more heat during drying and retain that heat longer. In general, the temperature of the paper in a commercial heatset web offset dryer may range from 110 to 200 °C. Let us demonstrate the effect of the temperature in a simple experiment. We will take a tube, 3 cm in diameter and roll-in the paper strip into it. We will vacuum dry the reference strip at room temperature. The rest of the samples will be dried in a convection oven (Fig. 9.31) so that the air passes through the tubes and removes the water.

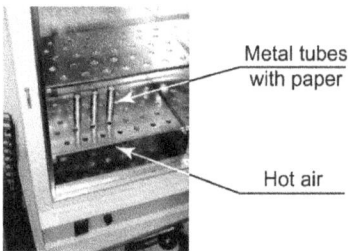

Metal tubes
with paper

Hot air

Fig. 9.31: Convection oven used to dry the paper inside the tubes.

The temperature will be gradually increased, and we will retrieve the tubes at different temperatures starting from 90 to 145 °C. At the time the first tube is retrieved, all the samples are fully dried so the only difference between them should be the temperature.

A comparison between the samples in Fig. 9.32 clearly shows that the temperature effectively increases the residual curvature. This means that paper accumulates more irreversible deformation under the same stress state at higher temperatures. Combining this conclusion with the fact that the image areas have a higher temperature than non-image areas gives us an explanation for the larger fluting amplitudes observed in the printed areas.

The fluting created during the convection drying is associated with certain stresses caused by bending. Although the fluting magnitude can the same in both inked and non-image areas, the inked areas have higher temperature. Higher temperatures soften paper and promote inelastic deformation. At a given waviness, the inelastic, unrecoverable deformations will be greater in the inked areas. This means, the residual waviness will also be larger in the image areas.

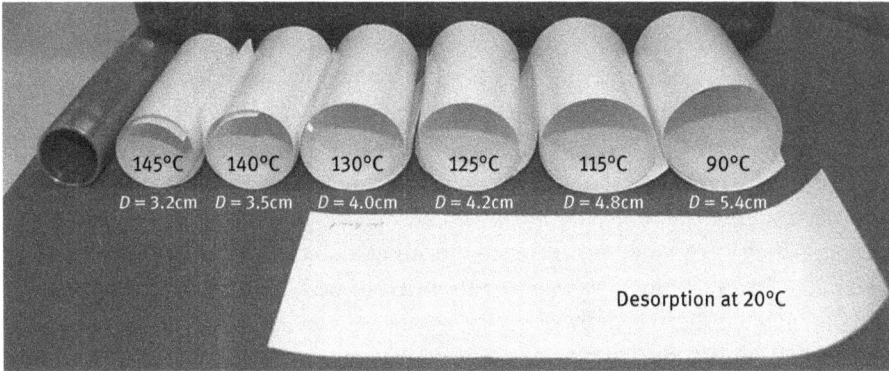

Fig. 9.32: Comparison of residual curvature in paper samples that were rolled into the metal tube and then dried in a convection oven to different temperatures. The permanent roll diameter *D* of the samples after reconditioning is shown. The reference sheet was dried at room temperature by desorption in vacuum. Reprinted from Kulachenko et al. (2007) with permission from Elsevier.

9.3.6 Interesting case: fluting without tension

It has been recognized that fluting is always associated with tension that triggers the instability in a form of a typical wavy pattern. There is an example of how the moisture can be applied in a way to create perfect fluting without any tension. This can be demonstrated with an inkjet printer. Take a used cartridge and fill it with water. Put it back into the printer and print a solid image with "high-quality print" settings. Usually, a lot of ink is supposed to be applied at a low speed with such a setting. Use an ordinary copy paper and you will see a beautifully fluted paper at the output tray of the printer (Fig. 9.33).

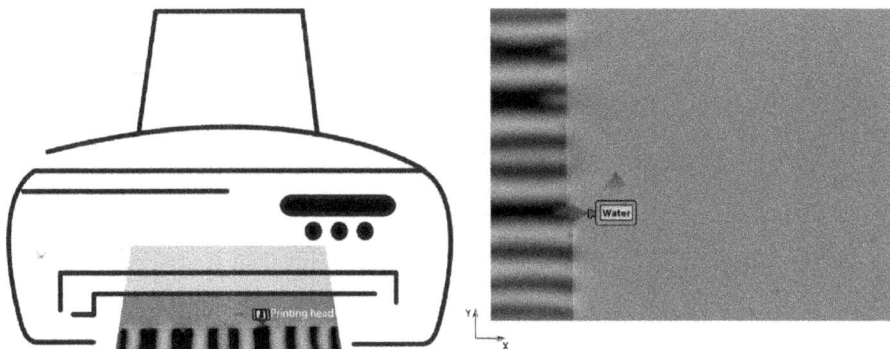

Fig. 9.33: Fluting created with an inkjet printer.

The reason for this is nothing more than the differential shrinkage combined with a specific loading history. The water is applied gradually, and the expansion cannot occur freely. The paper cannot curl globally because one of its boundaries is constrained in the Z-direction by the printer. The paper will inevitably escape compressive forces at the interface between the printed and unprinted areas through waviness.

The wavelength will depend on how fast you print. The faster the printing speed, the higher the wavelength. Once the waviness has started, a specific deformation history is created. For instance, buckling along the existing waves is more preferable than creating new waves with different wavelengths. By the end of the printing, you will have exactly the same wavelength as you have in the beginning. This confirms again that the loading history is very important, particularly when a lab test is used to mimic production-scale conditions.

9.4 Cockling

Cockling is a phenomenon that manifests itself as irregular small-scale out-of-plane distortion. Some authors use the words cockling and fluting interchangeably simply because it is sometimes difficult to discriminate these two phenomena. For example, Fig. 9.34 shows an unprinted A4 paper sample which would be normally characterized as cockled. It also contains, however, some irregular waviness with the wavelength typical to fluting.

Fig. 9.34: Example of cockled paper. Reprinted from Kulachenko et al. (2007) with permission from Elsevier.

There exists a commonly accepted view that some sort of small-scale in-plane strain variation is responsible for cockling. The incompatibility in the strain field creates negative stresses and they lead to local buckling. The main point of debate has been the reason for this variation. Things like fibre orientation (Leppänen et al., 2005), moisture variations during drying (Smith, 1950), and two-sidedness of the paper (Kajanto, 1993) have been named as potential candidates.

Small-scale strain variations can create both cockling and fluting. Both cockling and fluting are caused by local instabilities. Tension is normally required to produce fluting. Cockling may appear in a copy machine and during papermaking. In a copy machine, there is no tension, indeed, but in the paper machine, the web is constantly under tension. The paper sample in Fig. 9.34, for instance, is an example of cockling developed during papermaking. So, why do we see cockling and not regular fluting in the paper machine despite the fact that paper is under tension? A clue regarding this apparent inconstancy can be found in the experiment in which we dried the sheet by different drying techniques and attained dissimilar distortion patterns at the end of the drying (Fig. 9.29). The convection drying resulted in a fluting pattern and the IR drying did not. In the paper machine, water is predominantly removed through contact drying, in contrast to the offset printing press where convection drying is used. The contact drying creates fewer moisture variations than the convection drying (Hashemi and Douglas, 2003). Expecting less variation, we can return to our numerical experiment described in Fig. 9.27 and reduce the total coverage of the strain variation spots by a half. This will mimic the situation when fewer spots will have significant moisture differences. The resulting out-of-plane distortions after reducing the spot coverage are shown in Fig. 9.35.

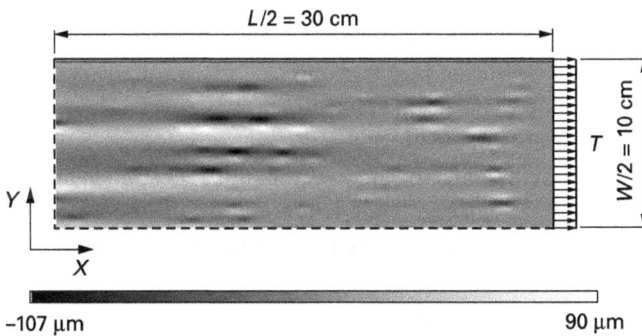

Fig. 9.35: Out-of-plane deformations are analogous to Fig. 9.28 but with a reduced coverage of the "wet" spots. Reprinted from Kulachenko et al. (2007) with permission from Elsevier.

Less spot coverage creates a less pronounced fluting pattern, and the individual "bumps" (or cockles) become more distinct. The same trend is observed when the spot coverage approaches 100% that is when the spots cover almost the entire paper area. Thus, *what is the cause of fluting in paper*? Unexpectedly, neither the tension nor the differential shrinkage could alone explain the characteristic fluting pattern with a typical wavelength of 1–2 cm. On the other hand, small-scale strain variations, such as those caused by in-plane moisture variations, could explain this typical pattern and are also consistent with the field observations. One of such key observation is that the faster convection drying, which involves interactions with the non-uniform paper structures, created the wrinkle wavelengths similar to fluting, in contrast to IR drying leading to moderate drying rate. Both cockling and fluting are caused by a similar phenomenon and the transition between them requires tension and enough coverage of the strain variation spots.

What exactly retains fluting on the printed area? It appears that it is not the ink layers. The estimated ink stiffness was unexpectedly small compared to the stiffness of paper. However, the inked areas attract more heat during drying and preserve it longer. Higher temperatures soften paper and promote inelastic deformation which do not relax over time retaining the waviness in the printed areas.

9.5 Concluding remarks

In this chapter, we studied the effect of moisture by investigating the physics behind three different problems associated with moisture change – curl, fluting, and cockling. Let us summarize the material by considering how these phenomena can be modelled.

We can solve the moisture transport and structural equations separately. The moisture history should be determined first. We can use diffusion equations for a slow rate or more advanced models in which the moisture and temperature are decoupled and the paper is considered as a multiphase material (Nilsson et al., 1993; Nilsson and Stenström, 1997; Hashemi and Douglas, 1999). When the moisture history is determined, we can apply it as an external time-dependent load in the structural equations of motion. Assuming we have all the governing equations and the boundary conditions are known, we still need to specify the following material properties:

1. Relation between moisture and hygroexpansion strain.
2. Elastic and inelastic properties as a function of moisture and position in the sheet.
3. Orientation of material axes if the material is anisotropic.

The last thing we need to account for is the previous load history. A commercial paper sheet is a product of a very complicated process, and it can inherit some stresses from it. For example, most of the drying process is done under constraints and tension. As we already know, this can lead to the development of dried-in stresses and strains. These stresses can be "released" during the first moistening and free drying cycle because the paper is more compliant when the water is added. The release of dried-in strains is manifested through irreversible shrinkage after the first moistening–drying cycle (Fig. 9.36). In a freely dried sheet, the irreversible shrinkage is insignificant (Fig. 9.3). Subsequent moisture changes happen without considerable irreversible changes in hygroexpansion.

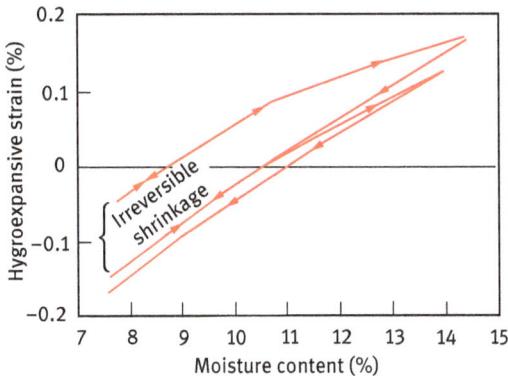

Fig. 9.36: Irreversible shrinkage during moisture cycling in a sheet dried under constraints (Uesaka et al., 1992). Reproduced with permission from Pulp and Paper Technical Association of Canada (PAPTAC).

In fact, these stresses alone can cause undesirable deformations, such as curl even under a slow and more or less uniform moisture change, if the distribution of dried-in strains in the thickness direction is non-uniform.

The appropriate modelling techniques include both the continuum (Nordström et al., 1998; Ketoja et al., 2001; Bortolin et al., 2006) and micromechanical models (Salminen et al., 2002; Sellén and Isaksson, 2014; Bosco et al., 2015; Motamedian and Kulachenko, 2019).

9.5.1 A note on hydroexpansion

You may realize by now that the paper–water interaction is a complex subject and accounting for all the factors may not always be possible. In printing, for example, ink and water are applied in a split of a second and under the nip pressure. Simulating such conditions in detail would be a formidable task. How can we study the response of the paper to fast moisture application experimentally?

In this case, we introduce the notion of hydroexpansion (Ketoja et al., 2001; Larsson 2010). Hydroexpansion is the dimensional change caused by liquid water in contrast to hygroexpansion that is measured as a response to slow humidity change. The water is applied in a way to reproduce the condition as close to reality as possible. The hydroexpansion is usually significantly lower than the hygroexpansion for the same paper because of the internal constraints in the thickness direction during the water absorption. Because the moisture history is so important, different set-ups for measuring hydroexpansivity may produce various incomparable results.

Literature references

Algar, W.H. (1965). Effect of Structure on the Mechanical Properties of Paper. In: Consolidation of the Paper Web, Bolam, F. (ed.), Proceedings of the 3rd Fundamental Research Symposium, 1965, Cambridge, vol. 2, 814–851.

Bortolin, G., Gutman, P.O. and Nilsson, B. (2006). On modelling of curl in multiply paperboard. J. Process Control. 16(4), 419–429.

Bosco, E., Bastawrous, M.V., Peerlings, R.H., Hoefnagels, J.P. and Geers, M.G. (2015). Bridging network properties to the effective hygro-expansivity of paper: Experiments and modelling. Philos. Mag. 95(28–30), 3385–3401.

Brunauer, S., Emmett, P.H. and Teller, E. (1938). Adsorption of gases in multimolecular layers. J. Am. Chem. Soc. 60(2), 309–319.

Christensen, P.K. and Giertz, H.W. (1965). The cellulose–water relationship. In: Consolidation of the Paper Web, Bolam, F. (ed.), Proceedings of the 3rd Fundamental Research Symposium, 1965, Cambridge, vol. 1.

Coffin, D. (2003). A Buckling Analysis Corresponding to the Fluting of Lightweight Coated Webs. In: Proceedings of the 2003 International Paper Physics Conference, Victoria, B.C., Canada.

Crank, J. (1956). The mathematics of diffusion, Oxford: Clarendon Press.

de Ruvo, A., Lundberg, R, Martin-Löf, S. and Sörenmark, C. (1973). Influence of Temperature and Humidity on the Elastic and Expansional Properties of Paper and the Constituent Fibre. In: The Fundamental Properties of Paper Related to Its Uses, Bolam, F. (ed.), Proceedings of the 5th Fundamental Research Symposium, Cambridge, UK.

Dunn, M.L., Yanhang, Z. and Bright, V.M. (2002). Deformation and structural stability of layered plate microstructures subjected to thermal loading. J. Microelectromech. Syst. 11(4), 372–384.

Falter, K.A. and Schmitt, U. (1987). Influence of Heat-Effects in Web Offset Printing on Strength Properties of the Paper and on Waviness of the Print. 19th International Conference of Printing Research Institutes, Eisenstadt, Austria.

Finot, M. and Suresh, S. (1996). Small and large deformation of thick and thin-film multi-layers: Effects of layer geometry, plasticity and compositional gradients. J. Mech. Phys. Solids 44(5), 683–721.

Fung, Y.C. and Tong, T. (2001). Classical and computational solid mechanics, Singapore: Word Scientific Publishing Co. Pte. Ltd.

Gates, E.R., Kenworthy, I.C. (1963). Effects of drying shrinkage and fibre orientation on some physical properties of paper. Pap. Technol. 4(5), 485–493.

Habeger, C.C. (1993). Tension wrinkling and the fluting of light-weight coated papers in web-offset printing. J. Pulp Pap. Sci. 19(5), 214–218.

Hailwood, A.J. and Horrobin, S. (1946). Absorption of water by polymers: Analysis in terms of a simple model. Trans. Faraday Soc. 42(B), 84–92.

Hashemi, S.J. and Douglas, W.J.M. (1999). Through drying of paper from mechanical and chemical pulp blends: Transport phenomena behavior. Dry. Technol. 17(10), 2183–2217.

Hashemi, S.J. and Douglas, W.J.M. (2003). Moisture nonuniformity in drying paper: Measurement and relation to process parameters. Dry. Technol. 21(2), 329–347.

Hirabayashi, T., Fujiwara, S., Fukui, T., Suzuki, Y., Tani, Y. and Watanabe, D. (2005). Development of New Printing Paper Called OK Non-Wrinkle. PTS Coating Symposium, Baden-Baden, Germany.

Hung, J.Y. (1984). Paper fluting – results of TEC studies, DePere, Wisconsin, USA: TEC Systems.

Hung, J.Y. (1986). Paper Fluting – Results of TEC Studies. Phase II, DePere, Wisconsin, USA: TEC Systems..

Kajanto, I. (1993). Finite Element Analysis of Paper Cockling. In: Products of Papermaking, Baker, C.F. (ed.), Proceedings of the 10th Fundamental Research Symposium, Oxford, vol. 1, 237–262.

Ketoja, J., Kananen, J., Niskanen, K. and Tattari, H. (2001). Sorption and web expansion mechanisms. In: The science of papermaking, Baker, C.F. (ed.), Proceedings of the 12th Fundamental Research Symposium, Oxford, vol. 2, 1357–1366.

Kulachenko, A., Gradin, P. and Uesaka, T. (2007). Basic mechanisms of fluting formation and retention in paper. Mech. Mater. 39(7), 643–663.

Larsson, P.A. (2010). Hygro-and hydroexpansion of paper: influence of fibre-joint formation and fibre sorptivity, Doctoral thesis, KTH Royal Institute of Technology, Stockholm, Sweden.

Leppänen, T., Sorvari, J., Erkkilä, A.-L. and Hämäläinen, J. (2005). Mathematical modelling of moisture induced out-of-plane deformation of a paper sheet. Model. Simul. Mat. Sci. Eng. 13, 841–850.

Lu, W. and Carlsson, L.A. (2001a). Influence of initial deflections on curl of paper. J. Pulp Pap. Sci. 27(11), 373–379.

Lu, W. and Carlsson, L.A. (2001b). Influence of viscoelastic behavior on curl of paper. Mech. Time-Depend. Mat. 5(1), 79–100.

MacPhee, J., Bellini, V., Blom, B.E., Cieri, A.D., Pinzone, V. and Potter, R.S. (2000). The Effect of Certain Variables on Fluting in Heatset Web Offset Printing, Web Offset Association, Affiliate of Printing Industries of America Inc.

Mäkelä, P. (2007). The effect of moisture ratio and drying restraint on the stress relaxation of paper. In: Proceedings of the 2007 International Paper Physics Conference, Gold Coast, Australia, Appita.

Mochizuki, S. and Aoyama, J. (1981). Effects of Fast Ink Drying Conditions on Multi-Colored Moving Web. TAGA Proceedings, 43–55.

Mochizuki, S. and Aoyama, J. (1982). "Multi-Colored Moving Web (II). TAGA Proceedings, 483–496.

Motamedian, H.R. and Kulachenko, A. (2019). Simulating the hygroexpansion of paper using a 3D beam network model and concurrent multiscale approach. Int. J. Solids Struct. 161, 23–41.

Nilsson, L. and Stenström, S. (1997). A study of the permeability of pulp and paper. Int. J. Multiphase Flow 23(1), 131–153.

Nilsson, L., Wilhelmsson, B. and Stenström, S. (1993). The diffusion of water vapor through pulp and paper. Dry. Technol. 11(6), 1205–1225.

Nissan, A.H. (1978). Water Effects on Young's Modulus of H-Bonded Solids. In: Fibre-Water Interactions in Papermaking. Fundamental Research Committee (ed.) Proceedings of 6th Fundamental Research Symposium, Oxford, UK, 1977, vol. 1, 609–640.

Nordström, A., Gudmundson, P. and Carlsson, L.A. (1998). Influence of sheet dimensions on curl of paper. J. Pulp Pap. Sci 24(1), 18–25.

Olsson, A.-M. and Salmén, L. (2001). Molecular mechanisms involved in creep phenomena of paper. J. Appl Polym. Sci. 79(9), 1590–1595.

Pulkkinen, I., Fiskari, J. and Alopaeus, V. (2009). The effect of hardwood fiber morphology on the hygroexpansivity of paper. BioResources 4(1), 126–141.

Roark, R.J. and Young, W.C. (1975). Formulas for stress and strain, New York, NY: McGraw-Hill Book Co., Inc.

Salmén, L., Fellers, C. and Htun, M. (1987). The development and release of dried-in stresses in paper. Nord. Pulp Paper Res. J. (2), 44–48.

Salminen, L., Alava, M., Heyden, S., Gustafsson, P.-J. and Niskanen, K. (2002). Simulation of network shrinkage. Nord. Pulp Pap.Res. J 17(2), 105–110.

Sellén, C. and Isaksson, P. (2014). A mechanical model for dimensional instability in moisture-sensitive fiber networks. J. Compos. Mater. 48(3), 277–289.

Simmons, S., Blom, B., Dreher, C., Dewildt, D. and Coffin, D. (2001). Parametric Evaluation of Web Offset Fluting. TAGA'S 53rd Annual Technical Conference, San Diego, CA, USA.

Smith, S.F. (1950). Dried-in strains in paper sheets and their relation to curling, cockling, and other phenomena. Paper-Maker Br. Pap. Trade J. 10(3), 185–192.

Strong, W. (1984). A study of fluting in coated papers printed by blanket-to-blanket offset presses, Chicago, USA: GATF Technical Forum.

Topgaard, D. and Söderman, O. (2002). Changes of cellulose fiber wall structure during drying investigated using NMR self-diffusion and relaxation experiments. Cellulose 9(2), 139–147.

Tydeman, P., Wembridge, D. and Page, D. (1966). Transverse shrinkage of individual fibres by micro-radiography. Consolidation of the Paper Web, 119–144.

Uesaka, T., Ishizawa, T., Kodaka, I. and Okushima, S. (1989). Curl in paper. 3. Numerical simulation of history-dependent curl. Japan Tappi J 43(7), 689–696.

Uesaka, T. (1991). Dimensional stability of paper: Upgrading paper performance in end use. J. Pulp Pap. Sci. 17(2), 39–46.

Uesaka, T. (1994). General formula for hygroexpansion of paper. J. Mater. Sci. 29(9), 2373–2377.

Uesaka, T. (2001). Dimensional Stability and Environmental Effects on Paper Properties. Handbook of Physical Testing of Paper. R.E. Mark, C.C. Habeger, J. Borch and B. Lyne. New York, Marcel Dekker, Inc. vol. 1, 115–171.

Uesaka, T. and Moss, C. (1997). Effects of fiber morphology on hygroexpansivity of paper-a micromechanics approach. In: The Fundamentals of Papermaking Materials, Baker, C.F. (ed.), Proceedings of the 11th Fundamental Research Symposium, Cambridge, UK, 663–679.

Uesaka, T., Moss, C. and Nanri, Y. (1992). The characterization of hygroexpansivity of paper. J. Pulp Pap. Sci. 18(1), J11–J16.

Uesaka, T. and Qi, D. (1994). Hygroexpansivity of paper: Effects of fiber-to-fiber bonding. J. Pulp Pap. Sci. 20(6), J175–J179.

Walker, J. (2006). Water in wood. Primary wood processing, Netherlands: Springer, 69–94.

Youngs, E.G. (1973). Fundamental Aspects of Fluid Flow through Porous Materials. In: The Fundamental Properties of Paper Related to Its Uses, Bolam, F. (ed.), Proceedings of the 5th Fundamental Research Symposium, Cambridge, UK, vol. 2, 452–463.

Tetsu Uesaka

10 Mechanics in printing nip for paper and board

10.1 Introduction

In impact printing processes, such as offset, gravure, letterpress, and flexographic printing, paper and board must go through a printing nip where ink is applied under compression (Fig. 10.1). Although the process is simple and straightforward, there are a number of important issues and questions related to the printing nip. The most important questions are how the ink is transferred to the paper and what kinds of forces are applied on the paper surface. These are directly related to the problems of print quality and printability, specifically print density, ink demand, print uniformity (micro and macro), linting, picking, and delamination. Another important, but not well-recognized, issue is an impact on runnability, such as web breaks and fan-out. Web breaks often occur in an open draw section, but field observations suggest that the ways by which the printing nip is operated and the materials used in the printing nip (e.g. offset blanket) have large impacts on web breaks and web tension (Uesaka, 2005; Uesaka et al., 2001). Fan-out is a phenomenon of sheet widening, normally considered as a result of hygroexpansion of paper which picks up water from ink and fountain solution during printing. However, it has been indicated that fan-out is also produced by the nip itself without any water.

A printing nip has been a black box for the industry and also for researchers. It is only recently that serious mechanistic studies started to disclose what is happening in the nip. In this chapter, we focus on those recent mechanistic studies. Readers

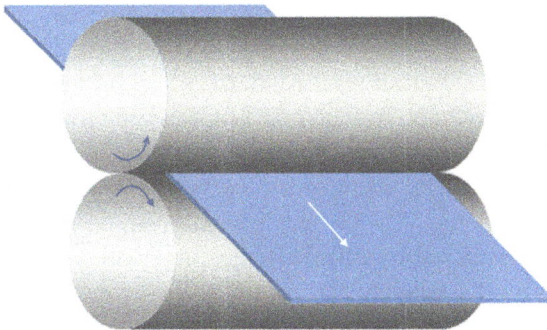

Fig. 10.1: Rolling contact in the nip of a web transport system.

Tetsu Uesaka, Mid Sweden University, Sundsvall, Sweden

https://doi.org/10.1515/9783110619386-010

who are interested in a broader background of printing and related problems are recommended to consult an appropriate textbook (e.g. Kipphan (2001)).

10.2 Nip mechanics in offset printing of paper

The problem of mechanics in a printing nip is often regarded as the simple compression of paper and board. However, looking at the condition in the nip in more details, we find that it is the problem of "rolling contact". For example (see Fig. 10.1), the paper is transported by the rotating rolls with friction through the nip. As the paper web goes through the nip, it is squeezed from the front end to the trailing end successively. This means that each position of the web receives time-varying compression and also shear stresses which reverse their signs in the nip, instead of simple compression. If the material, for example, paper, is an inelastic material, it could exhibit a different behaviour from the simple compression case, because the loading "history" (i.e. shear–compression–shear vs. simple compression) is different. Therefore, in nip mechanics, it is important to properly describe the stress–strain behaviour of all the materials used for printing rolls.

Print roll cover (outer surface) of the offset blanket is a rubbery material, and, in solid mechanics term, it is normally treated as a hyperelastic material (Fung and Tong, 2001), that is, stress is uniquely determined by strain, but the stress–strain relationship is not necessarily linear. Such materials usually undergo large deformation so that strain is characterized by finite strain, instead of infinitesimal strain that is commonly used for describing paper behaviour.

Within small strain, paper is treated as a linear elastic material where stress is a unique, linear function of strain, and it is characterized by nine (elastic) constants (see Chapter 2). Beyond a certain strain (yield strain, approximately 0.2%), paper starts showing inelastic behaviour, and it is often treated as an elastic–plastic material. For an isotropic material, the stress–strain relationships, called constitutive relation, for elastic–plastic materials are well established, for example, (Belytschko et al., 2000; Fung, 1965; Fung and Tong, 2001). However, in the case of orthotropic materials, the yield conditions and hardening rules must be chosen carefully both from the points of the frame indifference of the constitutive equations and also for computational efficiencies. The work is still in progress (Borgqvist et al., 2015; Mäkelä and Östlund, 2003).

The analysis of nip mechanics is based on satisfying the conservations laws (mass/momentum/energy), a constitutive relation, and boundary conditions. In addition to the generally non-linear constitutive relations, the boundary conditions of nip mechanics always involve contact problems in which, during the deformation process, the displacement-control conditions are suddenly switched to the traction conditions, creating another non-linearity (geometrical non-linearity). Therefore, all the nip mechanics analyses performed so far required numerical methods, particularly non-linear finite element methods (Belytschko et al., 2000).

Figure 10.2 shows the finite element model used for the rolling contact problem of an offset printing nip proposed by Wiberg (1999). As the paper enters the nip, it contacts with the roll with the nip width 2*a*. The offset blanket was modelled as a composite of three layers of materials consisting of hyperelastic (Mooney, 1940; Ogden, 1986) and linear orthotropic elastic materials. The paper was modelled as an elastic–plastic material using a phenomenological Karafillis–Boyce model in this case (Wiberg, 1999).

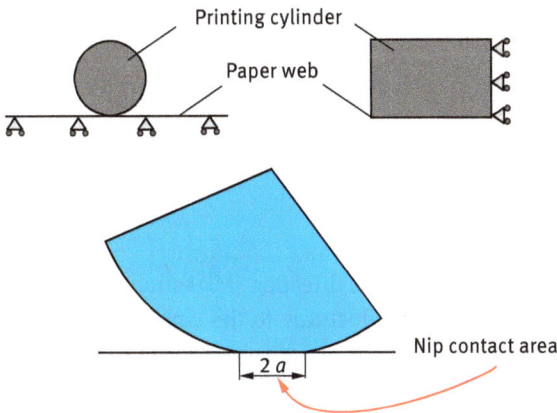

Fig. 10.2: Finite element model formulation for the rolling contact in offset printing nip (Wiberg, 1999). Reproduced with permission from the author.

The model was constructed into 3D with the *x*-direction in the travelling direction and the *y*-direction in the roll cylinder axis direction, as shown in Fig. 10.3. Because of the symmetry in the *y*-direction, the analysis was performed for half of the printing cylinder with the symmetry boundary condition.

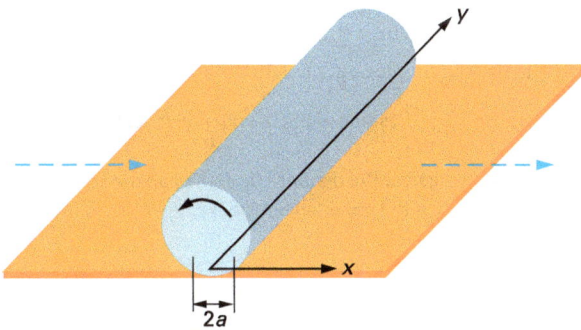

Fig. 10.3: Definition of the coordinates for printing nip problem, with 2*a* denoting the nip width.

Figure 10.4 shows the distribution of the cross-machine direction (CD) strain on the sheet plane (left graph) and along the cylinder axis (right graph). The CD strain is generally positive (sheet widening), and it peaked near the edge of the paper.

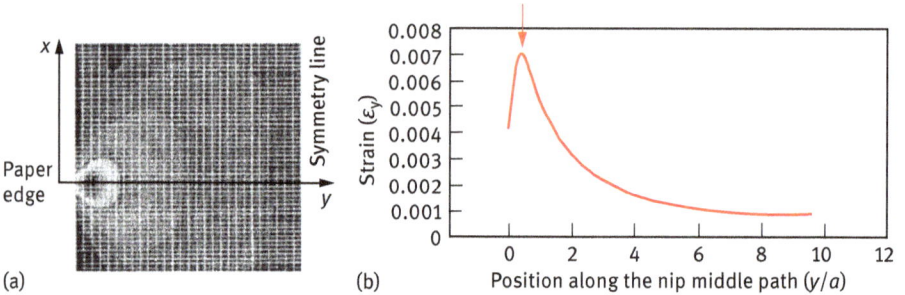

Fig. 10.4: (a) CD strain in the paper plane and (b)against the distance from the paper edge along the nip centre line. The CD strain has its maximum near the paper edge (Wiberg, 1999). Reproduced with permission from the author.

The similar distribution is seen also for the machine direction (MD) strain (Fig. 10.5). The positive strain means, again, the extrusion effect due to the compression. Note that the level of strain is very significant as compared with the typical strain (i.e. a draw) applied in the open draw section of printing presses, which is about 0.1–0.2% (Chapter 6). In the centre part of the printing cylinder, the paper is already subjected to about a 0.1% strain, and at the sheet edge, the strain increases up to 0.4%, twice as much as the typical yield strain of paper.

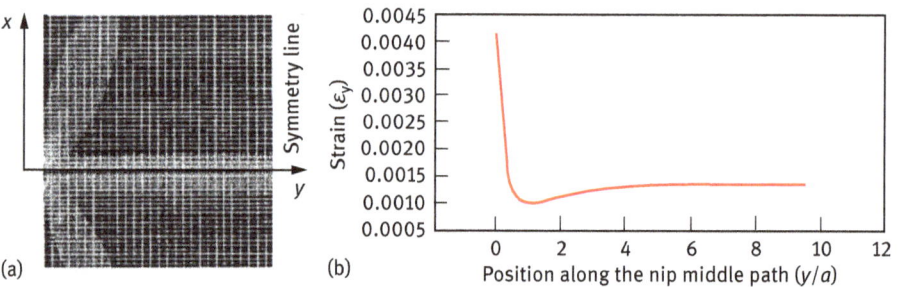

Fig. 10.5: (a) MD strain in the paper plane and (b) against the distance from the paper edge along the nip centre line. The MD strain has its maximum at the paper edge (Wiberg, 1999). Reproduced with permission from the author.

The implication of this non-uniform strain/stress distribution is significant. In a normal printing press operation, paper is subjected to a certain strain (draw) to transport from one section to the other, in addition to the strains created in the nip. Figure 10.6 illustrates this situation. The paper is subjected to a very high strain spike at the nip, in addition to the strain due to the draw, particularly at the sheet edges. The total strain could add up to the very significant level of strain-to-failure of some mechanical printing paper (≈1%). This straining of paper web in the nip is known to be dependent on the blanket layer compositions (Wiberg, 1999), which in turn can affect the feeding characteristics and thus web breaks in a printing press.

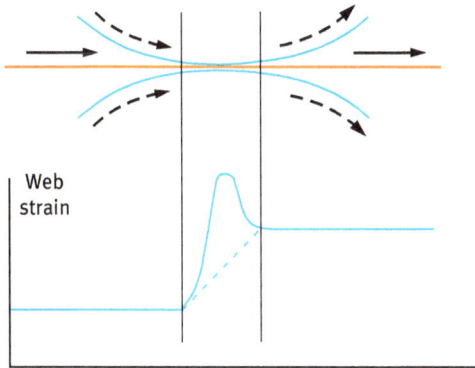

Fig. 10.6: Schematic of the MD strain of a web passing through a printing nip when draw is applied.

Printers are aware, by experience, that the web tension in an open draw section changes when different blanket materials are used in one nip and the other. This is due to the fact that the feeding characteristics of the nip are different depending on the types of blanket (and its compositions). Within the nip, the thickness direction compression deforms the blanket also in the feeding direction. By considering the simple mass balance of a one-dimensional material in the moving direction, we can show that the blanket deformation induces a (small) change in the peripheral velocity of the roll, Δv,

$$\Delta v = v_i \frac{\partial u}{\partial x},$$

(10.1)

where v_i is the incoming peripheral velocity of the roll in the moving direction x, and u is the surface displacement. Therefore, if the nip induces a tensile strain $((\partial u / \partial x) > 0)$, the roll accelerates, and if it induces a compressive strain$((\partial u / \partial x) < 0)$, the roll decelerates. The blanket that accelerates the roll is called a positive blanket, and the one that decelerates is called a negative blanket. This phenomenon has been known in the area of rolling contact, as, under different names, "rolling creep", "apparent slip", and "underdrive/overdrive" phenomenon. An excellent review and analyses in the context

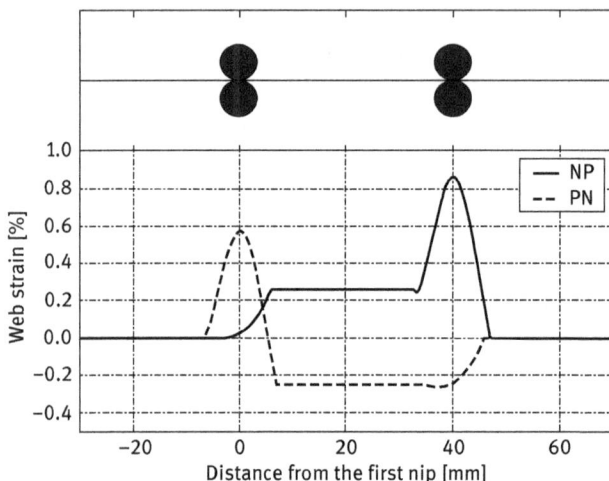

Fig. 10.7: Web strain developments. NP refers to the combination of the negative blanket at the first nip and the positive blanket at the second nip. PN refers to the positive and negative blanket combination.

of printing nips are given in the literature (Sorvari and Parola, 2014). Figure 10.7 illustrates how the web strain in a short open draw section changes, depending on the selection of the positive or negative blanket in each nip: For the negative–positive blanket combination (NP: solid line), the web strain increases in the open draw section, whereas the positive–negative blanket combination (PN: dotted line), the web strain decreases (Sorvari and Parola, 2014).

Figure 10.8 shows the distributions of the x-direction strain for the positive and negative blanket nips (Kariniemi et al., 2010). As expected from eq. (10.1), the positive blanket induces tensile strain in the centre nip zone, whereas the negative blanket induces compressive strain. Such tensile/compression difference in the x-direction strain is related to the deformation characteristics of the blanket: if the blanket is an "incompressible" material (i.e. Poisson's ratio is approximately 0.5, such as rubber), it induces tensile strain, but if the blanket is a more "compressible" material (i.e. Poisson's ratio approaching zero, such as foams), it induces compressive strain. Therefore, if these different types of blankets (positive, negative, and neutral) are used at different nips of the printing system (see Fig. 10.9), one can see unexpectedly high web tension in some open draw sections (Sorvari et al., 2016). Although in practice, it is rather rare to use different types of blankets at different nips in a printing house, one should be aware that the types of blanket can affect the web strains both in the open draw section and within the nip.

−0.007   0.031

−0.013   0.008

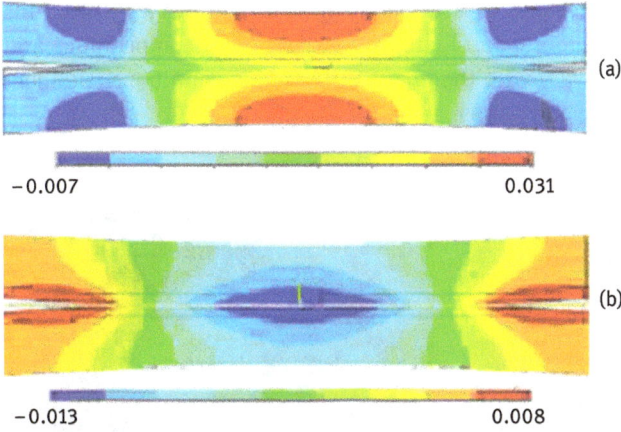

Fig. 10.8: Typical distributions of the *x*-direction strain for (a) the positive blanket and (b) the negative blanket. The figure is the courtesy of A. Kulachenko.

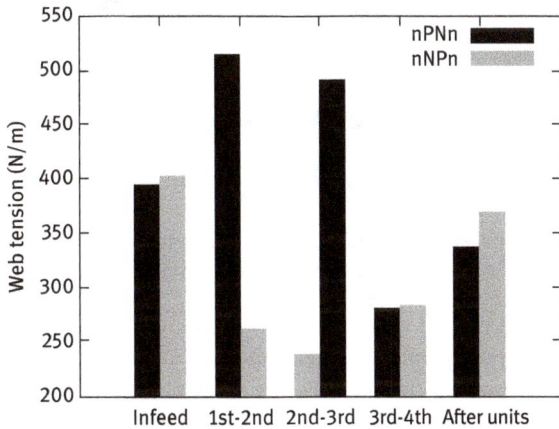

Fig. 10.9: Web tension at different units of the press system. "P", "N", and "n" refer to the positive, negative, and neutral blankets, respectively. Reprinted from Sorvari et al. (2016).

10.3 Nip mechanics in flexographic post-printing of corrugated board

Print nips accept not only a paper web but also a sheet of board with structures, such as corrugated board. Particularly, flexographic printing, the major printing method for packaging media, is widely used for printing both linerboard, called "pre-printing", and corrugated board, called "post-printing". In the latter, one of the outstanding print quality issues is striping, such as shown in Fig. 10.10. This print quality defect is related to the corrugated structures, but the appearance and severity of the defect greatly vary with the board and printing conditions (Hallberg et al., 2005; Netz, 1997; Zhang and Aspler, 1995).

Fig. 10.10: Example of print defect, striping in flexographic post-printing of corrugated board (Holmvall, 2010). Courtesy of M. Holmvall. The photos were taken by E.-K. Lindström.

An unique feature of flexographic post-printing is that the corrugated board is squeezed very lightly ("kissing" contact) in the nip in order to avoid any damages in the corrugated board structures. Therefore, the materials can be still assumed to be elastic, but the deformation can be still large (finite), involving contact problems, such as the case in offset printing.

Figure 10.11 shows a finite element model for investigating printing pressure variations over the top linerboard surface in flexographic post-printing (Holmvall and Uesaka, 2007, 2008a, 2008b). The model contains a portion of the corrugated board between the two peaks of the flute with the symmetric boundary condition on both sides. Linerboard and medium board were assumed to be linear elastic orthotropic materials, and the composite printing plate included two hyperelastic material models (Mooney, 1940; Rivlin, 1984).

Figure 10.12 shows printing pressure variations on the top linerboard between the two peaks of the flute. The elastic modulus of the photopolymer plate (the top cover of the printing plate) was changed by ±50% from a control value. As expected, the printing

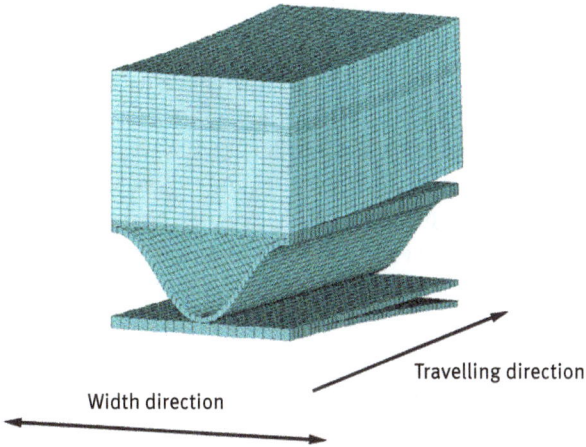

Fig. 10.11: Finite element model for the contact mechanics of a corrugated board in a flexographic post-printing nip (Holmvall et al., 2007).

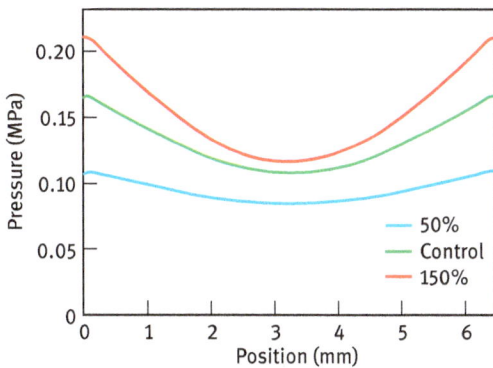

Fig. 10.12: Effect of the elastic modulus of the photopolymer cover of a flexographic printing plate on the printing pressure in the model of Fig. 10.10, with the flute tips located at the figure edges (Holmvall et al., 2007).

pressure varies considerably along the width direction (see also Fig. 10.11) due to the flute structure. Making the printing plate cover softer (or more flexible) alleviates this variation. However, it also reduces the overall printing pressure. Further parametric studies showed that compensating this overall reduction of printing pressure due to the use of softer plate by simply increasing print squeeze (more compression) reproduced similar non-uniformity of pressure. Therefore, using a softer plate does not provide a real solution for this non-uniform pressure distribution. This was also experienced in a converting plant. A material design solution for the printing plate construction was proposed, which included a hyperelastic material but with a yield stress, such as elastic foam materials (Holmvall and Uesaka, 2007).

In converting plants, the striping problem is often identified as "washboarding". Washboarding is a geometrical imperfection of corrugated board, in which linerboard sags between the flute peaks due to the excessive application of adhesives in the corrugating process or lower bending stiffness of linerboards. Figure 10.13 shows the effect of different degrees of washboarding on the printing pressure variation. Note that a 50 μm deflection is considered to be an excessive level of washboarding which is rarely seen in the fields. It is interesting that even with this level of large imperfection, the printing pressure variation changed rather modestly, except for the slight change in the average pressure. This means that washboarding is a rather minor contributor to the problem of striping. It was further shown that linerboard stiffness and its ink transfer characteristics affect the severity of striping, in addition to the printing plate properties. The nip mechanics analyses can quantify, systematically, the effects of different kinds of physical and geometrical imperfections of corrugated board.

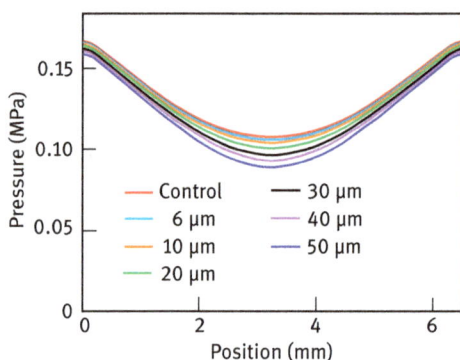

Fig. 10.13: Effect of different degrees of the washboarding of a corrugated board on the printing pressure in the flexographic post-printing model of Fig. 10.11 (Holmvall and Uesaka, 2007).

10.4 Ink–paper interactions

In a real printing nip, ink is present. Understanding ink–paper interactions in a printing nip, from the first principles, involves several challenges in paper mechanics and physics. First, paper is non-uniform and inhomogeneous, with surface pores of various sizes and geometries. We need to understand different length scales of interactions between ink and paper. Second, when analysing the problems, we need to consider both the deformations of solid (paper and printing cylinders) and the fluid (ink) deformation, simultaneously. Third, the entire process of printing includes the evolution of interfaces of two fluids (ink and air), such as the ink–film compression, penetration, retraction, and splitting. Fourth, the ink itself is a suspension of different fluids, pigments, and additives (complex fluid), and therefore, it is a non-Newtonian fluid. Nevertheless, there have been a few attempts to tackle this problem.

The analysis starts from the multi-phase flow of ink and air, since tracking the motion of interfaces is the essence of the ink–paper interaction problems. A diffused interface model, called a phase-field model, has been used (Dubé et al., 2008, 2009; Holmvall et al., 2011, 2010). In the diffused interface model, the interface is treated as a continuous transition of the phases (e.g. ink and air), rather than as a discontinuity, so that one can avoid the typical problem of the singularity of sharp interface models (Antanovskii, 1995; Jacqmin, 1999, 2000). The equilibrium position of the phases is determined by equilibrium thermodynamics, whereas the dynamics is governed by continuity and the Navier–Stokes equations, together with the convection–diffusion equation (called "Cahn–Hilliard" equation; Cahn and Hilliard, 1959). Therefore, the entire problem is to solve these equations for one continuous "binary" fluid with appropriate boundary conditions. The numerical method is based on the finite-difference projection method on a staggered grid. The boundary between the fluid and solid is the important analysis issue because it is where the solid mechanics problem and fluid mechanics problem are coupled. Currently, the complete two-way coupling has not been achieved, but the solid is treated as rigid, and its boundaries are treated as immersed moving boundaries (Holmvall et al., 2010).

An illustrative example of "printing" of a droplet on a model porous surface is given in Fig. 10.14 (Dubé et al., 2008). The droplet, for example, a half-tone dot, is pressed onto the surface of the model, and then retracted like the printing operation. The capillary systems investigated here are (1) a single vertical capillary, (2) a series of vertical capillaries with varying diameters and numbers, and (3) a combination of vertical and horizontal capillaries.

It was shown that, at a given porosity, smaller diameter capillaries tend to retain the fluid more effectively than larger ones (Fig. 10.15). When horizontal pores are present, such as shown in Fig. 10.14, they further increase fluid retention as well as fluid penetration. Note that the diameter and penetration length scales are less than a few micrometers.

When pore size is much larger than a few micrometers, the fluid droplet tends to split in the retraction stage without much retention in the pores. This phenomenon was observed in a similar simulation for fibre networks, where the structures were created by the fibre deposition/consolidation model (Drolet and Uesaka, 2005). Figure 10.16 shows the entire process of fluid transfer and splitting (Holmvall et al., 2011). As shown in both Figs. 10.14 and 10.16, the fluid–paper interactions in the nip are limited on a very top surface, less than a few micrometers in depth, and there is no extensive penetration into the sheet structure. This is understandable because the timescale in question is only in the millisecond order, and the fluid behaves like an elastic gel, rather than a "fluid", because of the surface tension and

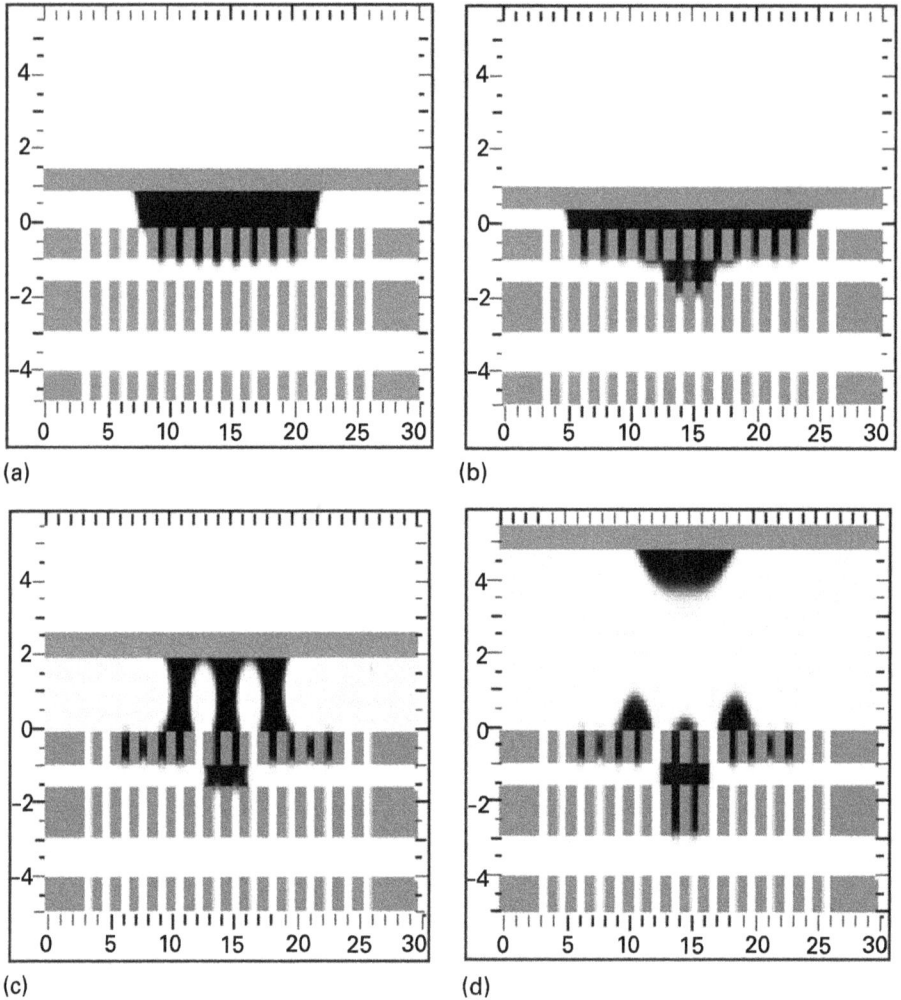

Fig. 10.14: Numerical simulation of an ink dot pressed onto a model network of capillaries. The initial diameter of the dot is 10 µm, and the vertical pore radius is 0.5 µm (Dubé et al., 2008). The acceleration of the plate is 1 m/s^2, surface tension is 40 N/m, and fluid viscosity is 0.4 Pa·s. Reproduced with permission from Pulp and Paper Technical Association of Canada (PAPTAC).

this short timescale. The penetration may happen only after passing through the nip in a much longer timescale. Another important insight from these numerical experiments is that the retention of the fluid, so-called ink holdout, is controlled by small pores less than a micrometre. In uncoated sheets, these pores correspond to the pores on the fibre surface, not those between fibres. For coated sheets, they are

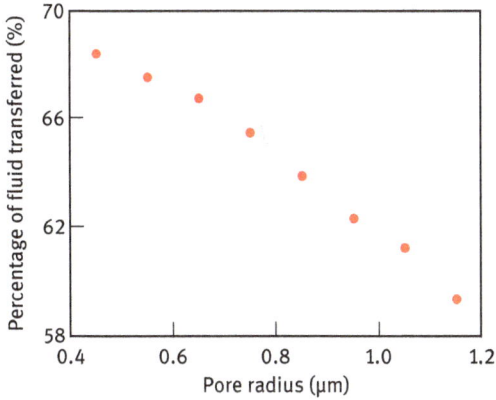

Fig. 10.15: Amount of fluid transferred against pore radius at a fixed pore volume fraction of 0.29 in the model capillary system of Fig. 10.14 (Dubé et al., 2008). Reproduced with permission from Pulp and Paper Technical Association of Canada (PAPTAC).

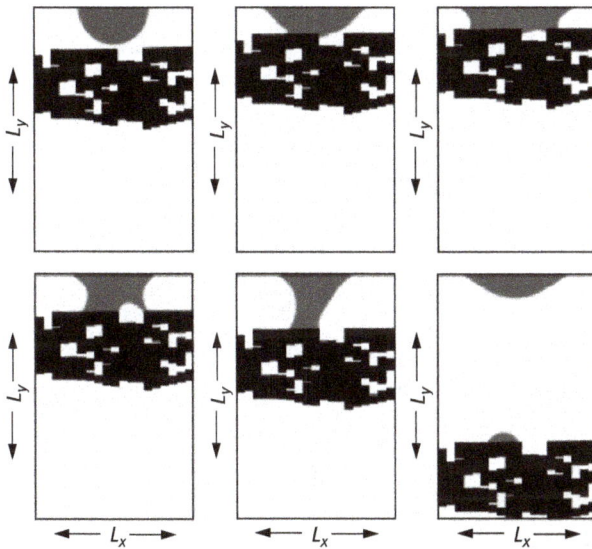

Fig. 10.16: Numerical simulation of an ink dot pressed onto a model fibre network in a flexographic printing system. The initial diameter of the dot is 22.3 μm. Reprinted from Holmvall et al. (2010) with permission from Elsevier.

between and within pigments. It is well known from the design practice of inkjet papers that a porous coating that creates nanoscale pores on the top surface is very effective in enhancing colour fidelity.

10.5 Delamination of paper and board in printing nip

As discussed in the previous section, the ink behaves very differently from what we perceive as a "fluid" in the timescale of the nip impression. It resembles an elastic gel, and the presence of the ink exerts additional forces on paper and board when passing through the nip. In the nip, the ink is compressed, decompressed, and then stretched until it splits (Fig. 10.16). At this last stage, the ink is under negative pressure, that is, under tension, and this tensile force, called "tack", sometimes can damage paper and board during printing. Delamination occurs when this tensile force exceeds the delamination strength of the paper and board. Therefore, the delamination issue is not only the problem of the inherent strength of paper and board but also the problem of ink–paper interactions affecting the magnitude of tensile force.

Inks are generally a (non-linear) viscoelastic fluid (Pangalos et al., 1985; Zang et al., 1991). The viscoelasticity comes mainly from polymers dispersed in the inks. In the interface regions of the ink, surface tension also plays a role, causing another elastic force component (microfluidic effect). The non-linear viscoelastic nature of the ink creates various (expected) effects on the tensile force growth, and thus delamination (Vaha-Nissi et al., 2010). For example, increasing printing speed increases the probability of delamination, partly due to the increased viscous stress effect. Increasing viscosity, particularly at the timescale relevant to nip impression, increases the tendency of delamination. The ink thickness, as affected by ink holdout of the paper substrate, also has an impact on the tack force: the higher the ink thickness, the higher the tack force (Zang et al., 1991). Therefore, the tack force is influenced not only by the printing conditions but also by ink setting properties, that is, ink–paper interactions.

Figure 10.17 shows the distribution of normal stress in the thickness direction of the board before the onset of delamination (Vaha-Nissi et al., 2010). As expected, the

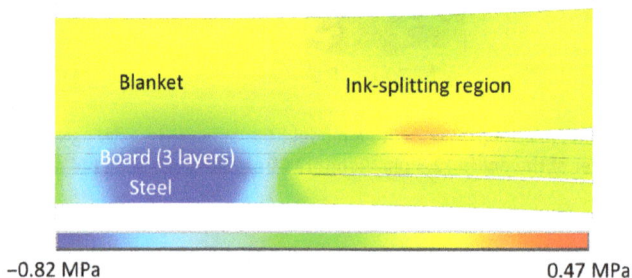

−0.82 MPa   0.47 MPa

Fig. 10.17: The distribution of normal stress in the thickness direction before delamination. The figure is the courtesy of A. Kulachenko.

top half of the board is subjected to tensile stress, which may cause the delamination if a weaker interface is present in the region of the board. The normal/shear strength does vary in the thickness direction of paper and board, as discussed in Chapter 4.

10.6 Concluding remarks

Significant insights have been accumulated over the years in the area of nip mechanics. For example, in an offset printing nip, the paper web is subjected to a significant level of strains both in the in-plane and thickness directions, which can easily damage the paper web, leading to web breaks and delaminations. Such strains are affected by the construction of an offset blanket. The sensitivity to the composite structure of printing cylinders was also demonstrated for flexographic post-printing. These studies suggest that there is an ample opportunity of designing printing materials and printing systems in order to avoid web breaks and to stabilize the process.

For the ink transfer problems, the still preliminary but the first principle approach has started drawing a new picture of ink–paper interactions. Ink in the printing nip is a microfluid where surface tension plays a significant role. Problems we often encounter are the dynamics of microfluids in the timescale of milliseconds. This notion explains many of the practical phenomena better than the traditional concepts based on the Lucas–Washburn theory or Stefan's law.

Literature references

Antanovskii, L.K. (1995). A phase field model of capillarity. Phys. Fluids. 7, 747–753.
Belytschko, T., Liu, W.K. and Moran, B. (2000). Nonlinear finite elements for continua and structures, Wiley.
Borgqvist, E., Wallin, M., Ristinmaa, M. and Tyding, J. (2015). An anisotropic in-plane and out-of-plane elastic-plastic continuum model for paperboard. Compos. Struct. 126, 184–195.
Cahn, J.W. and Hilliard, J.E. (1959). Free energy of a nonuniform system. III. Nucleation in a two-component incompressible fluid. J. Chem. Phys. 31, 688–699.
Drolet, F. and Uesaka, T. (2005). A stochastic structure model for predicting sheet consolidation and print uniformity, In: Baker, C.F. (ed.), Adv. Pap. Sci. Technol., The pulp and paper fundamental research society, U.K.: Lancashire, 1139–1154.
Dubé, M., Drolet, F., Daneault, C. and Mangin, P.J. (2008). Hydrodynamics of fluid transfer. J. Pulp Pap. Sci. 34(3), 174–181.
Dubé, M., Drolet, F., Daneault, C. and Mangin, P.J. (2009). Penetration of microscopic drops into paper structure, Proc. Papermak. Res. Symp., Kuopio, Finland.
Fung, Y.C. (1965). Foundations of solid mechanics, Englewood Cliffs, New Jersey, USA: Prentice-Hall.
Fung, Y.C. and Tong, P. (2001). Classical and computational solid mechanics, World Scientific Publishing Company.

Hallberg, E., Rättö, P., Lestelius, M., Thuvander, F. and Odeberg, G.A. (2005). Flexo print of corrugated board: Mechanical aspects of the plate and plate mounting materials. TAGA J. 2, 16–28.

Holmvall, M., Drolet, F., Uesaka, T. and Lindström, S. (2011). Transfer of a microfluid to a stochastic fiber network. J. Fluids Struct. 27(7), 937–946.

Holmvall, M., Lindström, S. and Uesaka, T. (2010). Simulation of two-phase flow with moving immersed boundaries. Int. J. Numer. Methods Fluids. 67(12), 2062–2080.

Holmvall, M. and Uesaka, T. (2007). Nip mechanics of flexo post-printing of corrugated board. J. Compos. Mater. 41(17), 2129–2145.

Holmvall, M. and Uesaka, T. (2008a). Print uniformity of corrugated board in flexo printing: Effects of corrugated board and halftone dot deformations. Packag. Technol. Sci. 21, 385–394.

Holmvall, M. and Uesaka, T. (2008b). Striping of corrugated board in full-tone flexo post-printing. J. Appita. 61(1), 35–40.

Jacqmin, D.. (1999). Calculation of two-phase Navier-Stokes flows using phase-filed modeling. J. Comput. Phys. 155, 96–127.

Jacqmin, D. (2000). Contact-line dynamics of a diffused fluid interface. J. Fluid Mech. 402, 57–88.

Kariniemi, M., Parola, M., Kulachenko, A., Sorvari, J. and von Hertzen, L.. (2010). Effect of blanket properties on web tension in offset printing. Adv. Print. Media Technol.

Kipphan, H. (2001). Handbook of print media: Technologies and production methods, Springer.

Mäkelä, P. and Östlund, S. (2003). Orthotropic elastic-plastic material model for paper materials. Int. J. Solids Struct. 40, 5599–5602.

Mooney, M. (1940). A theory of large elastic deformation. J. Appl. Phys. 11, 582–592.

Netz, E. (1997). Washboarding and print quality of corrugated board. Packag. Technol. Sci. 11(4), 145–167.

Ogden, R.W. (1986). Recent advances in phenomenological theory of rubber elasticity. Rubber Chem. Technol. 59(3), 361–383.

Pangalos, G., Dealy, J.M. and Lyne, M.B. (1985). Rheological properties of news ink. J. Rheol. 29(4), 471–491.

Rivlin, R.S. (1984). Forty years of non-linear continuum mechanics, Proc. 9th Int. Congr. Rheol., Mexico.

Sorvari, J. and Parola, M. (2014). Feeding in rolling contact of layered printing cylinders. J. Mech. Sci. 88, 82–92.

Sorvari, J., Parola, M., Kulachenko, A. and Leppänen, T. (2016). Effect of some printing nip variables on web tension. Nord. Pulp Pap. Res. J. 31(3), 491–498.

Uesaka, T. (2005). Principal factors controlling web breaks in pressrooms – Quantitative evaluation. Appita J. 58(6), 425–432.

Uesaka, T., Ferahi, M., Hristopulos, D., Deng, X. and Moss, C. (2001). Factors controlling pressroom runnability of paper, In: Baker, C.F. (ed.), Sci. Papermak., The Pulp and Paper Fundamental Research Society, Lancashire, UK, 1423–1440.

Vaha-Nissi, M., Kela, L., Kulachenko, A., Puuko, P. and Kariniemi, M. (2010). Effect of printing parameters on delamination of board in sheet-fed offset printing. Appita J. 63(4), 315–322.

Wiberg, A. (1999). Rolling contact of a paper web between layered cylinders with implications to offset printing, Department of Solid Mechanics, Stockholm, Sweden: KTH Royal Institute of Technology.

Zang, Y.H., Aspler, J.S., Boluk, M.Y. and De Grâce, J.H. (1991). Direct measurement of tensile stress ('tack") in thin ink films. J. Rheol. 35(3), 345–361.

Zhang, Y.H. and Aspler, J.S. (1995). Factors that affect flexographic printability of linerboard. Tappi. 78(10), 23–33.

Part IV: **Material properties**

Artem Kulachenko
11 Micromechanics

11.1 Introduction

The mechanical properties of paper, as well as many other materials, are originating from the microscale. The properties of fibres, their geometrical alignment, interconnectivity, and the bonding characteristics all have a certain impact on the bulk properties of the network. Researchers working with paper realized the importance of the microscale very early and many of the pioneering observations are dated back to the 1950s and 1960s as the development in microscopy enabled detailed observation of the microstructure. In this chapter, we will consider the theories connecting the micro- and macroscale properties, discuss the limitations, and present the main factors affecting the mechanical properties. We will, as a case in the discussion, use the differences observed between theoretical estimations of the elastic modulus and the experimental counterparts. This discussion paves the way for considering the strength of the network material from a micromechanical perspective, but here this topic will not be elaborated on.

Already in 1965, Algar (1965) published a review paper on the existing theories for the elastic modulus of paper. By then, eight different theories were reviewed, which is an indication that this matter is not simple.

11.1.1 Cox theory

We will consider the first theory on the list which was named Cox theory (Cox, 1952). Let us consider a planar fibre network subjected to uniaxial loading in the x-direction. Due to the uniaxial load applied, the network experiences strains. It extends in the direction of loading and typically contracts in the orthogonal directions (Fig. 11.1).

The strain state is described by a tensor (as discussed in Chapter 2). Similar to vector components, the tensor components vary with the chosen coordinate system. We will orient our coordinate system such that the x-axis will be aligned with the direction of loading. Considering paper as two-dimensional (2D) plane stress continuum, there are only two non-zero strain tensor components in case of uniaxial loading, namely ε_x and ε_y. The macroscopic shear in the chosen coordinate system γ_{xy} is zero in case of uniaxial tension. The network consists of many fibres oriented at different angles with respect to the loading direction. Let us consider one such fibre, oriented at the angle θ with respect to the loading direction (Fig. 11.2).

Artem Kulachenko, KTH Royal Institute of Technology, Stockholm, Sweden

https://doi.org/10.1515/9783110619386-011

Fig. 11.1: Fibre network under uniaxial loading. The dotted line shows the deformed state in response to the applied loading.

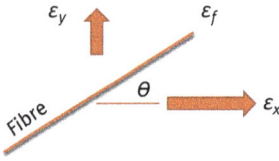

Fig. 11.2: Singled-out fibre oriented with an angle θ with respect to the x-direction of the applied load.

We will work under the assumption that this fibre is straight and has the same strain ε_f along its length. The first thing we do is relating the global strains ε_x and ε_y with the fibre strain with the help of the coordinate transformation of tensor components,

$$\varepsilon_f = \varepsilon_x \cos^2 \theta + \varepsilon_y \sin^2 \theta. \tag{11.1}$$

Knowing the strain in an arbitrarily oriented fibre, we can calculate the total force F_x projected on the x-axis from all the fibre by summing up the contribution from all the fibres in the following fashion:

$$F_x = N_f E_f \int_0^\pi \left(\varepsilon_x \cos^2 \theta + \varepsilon_y \sin^2 \theta\right) \cos \theta \cdot A_f \cos \theta \cdot f(\theta) d\theta. \tag{11.2}$$

Here, we introduce a number of new parameters where N_f is the number of fibres, E_f is the elastic modulus of the fibres, and A_f is the cross-sectional area of the fibres. Please note that A_f should also be projected and, therefore, multiplied with $\cos \theta$ similar to the force produced by a single fibre being equal to $EA_f\varepsilon_f$. This is the reason that $\cos \theta$ appears twice in the integrand. In eq. (11.2), $f(\theta)$ is the probability distribution function which tells about the fraction of the fibres oriented in a specific direction. Per definition, it obeys the relation

$$\int_0^\pi f(\theta) = 1. \tag{11.3}$$

By assuming an isotropic in-plane orientation of the fibres, which entails equal probability of fibres encountering in any direction meaning $f(\theta) = \text{const}$, we get $f(\theta) = 1/\pi$. We can now compute the integral in eq. (11.2), and it yields

$$F_x = K\left(\frac{3}{8}\varepsilon_x + \frac{1}{8}\varepsilon_y\right), \tag{11.4}$$

where $K = E_f N_f A_f$. Equation (11.4) relates the global strains with the force in the x-direction. Similar to the x-direction, we can compute the net force in the y-direction as

$$F_y = K\int_0^\pi \left(\varepsilon_x \cos^2\theta + \varepsilon_y \sin^2\theta\right)\sin^2\theta \cdot f(\theta)d\theta = K\left(\frac{1}{8}\varepsilon_x + \frac{3}{8}\varepsilon_y\right). \tag{11.5}$$

In the case of a uniaxial loading, the net force in y-direction should be zero. This enables us to relate ε_x and ε_y as

$$\varepsilon_y = -\frac{1}{3}\varepsilon_x. \tag{11.6}$$

It is a convenient relation that provides an interesting by-product, namely, Poisson's ratio, which is the coefficient connecting these two strains. It turns out that it is equal to 1/3, which is not too far from the classical 0.3 often taken as default for isotropic materials.

Combining eqs. (11.4) and (11.6), we can relate the force in the x-direction to the strain in this direction using the fibre properties:

$$F_x = K\left(\frac{3}{8}\varepsilon_x - \frac{1}{3}\cdot\frac{1}{8}\varepsilon_x\right) = K\frac{9-1}{24}\varepsilon_x = \frac{1}{3}N_f A_f E_f \varepsilon_x. \tag{11.7}$$

However, according to the definition, the same force can be constructed as

$$F_x = EA\varepsilon_x, \tag{11.8}$$

where E is the elastic modulus of the entire network and A is the nominal (without voids) cross-sectional area of the network. From eqs. (11.7) and (11.8), we can determine the relation between E and E_f as

$$E = \frac{1}{3}E_f N_f \frac{A_f}{A}. \tag{11.9}$$

The total number of the fibres in a given volume can be calculated by relating the total mass of the network to the mass of a fibre, which will give the relation,

$$N_f = \frac{DA}{\rho A_f},$$ (11.10)

for an arbitrary size of the network. Here, D is the density of the network and ρ is the mass density of the fibre. Combining eqs. (11.9) and (11.10) yields the required relation between the elastic moduli of the fibre and the entire network,

$$E = \frac{1}{3} E_f \frac{D}{\rho}.$$ (11.11)

Let us examine this relation using experimental evidence. We will start with the effect of density. The density of paper can be modified by two means: wet pressing and refining (see Sections 2.5.2 and 2.5.1). In the wet pressing method, we change the lateral pressure during the sheet preparation while the paper is still wet and the fibres compliant, and densification is achieved by fixing the fibres in the compressed state. In the approach with refining, we beat the fibres prior to sheet making. In this process, the fibres become more flexible and bonding is improved through creating fibrillar fines on the surface of the fibres. Those fibrillar fines contribute to the densification of the sheet, thereby increasing connections between the fibres and enhancing fibre joint regions through better bonding and mechanical interlocking (Motamedian et al., 2019). Figure 11.3 shows that the elastic modulus scales linearly with density increased by either method. Interestingly, the scaling in Fig. 11.3 starts with a density of around 200 kg/m^3. This density is associated with

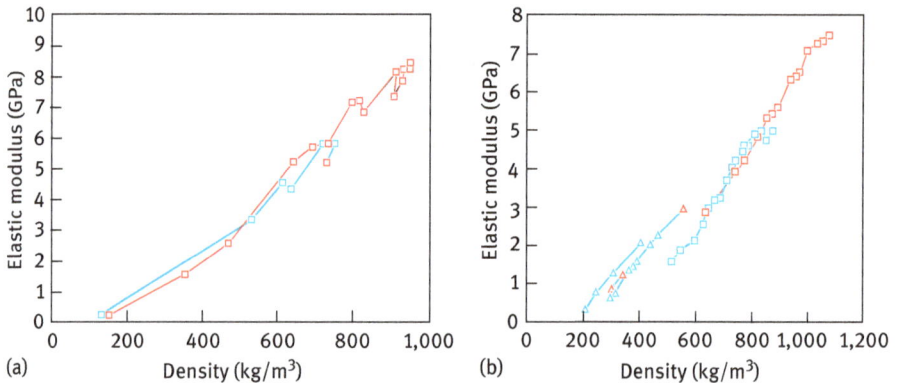

Fig. 11.3: Elastic modulus against density using data from Alexander and Marton (1968). (a) The density of laboratory sheets was increased by increasing the wet pressure. Data for three different chemical pulps are shown, each connected with a line. (b) The density of laboratory sheets was increased by increasing the refining. Data for two different chemical pulps (squares and triangles) and several wet pressures are shown, each refining series being connected with a line.

the percolation point, that is, the minimum density required to establish the connectivity across the network. We should have enough fibres to make sure that we can transfer the load from one boundary to another. Cox theory did not capture that as infinitely long fibres were assumed. Apart from this discrepancy, the linear scaling with the density is confirmed.

What about the actual predictive power of Cox equation? We have the values of the elastic modulus reported in Fig. 11.3. In order to test the formula and compare the results with these values, we need to know the density of the fibre and the elastic modulus. The density of the softwood fibres was reported at a level of $\rho = 1,500 \text{kg/m}^3$ (Hermans et al., 1946), while the elastic modulus of the pulp fibre having relatively low microfibril angle varied around 70 GPa (Page et al., 1977). Using these values and Cox equation (11.11), the elastic modulus of handsheets of mass density $D = 1,000 \text{ kg/m}^3$ should be equal to

$$E = \frac{1}{3} 70 \cdot \frac{1,000}{1,500} = 15.6 \text{ GPa},$$

while in practice, values attained in experiments with softwood pulp handsheets rarely exceed 8 GPa, which is almost 2 times lower than Cox equation would predict.

Let us now consider what the cause of this theoretical overestimation can be. In such cases, it is natural to revisit the assumptions made by Cox. One of them is assuming that the fibres are infinitely long, and the stress is constant along the fibre. In reality, however, the fibres have finite length and the stress should go down towards the free tangling ends of the fibre. Using simulations, (Åström et al., 1994) showed schematically how the stress along a fibre is distributed, demonstrating a decreasing stress towards the ends of the fibre (Fig. 11.4).

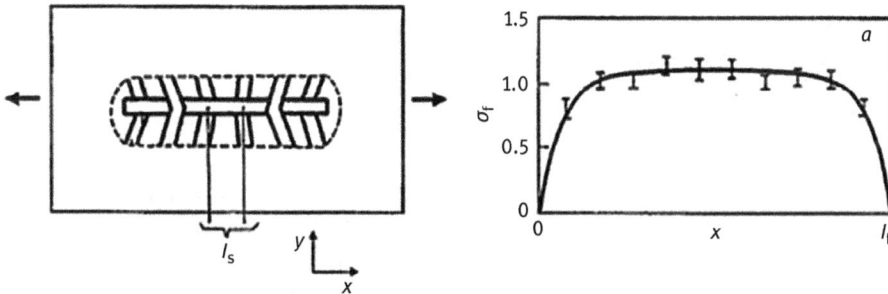

Fig. 11.4: Computed stress distribution along a fibre inside the network. Reproduced from (Åström et al., 1994), with the permission of AIP Publishing.

The direct consequence of stresses declining at the ends is the loss of the load-bearing capacity for a certain volume of the fibre. They contribute to the mass of the network but do not bear any load. This can be accounted for by so-called shear-lag correction similar to that in fibre composites, but having different nature, since there is no matrix in the fibre network we consider. The correction is expressed as

$$E = \frac{1}{3} E_f \frac{D}{\rho} \left(1 - \frac{l_{crit}}{l_f} \right),$$

(11.12)

where l_f is the average fibre length and l_{crit} is the averaged combined length of tangling ends along a fibre. How large should it be to be able to fit the results? By using 8 GPa as a target stiffness, we can determine that the fraction of unbonded length which would match the experimental stiffness at a sheet density of $D = 1,000 \, \text{kg/m}^3$ should be

$$\frac{l_{crit}}{l_f} = \left(1 - 3 \frac{E}{E_f} \frac{\rho}{D} \right) = 1 - 3 \frac{8}{70} \cdot \frac{1500}{1000} = 0.49.$$

This means that practically half of the fibre length should be unbonded in a dense sheet in order to meet the experimental results. Thinking about the possibility of the fibre to bond from both sides, this cannot be the single explanation. The use of shear lag theory will be discussed in the context of wood composites in Sections 12.3.3 and 12.3.4.

Let us list the assumption and the factors which can contribute to the overestimation.

1. Cox theory neglects the fibre joints and the finite elasticity of the joints. A large fraction of the fibre is bonded with other fibres. This means that a part of the elastic energy will be accumulated at the bonded regions. The fibre is a composite structure itself and it is not isotropic. With the load-bearing cellulose fibrils oriented along the fibre, it is actually more compliant in the transverse direction linked to the elasticity of the joints. Neglecting it can potentially make the predicted elastic modulus higher.

2. The fibres are assumed to be straight in contrast to how they are in reality (Fig. 11.5) and the contribution of bending is neglected. It is also a serious limitation as it eliminates a more compliant type of deformation, again making the response stiffer.

3. The measured fibre elastic modulus of 70 GPa was acquired from the fibres outside the network. During the papermaking, the fibres experience a stress history, they shrink and are deformed. This alone can introduce additional changes altering the mechanical properties of the fibre. The effect of papermaking can be easily illustrated by comparing the elastic properties of paper dried freely and under constraints. During restrained drying, the fibres cannot shrink freely, although they may still shrink locally and do so particularly at the joints. Partly inhibited shrinkage leads to positive stresses and results in straighter fibres. This process is called "activation" of the network (Niskanen, 2008) and is associated with increased stiffness. However, as the shrinkage still takes place, it decreases the

elastic modulus of the fibres through creating defects and dislocations (Page et al., 1985), reducing the microfibril-orientation angle and creating slack "inactivated" regions along the fibre.

4. The fibres are assumed homogeneous, in other words, have the same, length, width, and elastic properties. It is rarely so with pulp fibres. They show a large variation in geometrical and materials properties and they can vary not only between different fibres but also along a single fibre.

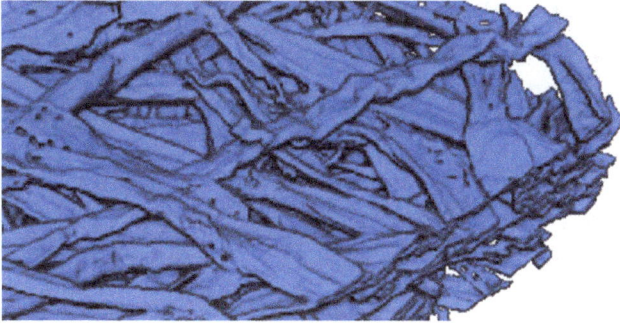

Fig. 11.5: Microtomography images of a fibre network.

Obviously, all of the abovementioned assumptions made by Cox theory increase the theoretical estimate, meaning that the theory provides an upper estimate for the elastic modulus of a randomly oriented fibre network as a function of elastic modulus of the constitutive fibres. Eliminating critical assumptions and yet preserve a simple analytical formula is a formidable task. However, it is possible to eliminate some of above-listed points by choosing another model system, namely, nanopaper. By doing this, we can at least see how the eliminated assumptions contribute to the overestimation.

The word "nanopaper" came from the cellulose nanofibres (CNFs) that are used to prepare the paper material instead of ordinary fibres. Despite the structural similarity with ordinary paper, nanopaper has superior mechanical properties, low porosity, and is naturally transparent (Henriksson et al., 2008). The CNFs are the smallest fibrous component of the plant cell wall. These fibres can be liberated by chemical (Saito et al., 2007) or enzymatic treatments (Henriksson et al., 2007) of plant materials with subsequent mechanical disintegration. Nanopaper is further discussed in Chapter 12.

The load-bearing components of nanopaper, the CNFs, are about 1,000 times shorter and thinner than the fibres used in the ordinary paper. Figure 11.6a shows a scanning electron microscope micrograph of the surface of nanopaper. The observed width of the nanofibres is rather uniform. At the same time, it is difficult to identify the length of nanofibres from such images or by any other established technique. Another remarkable feature of nanopaper is its layered structure, which can be seen in the image of the nanopaper cross section (Fig. 11.6b). Apparently, nano-fibres have a tendency to self-assemble into planar layers.

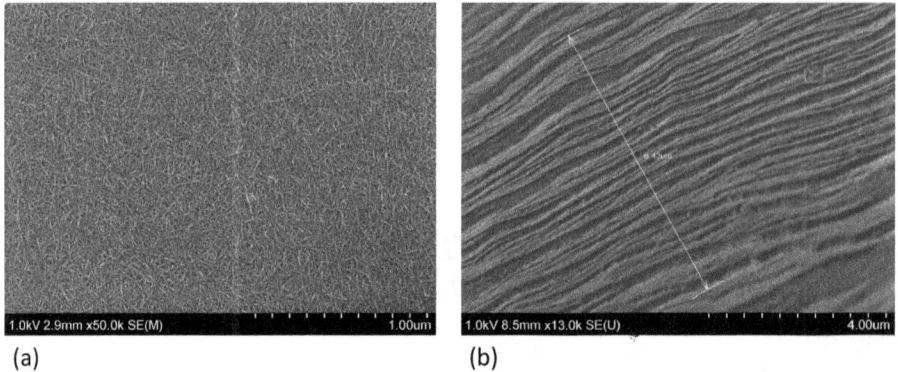

(a)   (b)

Fig. 11.6: SEM micrograph of: (a) the surface of nanopaper; (b) the cross section of nanopaper (the marked distance is 6.5 μm).

What are the advantages of using nanopaper as a model system relevant for us to try to find the source of the discrepancy between theory and practice?

1. Similar to conventional paper, nanopaper constitutes a network of fibres; however, the nanofibrils have more homogeneous morphology, which partly solves one of the problems of dealing with inhomogeneous dissimilar fibres and not being able to account for that reliably.
2. Nanopaper, being much denser, should intuitively be less susceptible to the effects of drying, since the segment length is much shorter, and the nanofibres are stiffer. Putting this in numbers, nanocellulose fibres, in the pure crystalline form, are reported to have an elastic modulus as high as 134 GPa (Sakurada et al., 1962; Tanaka and Iwata, 2006) in the longitudinal direction, which is almost twice as high as that of pulp fibres.
3. Individual nanofibrils have a simpler structure compared to ordinary fibres. They do not have a hollow structure and, therefore, are much closer to the type of fibres assumes in Cox theory.
4. The layered structure of the network makes each layer 2D, which is coherent to the basic assumption formulated by Cox.

Let us now examine the accuracy of analytical prediction by comparing it with experimental data. The reported elastic modulus of isotropic sheets made of nanocellulose varies between 12 and 19 GPa after renormalizing by multiplying with the density of the cellulose over the measured density of the sheet in order to account for the effect of the density (Henriksson et al., 2008; Syverud and Stenius, 2009). Most of the reported values of the elastic modulus are significantly lower than the theoretical maximum of 44 GPa ($\approx$134 GPa/3), computed using Cox equation. Even assuming an elastic modulus of nanocellulose equal to 80 GPa, which is the value measured on a single pulp fibre, the experiments still lag behind the theoretical prediction. A comparison of the measured experimental values by Henriksson et al. (2008) and Cox estimates (Fig. 11.7) shows that.

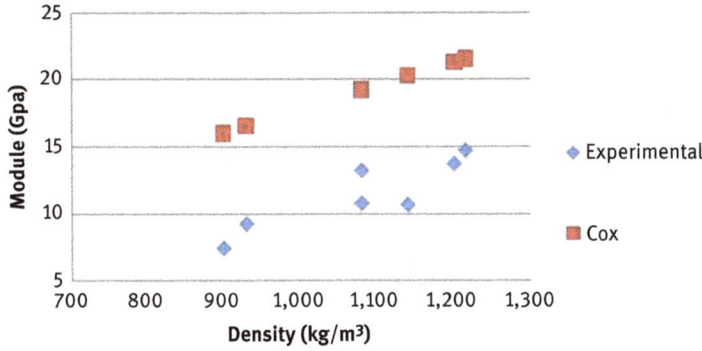

Fig. 11.7: Elastic modulus of nanopaper. Comparison of experimental data (Henriksson et al., 2008) with predictions using Cox equation, eq. (11.11).

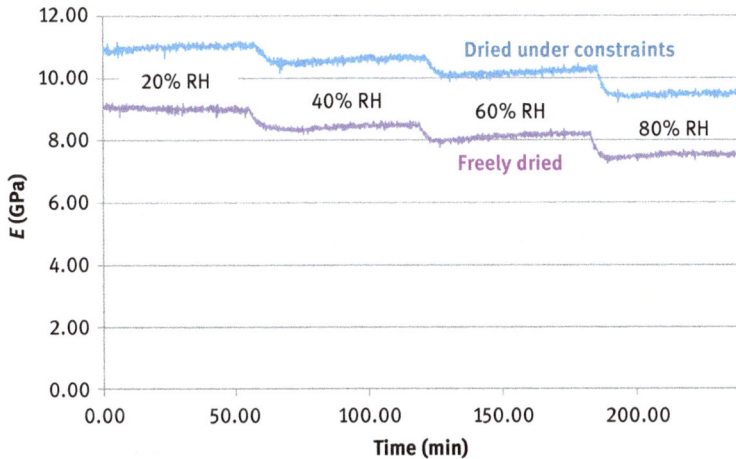

Fig. 11.8: Changes in elastic modulus during sorption for nanopaper dried freely and under constraints. Data from Kulachenko et al. (2012).

Perhaps we missed something important such as the effect of drying? Kulachenko et al. (2012) prepared isotropic nanopapers dried freely and under constraints. Figure 11.8 shows the elastic modulus of the two differently dried nanopapers under stepwise elevated relative humidity using a dynamic mechanical analyser. The two different films had a nearly identical density of 1,500 kg/m³, close to the theoretical density of cellulose.

The sample dried under constraints showed a 12% higher elastic modulus, but again, the difference was not as dramatic as in the case of sparse conventional paper, where it can exceed 50% (Rigdahl and Salmén, 1984).

These results show that drying constraints do not significantly alter the elastic properties of dense nanopapers and cannot fully explain the difference between theoretical predictions and experimental measurements of the elastic modulus. Furthermore, the obtained elastic modulus was not significantly greater than that of ordinary paper. What can explain this? As we have exhausted the possibilities offered by experimental methods in quantifying the factors contributing to the decreased elastic modulus, let us consider numerical methods instead.

11.2 Modelling a fibre network with the finite element method

With increased computing power, the limitations of the theoretical models have been overcome by numerical models (Åström et al., 1994; Heyden, 2000) in which the fibre is represented explicitly, as a chain of beam elements in the finite element analysis, for example. Heyden (2000) presents a comprehensive study of the effects of various fibre network parameters on the elastic properties. One of the limitations of early numerical models is their inability to consider dense fibre networks, partly because of limited computing power, but also because, in these models, the fibres have to share the nodes in the points of contact. Increasing the density requires a fine and irregular mesh to meet this condition. This limitation was avoided by using a bonded contact formulation to describe the bonds between the fibres (Motamedian and Kulachenko, 2018). In this contact formulation, the fibres represented with beam elements are pointwise connected in both translational and rotational degrees of freedom at the points of contact and do not need to share any node. This enabled us to consider a wide range of sheet densities in the analyses.

Beam elements that we use to represent the fibres have no through-thickness normal strains, which makes their cross section rigid against load in the normal direction. In reality, the fibres deform locally due to reciprocal forces in the bond regions. We account for the tangent and normal compliance in these regions through the contact stiffness in a penalty-based contact algorithm. In our case, when the cross sections of the fibres are rigid against pointwise loads, the contact stiffness alone is appropriate for representing the local deformations in the fibre joints.

The fibres were discretized with elastic Timoshenko beam elements. The fibres were assumed to have a square cross section with fibre width and height of 20 nm, corresponding to the size of nanofibril aggregates. The fibres may be curved. This fibre curl is quantified by the curl index C, which is the ratio between the straight

distance between the two ends of the fibre and the fibre length (Fig. 11.9). The outline of the model fibres traces a circular arc.

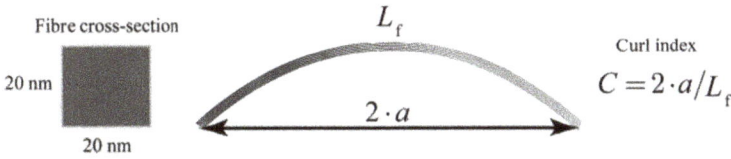

Fig. 11.9: Assumed fibre geometry with a controlled curl.

11.2.1 The geometry of the network model

We need to construct a randomly oriented rectangular fibre network to mimic the
structure of nanopaper. The network model is created by random deposition of fibres
onto a flat surface. Prior to deposition, all the fibres were assumed to be in a plane
parallel to the surface. In a 2D network model, the fibres are deposited onto the same
plane and, therefore, can intersect. The details of the three-dimensional (3D) deposition will be considered later in Section 11.2.10. During random positioning, parts of
the fibres can extend outside the rectangular cell. These fibres are clipped, and the
clipped parts are moved to the opposite side of the fibre network to avoid density
variations near the boundaries (Heyden, 2000). This type of approach enables creation of period structures. The schematics of the method is demonstrated in Fig. 11.10.
The fibres were deposited one by one until the target density was reached.

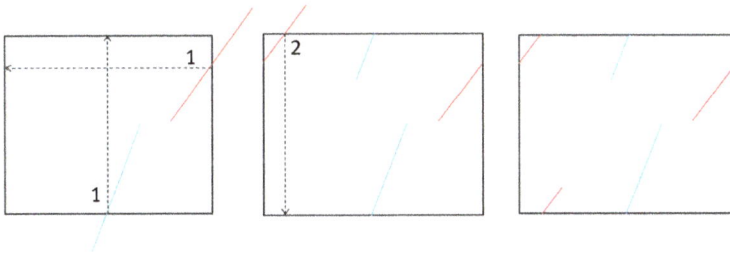

Fig. 11.10: Principle of forming a periodic network structure by cutting and transposing the fibre at
the boundary of a representative volume element.

11.2.2 Boundary conditions and loading

We tested the generated fibre network by applying a prescribed displacement in the x-
direction at one end and constraining the network at the opposite side (Figure 11.11a).

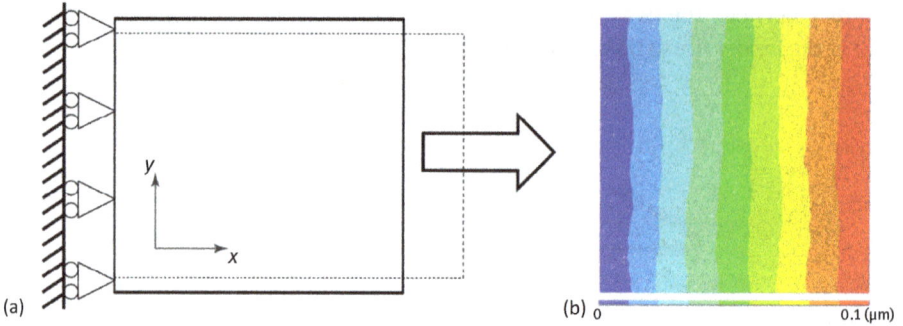

Fig. 11.11: (a) Loading conditions; (b) the displacement in the loading direction mapped on the fibres in the randomly oriented nanocellulose fibre network after 0.2% applied strain. The size of the fibre network is 50 × 50 μm². The average length of the fibres is 5 μm, and their width is 20 nm. The fibre network contains 30,250 fibres with an average number of contacts per fibre around 123.

At those two boundaries, zero traction in the y-direction was assumed. Periodic boundary conditions were applied across the sheet in the y-direction. Figure 11.11b shows the calculated displacement field mapped on the finite element mesh. The displacement field is macroscopically uniform across the fibre network, indicating that it is not noticeably affected by the boundary conditions.

11.2.3 Convergence with respect to mesh density

One of the first things one should do in the finite element method is testing the mesh convergence. The mesh resolution should not affect the results. We will do the mesh study for two different fibre network densities. Figure 11.12 shows the convergence of the elastic modulus for the fibre network densities of 500 kg/m³ and 1,500 kg/m³ with a decreasing length-to-width ratio of the beam elements. The fibre width in the simulations was 20 nm and the length was 5 μm. The size of the fibre network was 50 × 50 μm². We used a rather conservative estimation of nanofibril modulus of 80 GPa in most of our analyses. The geometry of the network model remained the same for each fixed density during testing. The dotted line in Fig. 11.12 shows the prediction by Cox theory.

The elastic modulus converges to a constant value with denser mesh. The convergence is smoother, and its rate is greater for the sparser fibre network. In a denser fibre network, the required number of elements to reach convergence is greater because of a shorter free segment length. The network model overestimates the stiffness if the free segments are insufficiently resolved with beam elements. However, the converged prediction is rather close to Cox theory, in particular for the dense mesh.

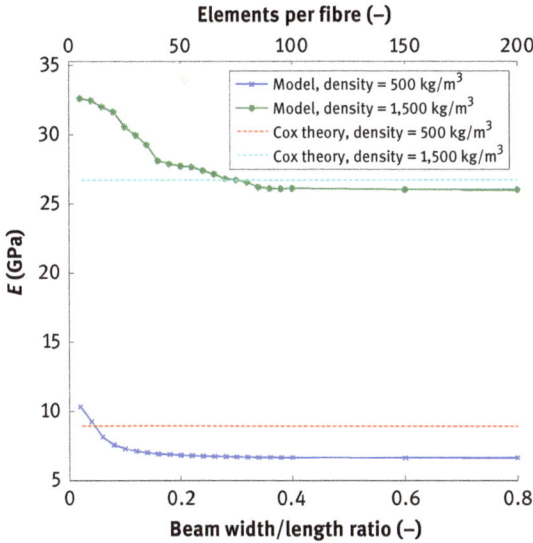

Fig. 11.12: Convergence of the numerical simulations with respect length-to-width ratio of the beams.

11.2.4 Representative fibre network size

Another important parameter is the representative size of the fibre network in the model with reference to the average fibre length. We varied the fibre network size while keeping the length of the fibre constant. For each size, we created 20 realizations of a 1,000 kg/m^3 fibre network and computed its elastic modulus. Figure 11.13a shows that the mean value of the elastic modulus stabilizes once the fibre network becomes approximately 6 times larger than the length of the fibres. At the same time, the standard deviation decreases and changes marginally after the fibre network becomes about 8 times larger than the fibre length.

Based on the mesh and size convergence studies, we used beam elements with a width-to-length ratio of 0.6 and a fibre network size 10 times the average fibre length throughout the simulations.

11.2.5 Effect of density

The elastic modulus of the fibre network scales linearly with density for different length-to-width ratio R of the fibres, as predicted by the Cox theory (Fig. 11.14). The elastic modulus decreases with a decreased length-to-width ratio of the fibres. This effect is attributed to the shear-lag phenomenon due to the axial stresses decreasing towards the ends of the fibres. The shorter fibres have more ends per given volume and therefore, smaller length to width ratio reduces the elastic modulus. Another

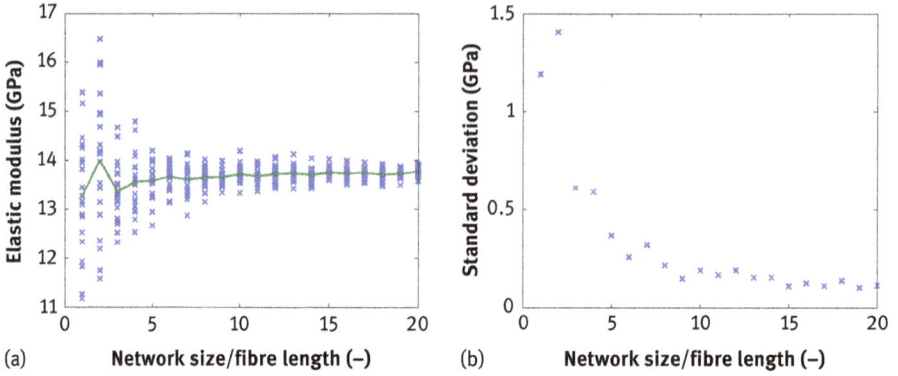

Fig. 11.13: (a) Elastic modulus and its variations for different realizations of the fibre network and (b) standard deviation of elastic modulus as functions of the ratio between fibre network size and fibre length.

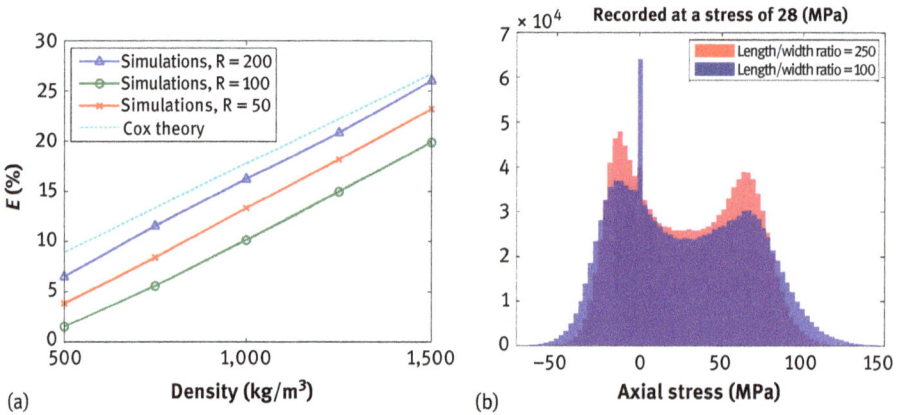

Fig. 11.14: The effect of mass density (a) and fibre aspect ratio (b) on the simulated elastic modulus of the fibre network.

way to demonstrate this is to sample the average axial stress from all beams and plot the histogram. Figure 11.14b shows the distributions of axial stresses in two networks composed of fibres having different length to width ratios. There are several interesting facts to be noted. First, the distinct peak at zero, especially for shorter fibres. It corresponds exactly to free ends. The next is the bimodality showing that there are parts of the fibre that are stressed positively (loaded in tension), unstressed, and even stressed negatively (loaded in compression).

Similar information can be presented in a polar plot with fibre oriented from 0 to 180 degrees with respect to the x-direction. The radial direction shows stress and the hoop direction shows the orientation of the segment. First, we map all the stresses with the dot (Fig. 11.15a) and then average them over a range to show the average stress in the specific direction (Fig. 11.15b).

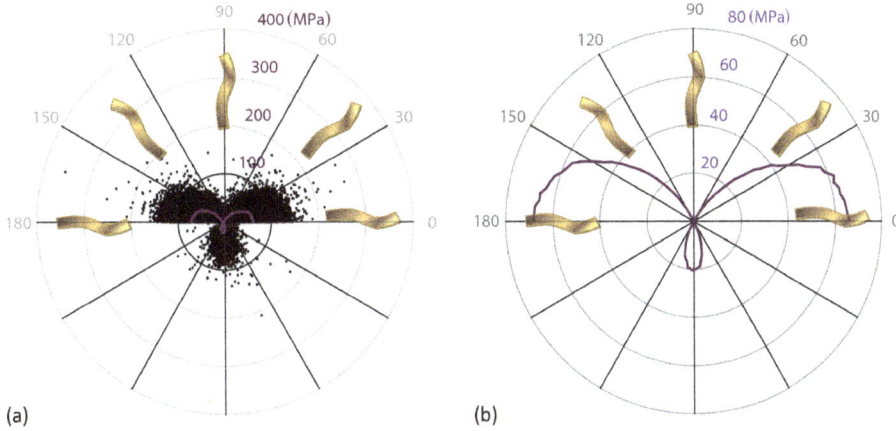

(a)

(b)

Fig. 11.15: Stress distribution in the fibre network as a function of fibre segment (beam) orientation: (a) all the stresses as a point cloud, and (b) average stresses.

Although Fig. 11.15a shows great variability, the average stress follows a smooth trend presented already in eq. (11.1). An interesting observation is a fact that the fibre segments located at 60 degrees to the loading directions have on average zero stress. It follows also from eq. (11.1) evaluated at 60 degrees that

$$\varepsilon_f(\theta) = \varepsilon_1 \cos^2 \theta - \frac{1}{3}\varepsilon_1 \sin^2 \theta \Rightarrow \varepsilon_f(60°) = 0. \tag{11.13}$$

Density also affects the stress variations along the fibre length. Figure 11.16a shows how the distribution of axial stresses changes with density. The stress distribution is bimodal as before. The contrast between the peaks becomes more pronounced at a higher density. Both the maximum and average stresses scale down with density. At a relatively low density (500 kg/m^3), the distribution of axial stresses is wider and has a very long tail. Figure 11.16b shows the effect of density on the tangent contact stresses, which were calculated by dividing the calculated contact forces with the contact area of orthogonally arranged fibres. Similar to fibre stresses, the magnitude of the contact stresses scales down with density. This is important when we talk about the strength of the network, which is often controlled by the bonds since the fibre strength is greater than the bond strength.

Density also increases the axial strain variations in the fibre network, which can be visualized after mapping, smoothing, and differentiating the displacement

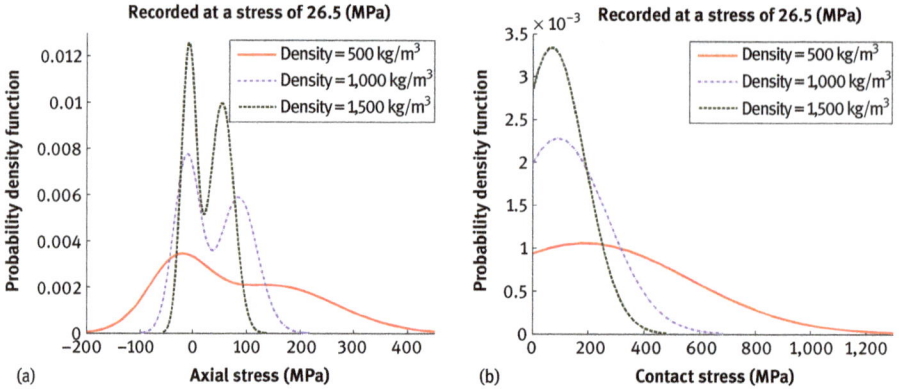

Fig. 11.16: Probability distribution functions showing the effect of mass density on: (a) axial stresses in the fibres and (b) tangent contact stresses in the joints.

field from the fibres onto a continuous surface. The strain field for different densities is visualized in Fig. 11.17. When comparing the scales, it becomes evident that sparse sheets have larger strain variations of smaller size. This is another aspect relevant to the strength of paper since the failure is due to localization and interaction between these variations.

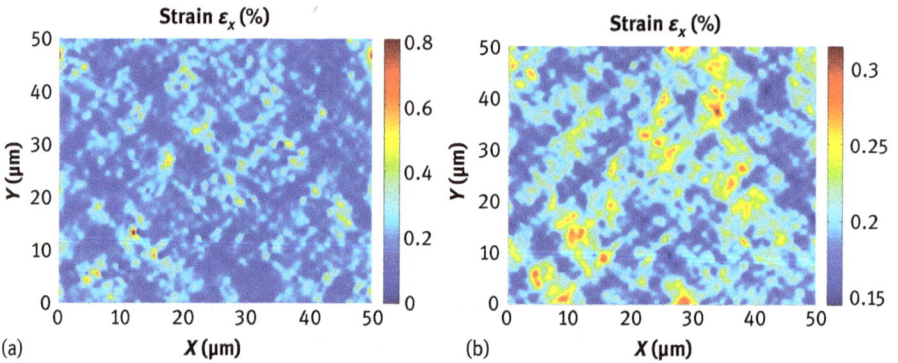

Fig. 11.17: Strains variations in the fibre network recorded at a strain of 0.2% and mass density of (a) 500 kg/m^3 and (b) 1,500 kg/m^3.

11.2.6 Effect of non-crystalline regions

For ordinary paper, fibre defects such as kinks and dislocations have been reported to affect the modulus (Page and Seth, 1980a, 1980b). Observing these defects is a formidable task in the case of nanofibres because of their small size. However, Nishiyama et al. (2003) reported that CNFs contain equidistant amorphous regions along their length, also referred to as "periodic disorders". This is schematically illustrated in Fig. 11.18. The reported length of the crystalline regions is 150 nm. For the reference case in our studies, the length of amorphous regions was selected to have a volume fraction of amorphous material around 30%, based on the measurements reported for TEMPO-oxidated nanocellulose (Saito et al., 2007). Higher values of the amorphous fraction for a similar type of material were reported elsewhere (Agarwal et al., 2010). The amorphous regions, however, can be located along the fibres and on their surfaces. The implication of this aspect will be discussed later in Section 11.2.10. A detailed discussion on the distribution of amorphous regions and their impact on the elastic modulus of cellulose microfibril was presented by Nishiyama (2009).

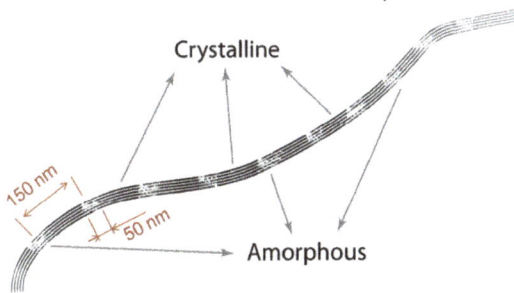

Fig. 11.18: Schematic representation of equidistant disordered regions along the fibre.

Such disordered regions can have a lower elastic modulus compared to crystalline regions. They also contain free OH groups, which can bind water molecules. Through this binding water makes cellulose softer, that is, decreases the stiffness even further. In addition, the amorphous regions can deform more accumulating residual strains upon shrinkage even in the case of constrained drying. We will investigate the effect of elastic modulus variations and initial strains separately by means of modelling.

11.2.7 Effect of residual/initial strains

Residual strains cannot be stored to any great extent within the cellulose crystal, but rather in the amorphous regions of a nanofibre since these regions are more compliant and more accessible to water during the sheet making. To estimate the effect of the residual strains on the modulus, positive initial strains, which correspond to slack, were introduced in equidistant amorphous regions along the length of the fibres, as depicted in Fig. 11.18. The simulations were performed in two steps. In the first step, the stresses caused by initial strains were equilibrated under constraints. In the second step, a prescribed displacement of 0.1 µm was applied to measure the effective modulus for different length-to-width ratios of the fibres. Figure 11.19 shows the simulation results. The non-linearity of the curve is owing to the shear-lag phenomenon mentioned earlier in Section 11.2.5.

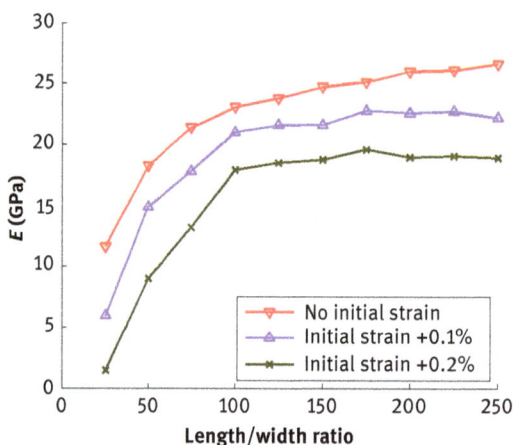

Fig. 11.19: Effect of initial strain on the elastic modulus of the fibre network. The initial strain indicated in the legend represents the initial strain within each amorphous region.

Initial strains contribute to a reduction of the elastic modulus. However, the effect is not sufficiently strong for explaining the discrepancy between theoretical predictions and observed elastic modulus of nanopaper. These numerical results are in harmony with our experimental finding that drying under constraints, which reduces slack, has a small enhancing effect on the elastic modulus. It is also observed that, when there are no initial strains, the modulus increases steadily with the fibre aspect ratio over the whole range of investigated fibre lengths. At the same time, when initial strains are introduced, the modulus reaches a plateau at an aspect ratio of about 100. This suggests that the fibres lose their ability to transmit stresses over long distances when initial strains are introduced.

11.2.8 Effect of elastic modulus variation

The repeated amorphous regions along the fibres are associated with a reduced modulus compared with the crystalline regions. To investigate the effect of these fibre modulus variations, we studied two cases. In the first case, we introduced equidistant compliant regions along the fibres, representing amorphous materials, having a lower elastic modulus of 30 GPa while the elastic modulus of 80 GPa was used for the remaining parts of the fibres, representing crystalline cellulose. In the second case, we distributed the compliant segments representing the amorphous regions randomly, that is, without any regularity in appearance along the fibre. In both cases, the mass fraction of amorphous cellulose was set to 30%. The first case is referred to as "Equidistant 70/30%", and the second case is denoted as "Irregular 70/30%". Furthermore, three different fibre networks without any modulus variations along the length of the fibres were computed for comparison: one consisting of only fibres with $E_f = 80$ GPa, one of only fibres with $E_f = 30$ GPa fibres and one with a 70/30% blend of 80 GPa and 30 GPa fibres.

The results in Fig. 11.20a show that the different blends of individually homogeneous fibres follow a simple mixing rule. Remarkably, this mixing rule does not apply when the modulus variations are introduced along the length of the fibre. The elastic modulus is significantly lower for the fibre networks composed of heterogeneous fibres as compared with the 70/30% blend of homogeneous fibres, which has the same relative amount of crystalline and amorphous material. The network with equidistant disorders (equidistant 70/30%) has a lower modulus than the network with randomly distributed disorders (irregular 70/30%). The difference diminishes as the fibre becomes longer. Furthermore, once initial strains were added to the more compliant segments, the elastic modulus was further reduced and the effect of the length-to-width ratio diminished, that is, the effects of initial strains were retained.

The amorphous regions can be located along the fibres and on their surfaces, which cannot be reliably distinguished with conventional crystallinity measurements. This means that the fraction of the disordered regions along the fibres can be smaller due to a surface contribution to the quantity of the non-crystalline regions. We quantified the effect of the amorphous fraction on the elastic modulus by reducing the number of amorphous regions along the fibre. The surface contribution reduces the elastic modulus of the nanofibre uniformly along the length and has implicitly been accounted for by taken a relatively conservative estimation of fibre elastic modulus of 80 GPa.

Figure 11.20b shows the effect of the ratio between crystalline and amorphous regions presented with the reference to the network with homogeneous fibres. For long fibres, the reduction in elastic modulus is proportional to the addition of amorphous regions.

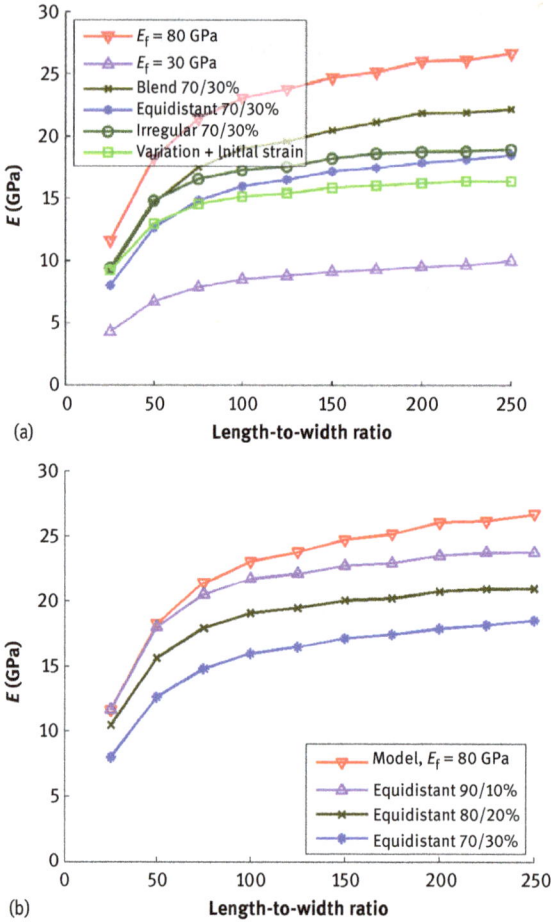

Fig. 11.20: Effect of varying elastic modulus along the fibres: (a) in comparison with constant elastic modulus and missing rule, and (b) effect of the fraction of amorphous regions.

11.2.9 Effect of fibre curl

Fibres are not perfectly straight and fibre curl or non-straightness in the plane of the sheet favours fibre bending, which is a more compliant form of deformation. Figure 11.21 shows the dependence of elastic modulus on curl. Denser networks are significantly less sensitive to changes in fibre curl, presumably due to the small ratio between the free fibre segment length and the radius of curvature. It should

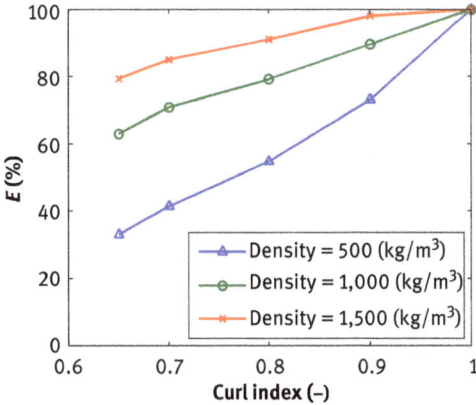

Fig. 11.21: Dependence of the elastic modulus on the curl index *C*.

be noted that the typical value for pulp fibre curl, expressed as a shape factor in Fig. 11.21, is above 0.9.

Increasing the curl index engages the fibres into bending deformation. Figure 11.22 shows how elastic energy is redistributed between different deformation modes in the fibres. For straight fibres, $C = 1$, the strain energy is dominated by the elongation of the fibres. The fraction of bending deformation is increased with decreased density of the fibre network. For curled fibres, $C = 0.8$, bending deformation dominates at a density of 500 kg/m^3. However, at a higher density of 1,500 kg/m^3, the difference between straight and curled fibres is insignificant.

11.2.10 Effect of three-dimensionality

For the 2D deposition method, the number of fibre bonds is identical to the number of intersections between the fibre backbones. During the preparation of nanopaper, however, a dispersion of nanocellulose fibres is poured onto a fine wire grid, and the fibres fall onto the underlying fibres and block parts of them from coming into contact with fibres deposited at a later point. As a consequence, the total number of bonds is reduced. To estimate the degree to which bonding was affected, we reconstructed the sheet forming mechanism numerically by depositing the fibres on top of each other. Figure 11.24 demonstrates different deposition methods. The method labelled "2D" is the method used for generating 2D fibre networks where the fibres are allowed to intersect. In Method 3D-1, the fibre is cut upon crossing the other fibre. This method yields the maximum packing, but it is obviously not realistic and will only be used for comparison. In Method 3D-2, the fibres wrap around the

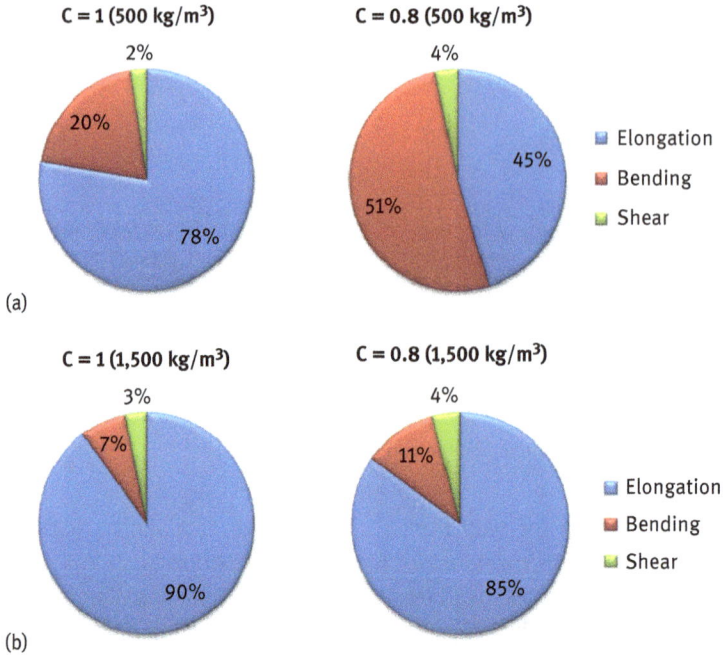

C = 1 (500 kg/m³) 2% 20% 78%

C = 0.8 (500 kg/m³) 4% 45% 51%

■ Elongation
■ Bending
■ Shear

(a)

C = 1 (1,500 kg/m³) 3% 7% 90%

C = 0.8 (1,500 kg/m³) 4% 11% 85%

■ Elongation
■ Bending
■ Shear

(b)

Fig. 11.22: Partition of the elastic energy stored in the fibres between elongation, bending and transverse shear strain energies for curl indices $C = 1$ and 0.8, respectively, and a fibre network density of: (a) 500 kg/m³; (b) 1,500 kg/m³.

underlying fibres, which is known to produce many 3D qualities of fibre network structures (Niskanen and Alava, 1994). The maximum angle at which a fibre can bend upon wrapping was kept below 85 degrees (Fig. 11.23).

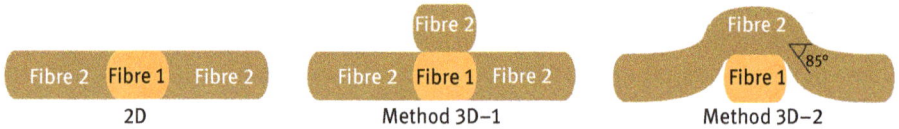

Fibre 2 Fibre 1 Fibre 2
2D

Fibre 2
Fibre 2 Fibre 1 Fibre 2
Method 3D–1

Fibre 2
85°
Fibre 1
Method 3D–2

Fig. 11.23: Illustration of different methods for modelling the fibre joint.

The difference between the 2D and 3D fibre networks is apparent in Fig. 11.24. In the 3D fibre network, the fibres do not overlap and stack on top of each which leads to a reduced number of contacts. The bonding region in the 3D fibre model has a local curvature.

Figure 11.25 shows how different packing methods affect the number of contacts in the fibre network. Method 3D-2 resulted in nearly half as many contacts

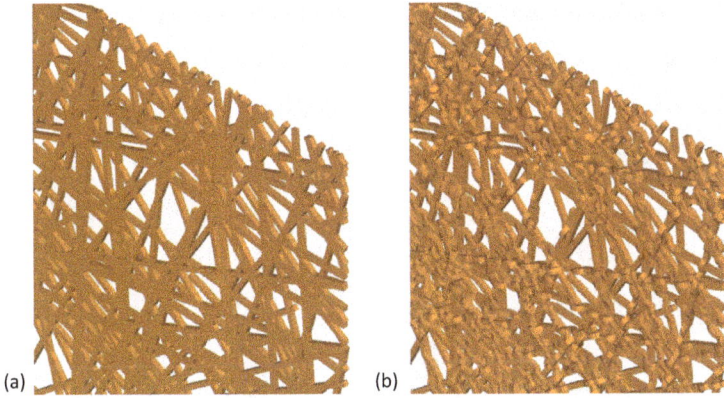

Fig. 11.24: Corner of a fibre network with a density of 1,000 kg/m^3: (a) 2D and (b) 3D.

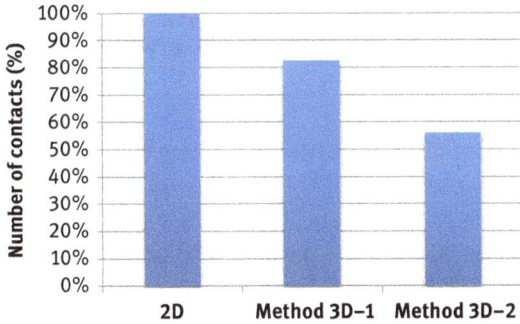

Fig. 11.25: Reduction in the number of fibre-fibre contacts due to 3D alignment of fibres.

compared to the 2D fibre network. It is not, however, equivalent to removing the bonds completely, because the fibres pile up on each other and remain interconnected through the fibres in-between.

We computed the stiffness of the fibre network created with Method 3D-2. The stiffness of the 3D fibre networks was lower than the stiffness of the 2D fibre network. A fraction of this reduction can be attributed to the lower density of 3D networks. The 3D deposition created a fibre network of greater thickness than in the 2D case. In order to quantify the reduction of the modulus due to three-dimensionality, we subjected the 3D fibre network to 2D loading constraints, that is, constraining out-of-plane displacement and rotations. Figure 11.26 shows that the elastic modulus of the 3D fibre network under 2D constraints (middle bar) is lower than for the proper 2D fibre network. Removing 2D constraints further reduced the modulus by 30% compared with the 3D fibre network under 2D constraints and by 40% compared with the 2D fibre network.

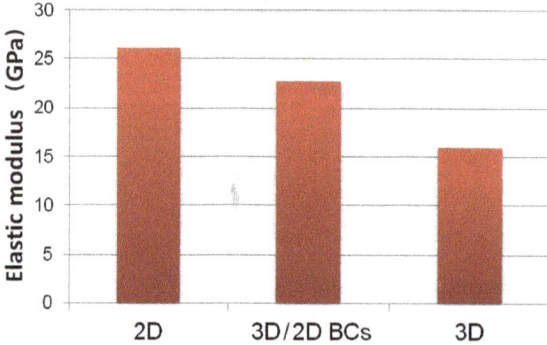

Fig. 11.26: The effect of three-dimensionality on the elastic modulus of the fibre network. The middle bar corresponds to the 3D fibre network with 2D constraints. The 3D fibre networks were created with Method 3D-2.

This reduction can be explained by observing the distribution of the elastic energy between different forms of deformation in the 2D and 3D fibre networks (Fig. 11.27). The drastic difference between these cases is that a large fraction of energy is stored in bending deformation in the 3D fibre network. Since the fibres are more compliant in bending, the 3D fibre network has lower stiffness.

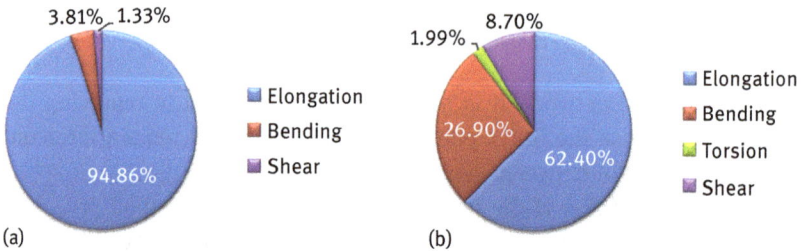

Fig. 11.27: Partition of the elastic energy stored in the fibres between different modes of deformation: (a) 2D fibre network and (b) 3D fibre network without 2D constraints.

11.2.11 Effect of bond density

The structure of nanopaper is layered in the thickness direction (Fig. 11.6b). During deposition, we assumed that all the fibres crossing in the plane form a bond, and all the bonds have the same load-bearing capacity. In reality, not all bonds can bear the load equally. Figure 11.28 demonstrates the development of the elastic modulus as the number of bonds is decreased by removing them randomly while maintaining the same density.

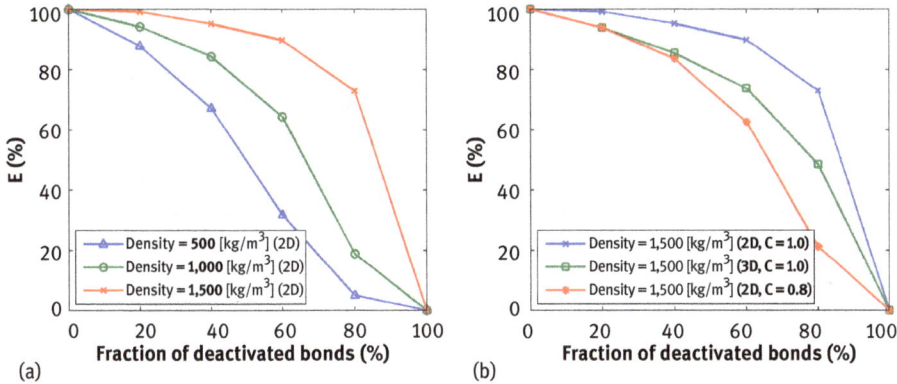

Fig. 11.28: Dependence of the elastic modulus on the fraction of active bonds in the fibre network. Results for: (a) different densities and (b) increased fibre curl index C and three-dimensionality of the fibre network.

In Fig. 11.28a, the results indicate that sparser fibre networks are more sensitive to the number of active bonds. Remarkably, however, the random elimination of 60% of the bonds caused only 10% decrease of the modulus in a dense planar fibre network with straight fibres. The 3D fibre network of the same density suffered a more considerable reduction (25%, Fig. 11.28b) since the removed bonds leave unconstrained curved regions along the fibres, which bend upon stretching and thus are more compliant in the tensile testing. A fibre network of curled fibres is also more sensitive to bond removal (Fig. 11.28b). Having fewer bonds means longer free fibre segments and, in the case of curled fibre, it also means that a larger fraction of the elastic energy is stored in bending deformation, as we already showed in Section 11.2.9. Bonds removal further increases the stress variation inside the fibre network, which has implications for the strength of the network.

11.2.12 Effect of bond stiffness

Finally, we varied the bond opening stiffness which depends on the local transverse properties of nanofibres. Beam-to-beam contact was a pointwise contact, in other words, the contact area is not resolved. This means that the actual stiffness is defined in the unit of N/m. The bond stiffness in all the analyses above was set to be equal to the elastic modulus (80 GPa) of the fibre times fibre width (20 nm), which gives the reference value for bond stiffness of 1,600 N/m. With this value, the elastic energy stored in the bonds turned out to be less than 1% of the total elastic energy. It is thus expected that the bond stiffness, when assigned a reasonable value, has a negligible effect on the fibre network stiffness.

Indeed, even when the bond stiffness is scaled by factors of 10 and 100, no significant increase can be seen in the elastic modulus of the fibre network, as shown in Fig. 11.29. On the other hand, it was possible to reduce the modulus of the fibre network by decreasing the bond stiffness. This effect is sensitive to the length-to-width ratio. At a given cross section, the elastic modulus of a fibre network with shorter fibres is more sensitive to a change in the bond stiffness. This can be explained by higher average bonding stress in the fibre network with shorter fibres, which increases the importance of bonds.

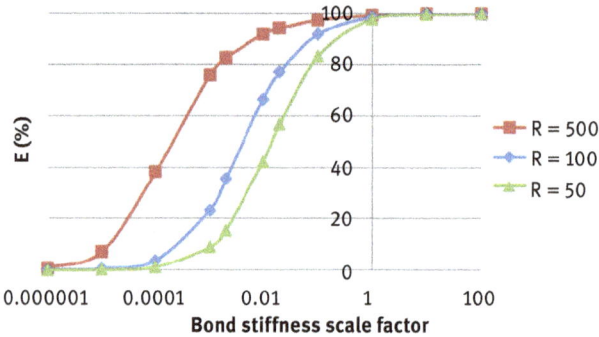

Fig. 11.29: Elastic modulus as a function of joint stiffness for different length-to-width ratios, R.

When the bond stiffness is reduced, the fraction of the elastic energy stored in the bonds increases. With the bond stiffness reduced by a factor of 100 from the reference level, Fig. 11.30 shows that the energy stored in the bonds increases dramatically from 1% to 20%. This increase is accompanied by a reduction of the fraction of elongation energy. The fraction of bending was not affected, and the fraction of shear was reduced since shear deformations predominantly take place near the bond regions, and, with more compliant bonds, the fibres were stressed less in the transverse direction.

Scale factor = 1

3% ⌐1%

7%

89%

Scale factor = 0.01

20%

1%

8%

71%

- ▣ Elongation
- ▣ Bending
- ▣ Shear
- ▣ Bonding

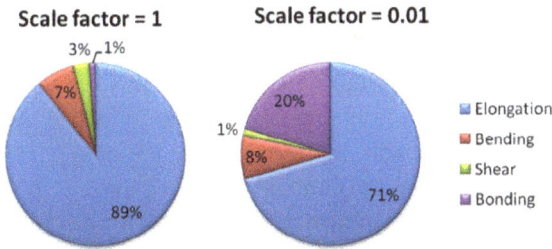

Fig. 11.30: Partition of the elastic energy stored in the entire fibre network between different modes of deformation for joint stiffness scale factors of 1 and 0.01.

11.3 Concluding remarks

We asked the question about what controls the stiffness of a fibre network and why it is lower than the theoretical predictions for a given density. We found that that there are a number of factors doing it although there is not a predominant one that can be blamed solely.

The details of the drying process are among the most influential ones affecting the fibre network through the activation process. As the fibre network shrinks without restraints, many of the fibre segments become slack, fibres are compressed accumulating initial strains and damage. It is particularly seen in conventional paper.

However, it is not the only contributor as we experimentally found with dense nanopaper in which the drying process had a lower effect because of having denser fibre networks with shorter free segments.

The length-to-width ratio of the fibre influences the elastic modulus, but its effect is limited for reasonably long fibres.

Fibre curl or non-straightness in the plane of the sheets promotes fibre bending during stretching of the network which is a more compliant form of deformation compared to the fibre stretching. It decreases the elastic modulus too.

Three-dimensionality of the bond geometry affects the elastic modulus through two factors: increasing a fraction of bending deformation in the contact regions, where the fibres wrap around the underlying fibres and reducing the number of bonds, although the effect from the number of bonds is non-linear and is not very significant for a relatively dense network. Removing 60% of the contacts from the fully bonded fibre network decreased the elastic modulus by 10% only in the 2D fibre network and by 25% in the 3D network.

Bond stiffness, which can also be related to transverse shear of the fibres, has a negligible impact if varied within a reasonable range in a dense nanopaper. The effect of the number of bonds and bonding compliance should be considered collectively. That is to say, in a fibre network with a low number of bonds per fibre, each contact site will bear greater force than in a better connected network and thus the effect of the bonding compliance will be more pronounced. This effect can be observed in sparse sheets or with bulkier hollow fibres.

The amorphous regions along the length of the nanofibres, which were modelled as variations of elastic modulus along the fibres, significantly influence the elastic modulus of nanopaper. They are probably the reasons why the nanopaper does have a dramatically greater stiffness than conventional paper.

The condensed list of factors reducing the stiffness of the fibre network can be presented as:

1. Finite length of the fibres (shear-lag effect)
2. Three-dimensionality of the network (decreased number of contact and induced non-straightness out of plane)
3. Drying (through network activation mechanism)
4. Finite compliance of the joints (more important with sparse sheets or hollow fibres)
5. Disorders along the fibres (contain softer regions)

11.4 Notes on the micromechanical aspects of network strength

In this chapter, we only considered the in-plane elasticity of the network. The mechanism of inelastic response and eventually failure are intricate and undoubtedly relevant to applications. In contrast to many other engineering materials, paper is often used beyond its elastic limit during forming and folding. Below is a list of some relevant questions regarding strength which can be addressed on the micromechanical level. Some of these questions are already discussed in previous chapters.

1. What is the dissipation mechanism during the plastic response of the fibre network? This can be bond failures and subsequent friction or plasticity of the fibre themselves. We saw that the largest fraction of the elastic energy in dry network is stored in the fibres. This information can help in answering this question.
2. What is the limiting factor for paper tensile strength: fibres or bonds? Every network will fail at some point as the strength of the fibres and bonds is limited. The fact is that in the majority of paper products the limiting factor is bonds, however, as paper ages, the fibres become embrittled and limit the strength. You can sometimes detect this by observing the fracture surface. In aged papers, it is straight without signs of fibre pull-out.
3. What are the precursors of the failure strain localization or bond failures? Multiple observations of strain fields using digital image correlation technique show that the strain localizes by bridging the areas of high strain shortly before failure. What is the factor controlling such localizations: bonds or inherited

inhomogeneities in paper such as density, thickness, or fibre orientation anisotropy? Using micromechanical simulation tools, it is possible to detect the precursor by disable debonding and comparing the strain field with the case with debonding enabled. Such observations show that strain localizations happen even without debonding and they are in the same place as with debonding, although they are not so sharp (Borodulina et al., 2012).

4. What is the fundamental difference between factors affecting compressive strength and tensile strength? The compressive strength is significantly lower than the tensile strength. During compression fibres may bend, buckle, develop plastic hinges, the network can delaminate, and these factors bring the strength down. Micromechanical simulations can help in detecting the basic mechanisms (Brandberg and Kulachenko, 2020).

5. How do fibre length, curl, coarseness, and their natural variations across the fibres affect the tensile and compressive properties of paper? We have learnt already how they affect the stiffness, but they also affect strength through similar mechanisms of redistributing the energy stored in different forms of deformation. These changes on both fibre and fibre joint level, respectively, are relevant to strength as well.

6. What is the role of secondary and primary fibrillar fines on strength? Refining is known to increase the strength of paper. Paper densifies in response to refining, but it is a part of the story. A large fraction of fibrillar content is generated in response to mechanical action applied to the surface of the fibres during refining. Theses fibrils reinforce the joints and they create layers connecting the fibres. These changes add other components to those brought by densification alone (Motamedian et al., 2019).

7. What is the influence of the drying constraints on strength? In this chapter, we learned that drying under constraints activated the network. It may affect the configuration of the fibres and fibre joints, which, in turn, will influence the strength.

8. What is the reason the mechanical properties of paper are affected by the loading rate? From Chapter 7, we saw that paper may exhibit rate dependency and intricate behaviour such as mechanosorptive creep. What are the mechanisms controlling it on the level of fibres and joints and below?

9. What is the fundamental difference between the wet and dry strength of paper? In the wet state, that fibres have not yet formed strong bonds and may slide with respect to each other. This is the reason why the wet strength is so much lower than the dry counterpart. The nature of the interactions is very different and as the balance is changed, the factors affecting the wet strength are very different compared to the dry strength (Kulachenko and Uesaka, 2012).

10. Why is the strength of the fibre-based product-size dependent, but the stiffness is not? From Chapter 8, we have learned that the strength of paper is size dependent. The dry strength scales follow the weak-link scaling law from a certain size. However, what is controlling the distribution at that size?

Literature references

Agarwal, U.P., Reiner, R.S., Filpponen, I. and Isogai, A. (2010). Argyropoulos DS crystallinities of nanocrystalline and nanofibrillated celluloses by FT-Raman spectroscopy. In: TAPPI International Conference on Nanotechnology for the Forest Product Industry, Helsinki, Tappi, p 7.

Alexander, S.D. and Marton, R. (1968). Effect of beating and wet pressing on fiber and sheet properties. II. Sheet properties. Tappi J. 51(6), 283–288.

Algar, W.H. (1965). Effect of structure on the mechanical properties of paper, In: Bolam, F. (ed.), Consolidation of the paper web, vol. 2, Cambridge: British Paper and Board Makers's Association, 814–851.

Åström, J., Saarinen, S., Niskanen, K. and Kurkijärvi, J. (1994). Microscopic mechanics of fiber networks. J. Appl. Phys. 75(5), 2383–2392.

Borodulina, S., Kulachenko, A., Galland, S. and Nygårds, M. (2012). Stress-strain curve of paper revisited. Nord. Pulp Pap. Res. J. 27(2), 318–328.

Brandberg, A. and Kulachenko, A. (2020). Compression failure in dense non-woven fiber networks. Cellulose 27(10), 6065–6082.

Cox, H.L. (1952). The elasticity and strength of paper and other fibrous materials. Br. J. Appl. Phys. 3(3), 72–79.

Henriksson, M., Berglund, L.A., Isaksson, P., Lindström, T. and Nishino, T. (2008). Cellulose nanopaper structures of high toughness. Biomacromolecules 9(6), 1579–1585.

Henriksson, M., Henriksson, G., Berglund, L.A. and Lindström, T. (2007). An environmentally friendly method for enzyme-assisted preparation of microfibrillated cellulose (MFC) nanofibers. Eur. Polym. 43(8), 3434–3441.

Hermans, P., Hermans, J. and Vermaas, D. (1946). Density of cellulose fibers. III. Density and refractivity of natural fibers and rayon. J. Polym. Sci. 1(3), 162–171.

Heyden, S. (2000). network modelling for the evaluation of mechanical properties of cellulose fluff, Lund: Lund University.

Kulachenko, A., Denoyelle, T., Galland, S. and Lindström, S.B. (2012). Elastic properties of cellulose nanopaper. Cellulose 19(3), 793–807.

Kulachenko, A. and Uesaka, T.. (2012). Direct simulations of fiber network deformation and failure. Mech. Mater. 51, 1–14.

Motamedian, H.R., Halilovic, A.E. and Kulachenko, A. (2019). Mechanisms of strength and stiffness improvement of paper after PFI refining with a focus on the effect of fines. Cellulose 26(6), 4099–4124.

Motamedian, H.R. and Kulachenko, A. (2018). Rotational constraint between beams in 3-d space. Mech. Sci. 9(2), 373–387.

Nishiyama, Y. (2009). Structure and properties of the cellulose microfibril. J. Wood Sci. 55(4), 241–249.

Nishiyama, Y., Kim, U.-J., Kim, D.-Y., Katsumata, K.S., May, R.P. and Langan, P. (2003). Periodic disorder along ramie cellulose microfibrils. Biomacromolecules 4(4), 1013–1017.

Niskanen, K.J. (2008). Paper physics. Papermaking science and technology, Vol. 16, 2nd, Helsinki, Finland: Fapet Oy.

Niskanen, K.J. and Alava, M.J. (1994). Planar random networks with flexible fibers. Phys. Rev. Lett. 73(25), 3475.

Page, D., Elhosseiny, F., Winkler, K. and Lancaster, A. (1977). Elastic modulus of single wood pulp fibres. Tappi J. 60(4), 114–117.

Page, D., Seth, R., Jordan, B. and Barbe, M. (1985). Curl, crimps, kinks and microcompressions in pulp fibres: Their origin, measurement and significance. In: Transactions of the 8th

Fundamental Research Symposium, 1985. Mechanical Engineering Publications Ltd., London, pp 183–227.

Page, D.H. and Seth, R.S. (1980a). The elastic modulus of paper. The effects of dislocations, microcompressions, curl, crimps, and kinks. Tappi J. 63(10), 99–102.

Page, D.H. and Seth, R.S. (1980b). The elastic modulus of paper. The importance of fibre modulus, bonding, and fibre length. Tappi J. 63(6), 113–116.

Rigdahl, M. and Salmén, N.L. (1984). Dynamic mechanical properties of paper: Effect of density and drying restraints. J. Mat. Sci. 19(9), 2955–2961.

Saito, T., Kimura, S., Nishiyama, Y. and Isogai, A. (2007). Cellulose nanofibers prepared by TEMPO-mediated oxidation of native cellulose. Biomacromolecules 8(8), 2485–2491.

Sakurada, I., Nukushina, Y. and Ito, T. (1962). Experimental determination of the elastic modulus of crystalline regions in oriented polymers. J. Polym. Sci. 57(165), 651–660.

Syverud, K. and Stenius, P. (2009). Strength and barrier properties of MFC films. Cellulose 16(1), 75–85.

Tanaka, F. and Iwata, T. (2006). Estimation of the elastic modulus of cellulose crystal by molecular mechanics simulation. Cellulose 13(5), 509–517.

Lars Berglund
12 Wood biocomposites and structural fibre materials

12.1 Introduction

The forest products industry relies heavily not only on paper, paperboard, and sawn timber, but also on a category of established products sometimes classified as "traditional wood composites": glulam, plywood, particle board, fibre board, and so forth. It is helpful to consider paper and paperboard products in the same context as wood composites because this puts a stronger emphasis on the engineering materials' nature of load-bearing materials. Traditional wood composites are often used in the building industry, where structures are subjected to significant static and dynamic loads. By using the term biocomposites in the heading of this chapter, it is underlined that the discussion will extend far beyond the traditional uses of wood composites.

A widening of the perspective for wood fibre-based materials by inclusion of polymer matrix composites and other new materials is of great interest since this may inspire new applications in large material volume areas such as the building, automotive, and packaging industries. Also, a context of composite materials rather than forest products is helpful because the science and engineering of materials puts a strong focus on the micro-scale structural organization of material constituents. Focus is on the relationships between processing and microstructure, and between microstructure and properties. Material components such as fibres and polymers are subjected to processing and combined into a material with a certain microstructure. During processing, the material is also given geometrical shape. Common examples in the context of structural mechanics include plates, beams, and cylinders. A material of a given shape can then serve simple or complex functions, such as transmitting loads, heat, and the ability to survive repeated folding or storing energy at minimum weight.

The term composite material does not have a unified definition accepted over all different categories of composites, but the following criteria have been presented (after Hull, 1981):

1. A composite material consists of two or more physically distinct and separable material components (constituents). Usually, the properties of different constituents are substantially different. Note that porous materials are composites.

Lars Berglund, Wallenberg Wood Science Center, KTH Royal Institute of Technology, Stockholm, Sweden

https://doi.org/10.1515/9783110619386-012

2. In order to optimize the properties, the composite can be prepared by mixing the constituents so that the structure, to some extent, can be controlled.
3. The properties are superior, and possibly unique, compared with properties of individual constituents.

Common constituents with potential for mechanical reinforcement include fibres, platelets, particles, and ribbons (fibre-like with rectangular cross section and substantially larger width than thickness). Air is also a constituent so that foams and porous networks (paper, fibreboard) can be classified as composites. Table 12.1 presents some examples of current material categories that can be classified as wood composites. We have made the classification according to the micro-structural characteristics and the scales of constituent size or constituent type. Some application examples are also presented.

Table 12.1: Material categories that may be classified as wood composites.

Wood composite category	Specific wood composite material	Description	Example of applications
Polymer-modified wood	Impreg	Wood is impregnated by monomers, which are polymerized	Flooring
Laminated wood	Plywood	Veneer layers are laminated and bonded with a certain veneer orientation distribution	Building industry, furniture
	Laminated veneer lumber	Veneer layers are stacked to form laminated beams	High-strength beams for building industry
	Glulam	Board layers are stacked to form beams	Beams for building industry
Strands or particles with adhesive	Particle board	Large wood particles are coated by adhesive and hot-pressed to porous particleboard	Furniture, building industry
	Oriented strand board	Anisotropic strands are coated by adhesive and compressed to oriented high-density boards	Competes with plywood at lower cost
Porous wood fibre networks	High-density fibreboard ($850-1,100$ kg/m^3)	Mechanical or Masonite pulps are hot-pressed and bonded by lignin or adhesive	Flooring, siding, wall panels, furniture

Table 12.1 (continued)

Wood composite category	Specific wood composite material	Description	Example of applications
	Medium-density fibreboard (600–800 kg/m³)	Mechanical pulp is combined with an adhesive and hot-pressed	Furniture, cupboards, doors flooring
	Paper (kraft paper 600–800 kg/m³)	Wood pulp is filtrated and dried into network	Printing, packaging
	Paperboard	Typically thicker than about 0.25 mm (ISO definition: >224 g/m²)	Packaging
Impregnated wood fibre networks	Paper laminates	Paper is impregnated by resin and polymerized	Electric insulation boards, flooring
Short, discrete wood fibres in polymer matrix	Wood plastics	Saw dust or wood pulp is mixed with thermoplastic or resin and is extruded, injection moulded, or foamed	Decking, building industry, furniture, automotive

The orientation distribution of the reinforcement component as well as its size is important for the mechanical properties of the composite. Larger constituents tend to result in materials with larger defect size and, therefore, lower strength. Oriented reinforcement provides higher strength in the orientation direction, and this is one of the main advantages of composite materials. It makes it possible to tailor the anisotropy (orientation dependency) of the material properties.

The typical densities and mechanical properties of different wood composites are listed in Table 12.2. Density is important to consider in material comparisons because mechanical properties show strong dependency on density. Because wood composites tend to be porous, the correct parameter in a micromechanics context is the relative density or volume fraction, as will be discussed later.

Table 12.2: Typical densities and mechanical properties of different wood composite materials.

Wood composite	Density (kg/m³)	Elastic modulus (GPa)	Bending strength (MPa)	Tensile strength (MPa)
Spruce board	400	11	9.3	n.a
Spruce glulam	400	12.4	16.6	n.a

Table 12.2 (continued)

Wood composite	Density (kg/m³)	Elastic modulus (GPa)	Bending strength (MPa)	Tensile strength (MPa)
Spruce laminated veneer lumber	≈550	14	51	42
Spruce plywood	≈550	6.9–13	21–48	6.9–13
Oriented strand board	≈550	4.8–8.3	21–28	6.9–10.3
High-density fibreboard	850–1,100	2.8–5.5	31	15
Medium-density fibreboard	600–800	2	20	n.a
Particleboard	550–750	2–4 (from bending)	15–25	n.a
Kraft linerboard Hot-pressed holocellulose fibres*	600–800 1,180	2.8–4.1 18	n.a n.a.	25 (4% failure strain) 195
40 wt% wood fibre/ polypropylene and maleic anhydride–polypropylene	1,030	4.2	n.a	52 (3.2% failure strain)
Nanopaper	1,300	13–17	n.a	200–300 (10% failure strain)

* Yang and Berglund (2020).

Comparing the different materials in Table 12.2, spruce wood has good mechanical properties but is limited by the restricted geometric shape. Complex machining operations are required. The anisotropy and high porosity also results in locally weak regions in machined structures of complex shape. Laminated structures such as laminated veneer lumber beams and plywood sheets often show high strength due to the thin lamellae. In addition, the orientation distribution of the lamellae can be controlled. Again, there is little freedom with respect to shape. The comparison between high-density fibreboard and linerboard is of some interest. Kraft linerboard is a paperboard made of chemical pulp fibres (Section 2.5.1) and used as the surface ply in corrugated boards (Section 3.2.1). It has higher strength than the fibreboard, despite lower density. Hot-pressed holocellulose fibres are a new material concept, where high-strength wood fibres are hot-pressed to form a dense material of high modulus and strength (Yang and Berglund, 2020). The wood fibre/polypropylene composite (an example of wood plastics) is also interesting. This material category is very successful in North America for decking applications, replacing impregnated wooden boards. It can also be injection moulded into complex geometrical shapes. The elastic modulus is quite high, and the strength is respectable compared with many other materials. The main structural advantage of wood plastics is low porosity. This is interesting because it indicates the potential of new types of composites based on wood fibres.

The wood fibre itself has attractive characteristics including high aspect ratio (length-to-diameter ratio ≈100 for softwoods), high axial strength and elastic modulus in the fibre wall, as well as favourable fibre network forming characteristics. Networks made of strong wood fibres or chemically tailored fibres can be used in new fibre architectures of designed orientation distributions and combined with new polymer matrices, foams, or other porous materials to form new types of wood composites. Interesting functions include thermal insulation and mechanical performance. Wood fibre composites could also provide new opportunities with respect to melding of intricate geometrical shapes. The "nanopaper" material in Table 12.2 represents some of the advantages that can be obtained with cellulose nanofibrils (CNFs). Their dimensions are three orders of magnitude smaller than regular wood fibres. The elastic modulus is 13–17 GPa, and the strength in tension exceeds 200 MPa due to the fine structure of the material. This will be discussed in Section 12.4.

The development of new wood composites is motivated by new applications for wood-based materials. Property comparisons with other categories of materials will then be helpful and can be pedagogically made by the use of so-called property charts, such as Fig. 12.1, that have been introduced by Michael Ashby and co-workers (Ashby, 1999). An interesting feature of these charts is that the material efficiency of specific structures can be compared for different classes of materials. For instance, the efficiency of tensile members ("stiff ties") is given by the ratio E/ρ of elastic modulus and density. A higher value for this index gives a lower-weight tie for the same stiffness. The efficiency of a beam loaded in bending is measured by the index $E^{1/2}/\rho$. The corresponding index for flat plates loaded in bending is $E^{1/3}/\rho$. Three stiffness guidelines corresponding to these indices are plotted in Fig. 12.1 together with the elastic modulus and density data for different material categories, all on log–log scales to facilitate comparison.

Different values for the stiffness indices appear as a set of parallel lines in Fig. 12.1 so that one can compare material categories. The ratio E/ρ for wood parallel to grain is almost as high as for steel. The corresponding values of $E^{1/2}/\rho$ and $E^{1/3}/\rho$ are much higher than for steel, which means that the efficiency in bending of beams and plates is much higher for wood than for steel. Wood yields lighter-weight structures for a given stiffness. It is also interesting to note that in bending wood is stiffer than many man-made synthetic fibre composites. Related property charts can be worked out for other properties and characteristics in order to identify if there is "free" property space in the charts, where novel wood composites may be designed to give new combinations of elastic modulus and density. Some particularly interesting new wood composites are those that extend the current range of properties, as exemplified by the nanopaper material in Table 12.2.

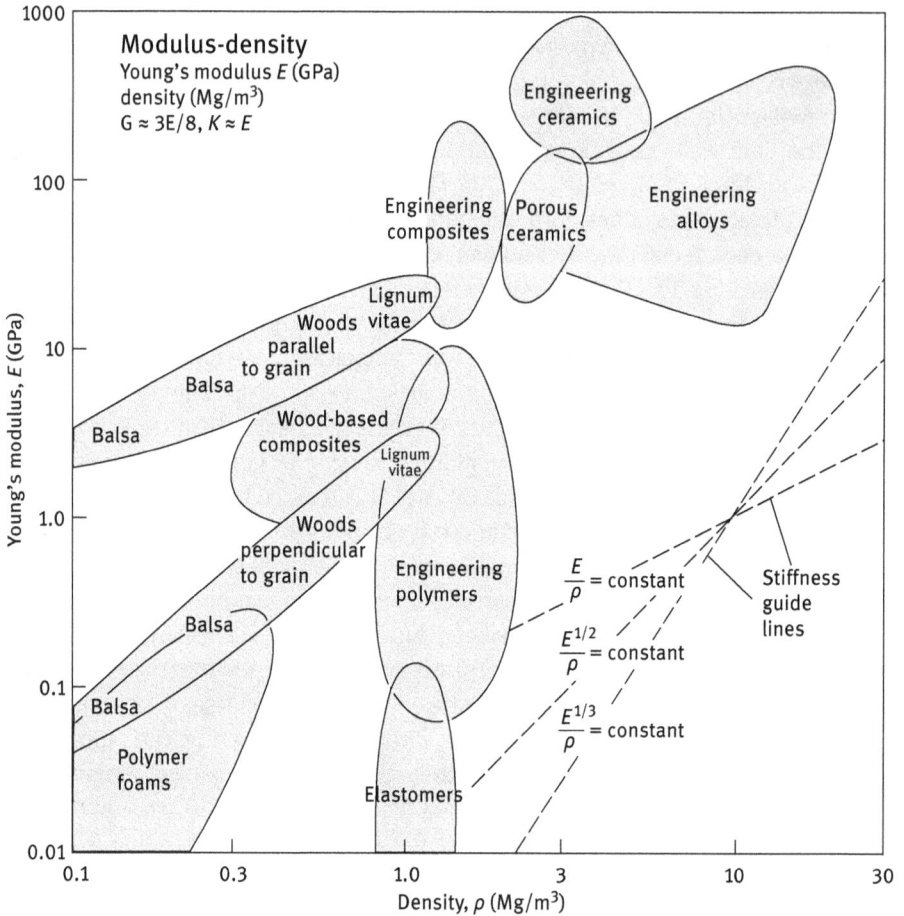

Fig. 12.1: Elastic modulus as a function of density for different material categories, plotted on log–log scale. Note the materials efficiency guidelines explained in the text (Wegst and Ashby, 2004).

12.2 Material components: fibres and polymers

12.2.1 Plant fibre structure

Even in an engineering context, it is helpful to consider that the plants, from which we derive wood and other fibres, are biological organisms. Fibres have specific functions in the plant organism, and these functions, together with growth aspects, explain the structure of plants. Plant fibres in trees and grasses (i.e. annual plants such as flax) are single cells. The fibrous shape indicates that they have a mechanical function in the plant. Fibre geometry provides anisotropy, and the load-carrying ability is high in the direction of fibre axis. In addition, the fibre cell wall itself is anisotropic due to the organization of its components. The cell wall is much stiffer in its length direction than in the transverse directions.

Figure 12.2 presents a micrograph of a plant fibre or cell. Softwoods (coniferous trees such as pine and spruce) are most relevant in the context of composites because they consist of long fibres called tracheids. Their length is typically 2–4 mm and diameter 20–40 μm, and thus the aspect ratio (length/diameter) is around 100. The lumen is the empty space at the centre of the cell. The fibre cell wall consists of a thin primary wall layer and a thick secondary wall layer. The latter divides further into thin S1 and S3 layers and a thick S2 layer that occupies about 85% of the fibre cell wall thickness. An important structural feature is that the cell wall is composed by cellulose microfibrils with a diameter of 5–15 nm, depending on the plant. Cellulose microfibrils reinforce the cell wall and are oriented at a certain angle to the fibre direction (microfibril angle, MFA). The smaller the MFA, the stiffer and stronger is the fibre. The S2 layer typically has MFA = 10°–30°, but at chest height of the trunk the outer part in a mature coniferous tree contains tracheids (fibres) of very small MFA. This is the region of the tree that is subjected to high stresses when the tree trunk is bent by heavy winds.

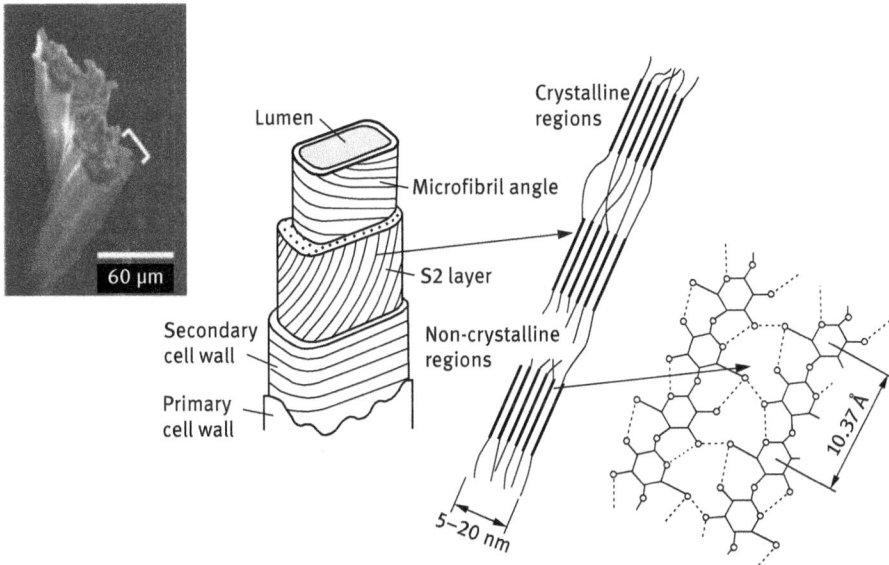

Fig. 12.2: Micrograph of a plant fibre cell (left), the cell wall structure consists of the primary cell wall, and the secondary cell wall with the S1, S2, and S3 layers of different organization and microfibril angle of the cellulose(centre), and the cellulose microfibril with ordered and less ordered regions (right); courtesy of Prof. T. Nishino, Kobe University.

The cell wall has an organization of a laminated composite material with cellulose microfibrils in a matrix of highly hydrated lignin–hemicellulose complex. Softwoods typically have around 42% cellulose, 27% hemicelluloses, and 28% lignin out of the dry matter and a few per cent extractives (fatty acids and phenolics). The water content in a native wood fibre is around 30%. Most likely, the hydrated lignin–hemicellulose matrix is strongly associated with the cellulosic microfibril by physical adsorption of hemicelluloses. Lignin and hemicelluloses are mixed and form a polymer matrix for reinforcing cellulose fibrils. The mechanical behaviour of a wet wood cell wall is poorly understood. It shows interesting features, including an impressive combination of strength, stiffness, and toughness despite its hydrated state.

In a mechanical property sense, cellulose microfibrils need to be stiff and strong because their main function in plant cell walls is to provide tensile performance. Obviously, wood tracheids also carry compressive stresses, but the tensile material function is still the most important one for cellulose in plant organisms. The estimated axial elastic modulus of the cellulose crystal is 134 GPa. It arises from the extended chain conformation of cellulose molecules, giving high density of strong inter-molecular covalent bonds, and strong intra-molecular hydrogen bonds that stiffen the molecule. The hydrated lignin–hemicellulose network, which constitutes the cell wall matrix, is amorphous in nature. In spite of some degree of hemicellulose orientation, the elastic modulus of the amorphous network is unlikely to exceed 1 GPa.

Geometric features of some plant fibres, including the MFA and cellulose degree of polymerization (DP) (a measure of the molar mass or length of cellulose molecules) are presented in Table 12.3. However, during the cooking of wood fibres into a chemical pulp, some degradation of cellulose DP takes place, and this will decrease the strength of cellulose. It is interesting to note that hemp, jute, flax, and ramie have very low MFAs. Although it is tempting to relate this to plant stem function, we have to keep in mind that these are plants used for thousands of years as fibre sources for textiles. As a consequence, the breeding of these grasses has been strongly focused on the production of stronger fibres, probably selecting for small MFA and high cellulose content. It is also interesting to note the very large aspect ratio of grass fibres. In a practical context, it is difficult to process fibres that are 15 cm long (ramie). Usually the fibres used in textiles are made by spinning so that the plant fibre cells are intertwined into a continuous thread of larger diameter. Cotton is not a tensile material but has a seed hair function. Wood is very interesting due to its comparably low cost and the considerable infrastructure for harvesting and processing that is present in many regions of the world. The aspect ratio (length to diameter) is typically more than sufficient for biocomposite requirements, although continuous wood fibre products (such as yarns) are not readily available.

The mechanical properties of plant fibres are interesting for estimates of the property potential of biocomposites. In Table 12.4, some data are provided. It is interesting to note that the dry elastic modulus E of plant fibres compares well with glass fibre ($E_{glass} = 70$ GPa). As a consequence, specific stiffness (elastic modulus

Table 12.3: Geometrical parameters, microfibril angle (MFA), and cellulose degree of polymerization (DP) in some plant fibres (after Wainwright et al., 1982).

Fibre	Width (µm)	Length (cm)	Aspect ratio	MFA	DP
Hemp	16–50	1.2–2.5	500–700	2.3°	9,300
Jute	20–25	0.15–0.5	75–200	7.9°	n.a
Flax	11–20	1–3	900–1,500	6°	7,000–8,000
Ramie	20–75	12–15	2,000–6,000	6°	10,800
Cotton	15–30	2.5–6	1,700–2,000	28°–44°	15,300
Wood	20–30	0.1–0.3	33–150	10°–40°	10,000

Table 12.4: Elastic modulus, tensile strength, and breaking strain of some plant fibres in the wet and dry state (after Wainwright et al., 1982).

Fibre	Elastic modulus E (GPa)		Elastic modulus ratio E_{dry}/E_{wet}	Tensile strength σ^{max} (MPa)		Tensile strength ratio $\sigma^{max}_{dry}/\sigma^{max}_{wet}$	Breaking strain (%)	
	Wet	Dry		Wet	Dry		Wet	Dry
Hemp	35	70	2	n.a	920	n.a	n.a	1.7%
Jute	n.a	60	n.a	n.a	860	n.a	n.a	3.0%
Flax	27	80–110	3–3.7	880	840	0.96	2.2%	1.8%
Ramie	19	80	4.2	1,080	900	0.89	2.4%	2.3%
Wood	n.a.	35	2	82	120	1.5	Wet	Dry

divided with density) for plant fibre biocomposites compares well with glass fibre composites. The elastic modulus of the wood cell wall is around 30 GPa, which is about one order of magnitude higher than in glassy polymers. The elastic modulus of different fibres is controlled by the cellulose content and MFA (higher cellulose content and lower MFA increase the elastic modulus). All plant fibres have significantly reduced elastic modulus in the completely wet state. However, the tensile strength is not significantly reduced because moisture increases the plasticity and toughness of the cell wall so that the fibres become less sensitive to defects. The breaking strain ε^{max} of fibres is not very high because the cellulose molecules in microfibrils are in extended chain conformation and aligned with the fibre axis and cannot stretch very much without failing. Wood fibres with high MFA (occurring in so-called compression wood) are an interesting exception together with coir fibres because at high MFA the microfibrils can slide and reorient relative to one another, providing high ductility to the cell wall.

The wood products industry supplies many different types of wood fibres for different applications. It is interesting to consider the suitability of different fibres for new types of wood fibre biocomposites. Table 12.5 lists the most common types of wood fibres, with comments on their characteristics. Due to the low cost, saw dust and other particles from saw mills and machining of wood are in widespread use in particle-boards and melt-processed thermoplastic wood composites. However, because the typical aspect ratio of these particles is low (≤10), the reinforcement potential of the stiff wood fibre is not utilized. In contrast, mechanical pulps such as thermomechanical pulp (TMP) and chemi-thermomechanical pulp (CTMP) (Section 2.5.1) have much higher aspect ratio and have better reinforcement efficiency in biocomposites. The chemical composition in TMP and CTMP is fairly similar to wood. This may cause problems with odour (thermal degradation of hemicelluloses and lignin) or discoloration (extractives) problems in biocomposites that are processed at high temperature (e.g. melt processing or compression melding). Bleached kraft pulps are more stable. They can also have high molar mass of cellulose, which gives good fibre strength. The hemi-cellulose content is usually significant since it may be hydrolysed in high-temperature processing. Sulphite pulps can have very low hemicellulose content, but at the same time, the DP of the cellulose is lower than in kraft pulps.

Table 12.5: Chemical composition and characteristics of wood reinforcement particles.

Wood fibre type	Chemical composition	Characteristics
Saw dust, wood flour	Similar to wood	Short aspect ratio, large size
Mechanical wood pulp	Similar to wood (high yield)	Individualized fibres 20–30 µm in width and 1 mm (hardwood) to 3 mm (softwood) in initial length, mechanically cut or damaged, little fibre collapse
Chemi-thermomechanical pulp	Low extractives content, otherwise similar to wood (yield 90–93%)	Less mechanical damage compared with thermomechanical pulp
Kraft pulp	Lignin, 2–12%, hemicellulose 10–15%, cellulose 75–85%	Individualized, mechanically intact, collapsed after refining
Sulphite pulp	Low lignin conc., 4–15% hemicelluloses, 85–96% cellulose	Individualized, mechanically intact, collapsed after pulp refining, lower cellulose molar mass than kraft pulp (usually)
Microfibrillated cellulose	Typically 4–15% hemicellulose, depending on pulp source	10–500 nm in width, large size distribution, large aspect ratio
Cellulose nanofibrils	Typically 4–15% hemicellulose, depending on pulp source	5–15 nm in width and several micrometres in length

CNF is a new component for biocomposites. It is obtained by mechanical disintegration of wood pulp fibres. Usually, the source is a chemical sulphite pulp or bleached kraft pulp. A chemical or enzymatic pre-treatment step is often used to reduce the energy consumption so that the manufacturing cost becomes reasonable. The dimensions of CNF are three orders of magnitude lower than for plant fibres (5–30 nm in diameter and 5–10 μm in length). As a consequence, biocomposites of very fine structure can be prepared. CNF-based composites have much greater strength and ductility than plant fibre-based biocomposites because defect size is governed by the particle size of the reinforcement phase. Industrial grades of nanocellulose are often microfibrillated cellulose (MFC) with much larger fibril size distribution.

12.2.2 Polymer matrices and binders

The polymer matrix improves the properties of composites. Compared with a fibre network without additional polymers, inter-fibre bonding and interaction can be better controlled. Also, the density range can be extended. The term binder refers to when the wood composite is a porous fibre network and the polymer binder is an adhesive, which primarily improves the inter-fibre bonding but does not completely fill the network.

The different categories of polymer matrices are listed in Table 12.6. The most common thermoplastics for melt-processed biocomposites are polypropylene (PP) and polyethylene (PE), which are widely used in large-volumed products, such as the decking market in North America. There is a substantial market for recycled PE and PP, making them readily available. Their disadvantage is low strength, brittleness, odour, and surface appearance problems. Polylactic acid (or polylactide, PLA) is an interesting thermoplastic with biomass origin, but it is quite brittle unless plasticized and is also quite expensive.

Water-soluble thermoplastics are used in paper manufacturing and converting (such as the manufacture of corrugated boards). They include cellulose derivatives and starch. Starches are obtained from potato, corn, and other food sources, and this is a disadvantage. Amylopectin-rich starch is quite brittle without plasticizer, but the main problem is still the high water solubility. Although advantageous in processing, this limits the durability to the biocomposite material in humid conditions. Cellulose derivatives are interesting because they are made from sustainable resources. For example, hydroxyethyl cellulose (HEC) is a very tough polymer. However, as with starches, biocomposites based on cellulose derivatives need either protection against water or additional chemical modification to reduce water solubility.

Thermoset resins are chemically cross-linked network polymers, which cannot be dissolved or melted. They are prepared by chemical reactions that initiate either by the mixing of two components or by heating. They are widely used for wood

Table 12.6: Categories of polymers used as polymer matrix or binder in biocomposites.

Polymer category	Polymer	Characteristics
1. Thermoplastics (melt processed) polylactide	Polypropylene (PP), polyethylene (PE, HDPE, LDPE)	Petroleum based (not ethanol-PE), wide industrial use, ductile, low cellulose compatibility
	Polylactic acid, or polylactide	Corn-based (starch), increasing industrial use, somewhat brittle
2. Water-soluble thermoplastics	Starch	Cheap, brittle without plasticizer, available as melt-processable grades and blends with other thermoplastics
	Cellulose derivatives	Examples: carboxymethyl cellulose, hydroxyethyl cellulose, good cellulose compatibility, interesting binder with cellulosic fibres
3. Thermoset resins	Formaldehyde cured resins	Currently used as binders in wood composites; phenol–formaldehyde, melamine–formaldehyde, urea–formaldehyde
	Epoxies, unsaturated polyesters	Dominating resins in current glass fibre composites, low viscosity, good mechanical properties, insensitive to moisture, petroleum-based although new grades are partly of bio-origin
	Furan resins	Bioresin based on sugar cane, good mechanical properties, complex curing chemistry, dark colour, small current market
4. Wood cell wall matrix	Lignin–hemicellulose complex	Hydrated lignin–hemicellulose network. A unique feature is its highly hydrated state and strong association with cellulose microfibrils

composites in the form of phenol–formaldehyde (PF), melamine–formaldehyde (MF), and urea–formaldehyde (UF) resins or adhesives. PF has good hygrothermal stability, whereas MF is not as good. MF can be optically transparent. UF is of low cost but sensitive to hydrolysis when water is present. There is strong interest in replacing formaldehyde resins with more environmentally friendly alternatives, but this is a challenge because of associated cost increases which are difficult to accept in commercial applications.

Epoxies and unsaturated polyesters (UPs) are well established as polymer matrices in the composites industry. Several bio-based resins are available that typically have an increased proportion of molecules from renewable resources. For instance, polyesters can be made partly from soybean oil that is chemically modified (epoxidized or acrylated) so that conventional curing chemistry can be used. Epoxies have high hygromechanical durability, but they are much more expensive than polyester resins.

Recently, there has been an increasing interest in furan resins. They usually derive from sugar cane and are prepared from sugars. However, complex chemistry is needed, and the cured thermoset resin is often brittle, has pores, and is black in colour. Still, there are strong development efforts underway to improve the characteristics of furan resins.

The final row in Table 12.6 describes the characteristics of the hydrated lignin–hemicellulose matrix complex in the wood cell wall. Although this complex is unlikely to be directly useful as a polymer matrix, we can learn from its function in the cell wall. The first interesting characteristic is that it is strongly associated with the cellulose microfibrils. Possibly, this intimate association or bonding is why the wood cell wall shows such high stiffness and strength, in spite of the hydration of the matrix at the 30% water content of the cell wall. In conventional micro-scale biocomposite materials, the fibre–matrix interface is likely to debond when the material adsorbs moisture and expands. In contrast, the wood cell wall is molecularly designed to perform in the wet state. The second interesting characteristic is the significant toughness of the wood cell wall. Local strain in the cell wall can exceed 20%, and the cell wall still has considerable load-carrying capacity. Most likely, this is related to the fine structure of the reinforcing microfibrils in combination with some mechanism whereby the microfibril–matrix interaction holds despite the high strain. This has not yet been achieved with micro-scale fibre composites, where the fibre–matrix interface will crack and debond at high strain.

12.3 Micromechanics of fibre composites

In this chapter, the potential of wood biocomposite materials is in focus. In order to appreciate this potential, it is important to understand the basic stiffening mechanisms of fibre reinforcement. Here, the micromechanics of fibre reinforcement of a polymer matrix is helpful. Let us consider the case of continuous cylindrical fibres embedded in a matrix (Fig. 12.3). An important effect of stiff and strong fibres added to a soft polymer matrix is an improvement of the mechanical properties that depend on the relative proportions of the matrix and fibre phase. It is the constituent volume fraction, not weight fraction, which determines the reinforcement effect of a certain amount of fibres. This is easily understood by considering the hypothetical comparison of lead "fibres" and glass fibres in a polymer matrix. Even if the two fibre types would have the same elastic moduli and geometric dimensions, a given weight fraction of lead fibres in the composite would give much lower elastic modulus than the same weight fraction of glass fibres because the number of lead fibres in the composite would be much smaller.

12.3.1 Weight fraction and volume fraction

Let us consider a composite material consisting of a volume fraction V_{fc} of fibres and V_{mc} of matrix material. The corresponding weight fractions are W_{fc} and W_{mc}, so by construction

$$V_{fc} + V_{mc} = 1 \quad \text{and} \quad W_{fc} + W_{mc} = 1. \tag{12.1}$$

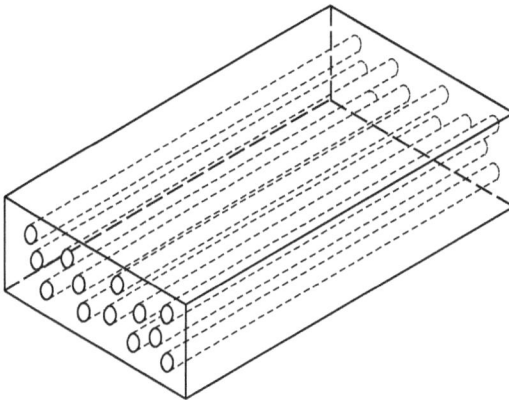

Fig. 12.3: The architecture of a unidirectional fibre composite. Cylindrical fibres are continuous, oriented parallel to each other in one direction, and surrounded by a matrix phase.

Weight fractions are easily estimated by weighing the constituents. However, theoretical analysis of composite properties (Agarwal and Broutman, 1990) is based on the knowledge of volume fractions, since they control physical properties. A very important relationship expresses how the volume fractions of fibre and matrix are calculated from the weight fractions and the densities (ρ_{fibre} and ρ_{matrix}) of the constituents, and the density of the composite in the absence of voids ($\rho_{void\text{-}free}$):

$$V_{fc} = \left(\rho_{void-free}/\rho_{fibre}\right)W_{fc} \quad \text{and} \quad V_{mc} = \left(\rho_{void-free}/\rho_{matrix}\right)W_{mc}. \tag{12.2}$$

The density of a void-free composite is

$$\rho_{\text{void-free}} = \frac{1}{(W_{\text{fc}}/\rho_{\text{fc}}) + (W_{\text{mc}}/\rho_{\text{mc}})}. \tag{12.3}$$

The volume fraction of voids Φ in a real composite is also important since it has a strong influence on the mechanical properties. It requires the experimental determination of the composite density ρ_{comp}:

$$\Phi = \left(\rho_{\text{void-free}} - \rho_{\text{comp}}\right) / \rho_{\text{void-free}}. \tag{12.4}$$

12.3.2 Elastic properties in unidirectional composites

Fibre composites have usually high elastic modulus combined with low density. The reinforcing mechanism of stiff fibres in a polymer matrix can be understood by a micromechanics analysis of the stress and strain distributions at the scale of fibres. Composites are complex materials from the point of view of stress analysis. For this reason, a simple composite structure is a good starting point that provides an insight into reinforcement mechanisms.

We start from the unidirectional composite in Fig. 12.3, where the longitudinal and transverse directions are defined as parallel and perpendicular to the fibre direction. The fibres are perfectly bonded to the polymer matrix, so that the in the longitudinal direction the fibre strain and matrix strain are identical with the composite strain, and the elastic modulus E_{L} of the composite in the longitudinal direction is

$$E_{\text{L}} = E_{\text{fibre,L}} V_{\text{fc}} + E_{\text{matrix}} V_{\text{mc}}, \tag{12.5}$$

where $E_{\text{fibre,L}}$ is the longitudinal (or axial) elastic modulus of the fibres, and E_{matrix} is the elastic modulus of the matrix that can be assumed isotropic. For stiff fibre composites, the axial elastic modulus of fibres is typically much higher than the elastic modulus of the matrix, and therefore the stress carried by the fibre phase is higher. This is the main reinforcement principle for fibre composite materials. For instance, the longitudinal cell wall modulus of wood fibres is typically 30 GPa, whereas glassy polymers have an elastic modulus of about 3 GPa.

In a void-free composite $V_{\text{mc}} = 1 - V_{\text{fc}}$, and we can write

$$E_{\text{L}} = E_{\text{fibre,L}} V_{\text{fc}} + E_{\text{matrix}}(1 - V_{\text{fc}}). \tag{12.6}$$

This equation is usually called the rule of mixtures. Its agreement with experimental data is very good when the constituents are elastic. The reason for the good agreement is that the real structure is very similar to the structure assumed (unidirectional fibres in the loading direction) and the fibre–matrix bonding is good at low strains.

A model for transverse elastic modulus E_T can be derived assuming constant stress in the matrix and the fibres. The final expression is

$$E_T = \frac{E_{\text{fibre, T}} E_{\text{matrix}}}{E_{\text{fibre, T}}(1 - V_{\text{fc}}) + E_{\text{matrix}} V_{\text{fc}}},\tag{12.7}$$

where the correct fibre elastic modulus $E_{\text{fibre,T}}$ is in the transverse direction. The agreement between predictions based on the above expression and experimental data is poor, especially at high volume fractions of stiff fibres. The constant stress assumption is not very good in that case because the stress state in the matrix is highly non-uniform, and often very different from the fibre stress. Instead, the Halpin–Tsai model is better suited as a predictive tool (Agarwal and Broutman, 1990; Hull, 1981). For cylindrical fibres, the model is

$$\frac{E_T}{E_{\text{matrix}}} = \frac{1 + 2\eta_T V_{\text{fc}}}{1 - \eta_T V_{\text{fc}}} \quad \text{with} \quad \eta_T = \frac{(E_{\text{fibre, T}}/E_{\text{matrix}}) - 1}{(E_{\text{fibre, T}}/E_{\text{matrix}}) + 2}.\tag{12.8}$$

This is an approximation based on finite element modelling results. The predictions agree quite well with experimental data, and the model is theoretically sound.

12.3.3 Elastic properties in short-fibre composites

In the context of fibre composites, wood fibres are termed short fibres because they have a finite length. As may be expected, short fibres have lower reinforcement ability than continuous fibres. When the composite material is subjected to mechanical load, the externally applied stress is transferred from the matrix to the stiff fibres. The situation is similar to the fibre network of paper discussed in the preceding chapter.

In order to understand the nature of the stress distribution in a short-fibre composite, consider a single fibre oriented parallel to a uniaxial external load similarly to the analysis in Section 11.1.1. The fibre has a finite length l_{fibre} and is embedded in a matrix. Figure 12.4a illustrates the strain distribution along the fibre embedded by showing deformation lines of the matrix. Close to the fibre ends, the fibre strain is lower and matrix strain larger than elsewhere because the fibre is stiffer than the matrix. The tensile stress in the fibre (Fig. 12.4b) is highest in the middle of the fibre and decreases at the ends, where in turn the interfacial (fibre/matrix) shear stress is high. The length fraction at a fibre end where tensile stress is reduced is $l_{\text{crit}}/2$. Figure 12.4 illustrates the importance of the aspect ratio of fibres $l_{\text{fibre}}/d_{\text{fibre}}$, where l-$_{\text{fibre}}$ is fibre length and d_{fibre} is fibre diameter. High aspect ratio means that only small proportions of the fibre ends carry lower stress. For example, in saw dust composites the reinforcing particles can have an aspect ratio below, which gives very low reinforcement efficiency.

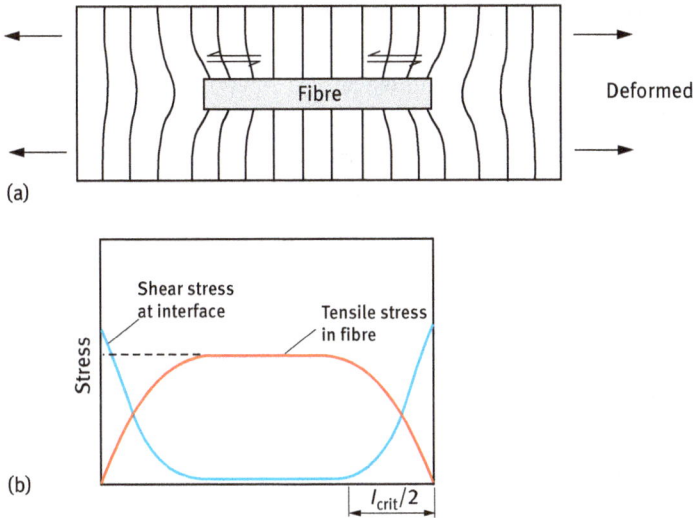

Fig. 12.4: Displacement field created by a stiff fibre of finite length that is perfectly bonded in a continuous matrix when tensile stress is applied to the fibre–matrix assembly in the direction parallel to the fibre axis (a), and the corresponding distribution of tensile stress in the fibre and shear stress at the fibre–matrix interface. Half the critical length $l_{crit}/2$ denote the section where the tensile stress is lower than in the centre.

Prediction of the elastic modulus of real short-fibre composites is a complex problem, since the elastic modulus depends on the distributions of fibre length and fibre orientation. A simple case can help us to understand the important parameters. Consider the elastic modulus of a short-fibre composite where all the fibres are aligned. The Halpin–Tsai equation has been extended to the longitudinal elastic modulus of short-fibre composites

$$\frac{E_L}{E_{matrix}} = \frac{1 + 2\frac{l_{fibre}}{d_{fibre}}\eta_L V_{fc}}{1 - \eta_L V_{fc}} \quad \text{with} \quad \eta_L = \frac{(E_{fibre,L}/E_{matrix}) - 1}{(E_{fibre,L}/E_{matrix}) + 2\frac{l_{fibre}}{d_{fibre}}}. \tag{12.9}$$

This expression is at best empirical. Note the similarity with eq. (12.8). For E_T the fibre aspect ratio l/d is simply 1, since a cylindrical fibre loaded in the transverse direction is considered. The orientation distribution of the fibres will obviously also influence the moduli (see Hull, 1981). For the practically important case of isotropic fibre orientation distribution, a simple empirical expression can be used for the elastic modulus of the short-fibre composite:

$$\overline{E} = \frac{3}{8}E_L + \frac{5}{8}E_T, \tag{12.10}$$

where E_L and E_T are obtained from the Halpin–Tsai expressions (12.8) and (12.9).

12.3.4 Interfacial strength in short-fibre composites

The stress distribution along a short fibre influences the composite strength. For a fibre with short aspect ratio, the stress in the fibre never reaches the fibre strength, and when the composite fails intact fibres are pulled out of the matrix. Some minimum fibre length is therefore needed in order to utilize the fibre strength. This gives a definition for the critical fibre length l_{crit} illustrated in Fig. 12.4. It is the shortest length of fibres in a given composite that allows the fibres to be loaded maximally so that they start failing. Consider a short cylindrical fibre embedded in a block of the matrix material. As the length of the fibre is increased, the critical fibre length l_{crit} is reached where the fibre stress at composite failure is only marginally smaller than the fibre strength σ_{fibre}^{max}. The force balance between the tensile force in the fibre and the mean shear force at the fibre–matrix interface can be analysed in a straightforward manner, resulting in the following expression for the critical aspect ratio:

$$\frac{l_{crit}}{d_{fibre}} = \frac{\sigma_{fibre}^{max}}{2\tau^{max}}, \tag{12.11}$$

where τ^{max} is the shear strength of the fibre–matrix interface or the yield strength of the matrix, whichever fails first. If we assume that the wood fibre cell wall strength is 300 MPa and the interfacial shear strength is 10 MPa, we end up with an aspect ratio of 15 in order to ensure fibre fracture. For a 30 μm diameter wood fibre this is well below 1 mm, the typical length of hardwood fibres.

We can conclude that if the interfacial strength or matrix strength is low, the minimum aspect ratio (12.11) is high. Low fibre strength has the opposite effect. If the strength of the short-fibre composite is considered, higher interfacial shear strength increases the composite strength, as will become apparent from the experimental data in the next section. Note also that the discussion here only considered the conditions for barely reaching a stress in the fibre that corresponds to the fibre strength. Much higher aspect ratio than in (12.11) is required in order to have the load-carrying efficiency of short fibres approaching continuous fibres. More advanced treatments also consider the statistical distribution in fibre strength.

12.4 Composites data: wood fibre/thermoplastic

Thermoplastic composites are of great interest to industry. An important reason is that thermoplastic biocomposites can be melt-processed, for instance by continuous extrusion of profiles such as sheets or rods. In addition, injection moulding of three-dimensional components can be carried out at great production speed for use in automotive applications, including interior components. The most well-known example of thermoplastic biocomposites in the forest products sector is wood flour composites with PE or PP as thermoplastic matrix. In North America, PE-based materials have been widely used in the large decking market, where the "wood plastics" provide an alternative to impregnated wood. IKEA has launched a few injection-moulded furniture products based on a wood flour/PP composite.

Table 12.7 shows the tensile strength and elastic modulus for different compositions of the wood/PP composites. Using 40 wt% of wood flour reinforcement, the strength is virtually unchanged, whereas elastic modulus is increased from 1.5 to 3.9 GPa. Interestingly, the strength increases considerably if a little maleic-anhydride-modified polypropylene (MA-PP) is added. The MA-PP serves as a compatibilizer that improves the interfacial strength, τ^{max}, and also helps the dispersion of wood flour particles, which often improves the strength of the composite. In terms of eq. (12.11), the increase in the interfacial shear strength increases load transfer efficiency of wood flour particles by making the critical aspect ratio smaller. Similar effects are seen with wood fibres. Without the MA-PP addition, wood fibres influence only the elastic modulus but not the strength of the composite. When the MA-PA is added, also the tensile strength increases compared to the unreinforced PP. With 60 wt% of wood fibres and 3% of MA-PP, the elastic modulus is close to 7 GPa and the material is quite brittle.

Table 12.7: Elastic modulus and tensile strength of polypropylene (PP)/wood flour and PP/wood fibre composites for different mixing ratios, with and without added maleic anhydride (MA)–PP compatibilizer (Stark and Rowlands, 2003).

Composite	Elastic modulus (GPa)	Tensile strength (MPa)
PP	1.5	28.5
40 WPP wood flour	3.9	25
40 WPP wood flour + 3% MA-PP	4.1	32
40 WPP wood fibre	4.2	28
40 WPP wood fibre + 3% MA-PP	4.2	52
50 WPP wood fibre	5.8	28
50 WPP wood fibre + 3% MA-PP	6.7	53

Polyamide-6 (PA6) is an interesting alternative to PP, as demonstrated in Table 12.8. The composition of PA6 with 33 wt% of sulphite pulp reinforcement can have an elastic modulus of 5.7 GPa and a tensile strength of 87 MPa. This is quite encouraging since the values are close to those for PA6 reinforced with glass fibres, but with a lower density. One reason for better performance of PA6/wood composites compared with PP/wood composites is that PA6 itself has better properties. The interaction between PA6 and cellulose is more favourable and the interfacial shear strength is likely to be higher. The sulphite pulp with high cellulose content (95%) was chosen

Table 12.8: Melt-processed thermoplastic composites where chemical sulphite pulp fibres (≈95% cellulose content) made of different wood materials are combined with polypropylene (PP) or polyamide-6 (PA6). A glass fibre/PP composite is given as a reference (Jacobson et al., 2002).

Composite	Elastic modulus (GPa)	Tensile strength (MPa)
PP	1.39	27.6
PP/33 WF	3.38	33.1
PA6	2.75	60.2
PA6, 33% WF, HW	5.71	86.5
PA6, 33% WF, SW	5.35	81.9
PA6, 33% GF	8.01	111

because of the high processing temperature needed with PA6, which would lead to thermal degradation effects if the fibres contained hemicelluloses.

Other polymer matrices such as PLA and starch can also be used for thermoplastic composites. PLA tends to be brittle if plasticizers are not used. On the other hand, plasticizers reduce the thermal stability and strength of the composite. Thermoplastic starch is used in some packaging applications. Amylopectin-rich starch from potato can be used and mixed with glycerol plasticizer or another thermoplastic. The use of thermoplastic starches combined with plant fibres is hampered by its high solubility in water, which often causes problems with the long-term stability and durability of biocomposites.

12.5 Composites data: wood fibre/thermoset

Wood fibre composites with thermoset matrix are already used industrially in mouldable compounds with polyphenol–formaldehyde (PF) or melamine–formaldehyde (MF) resin as the matrix. The first black telephones were made of wood fibre/PF compounds, and white wood fibre/MF compounds are used in electrical insulation components (wall sockets, plugs). The compound is premixed and cured under pressure and elevated temperature in the injection moulding or compression moulding process. Another fabrication route is to impregnate paper sheets with a water solution of the resin, dry the impregnated paper, and then laminate several layers for curing under pressure. Such laminates are used in electronic circuit boards but also in flooring and kitchen surfaces. The main problem with these wood fibre composites is that the use of formaldehyde requires careful safety precautions. Formaldehyde-free wood fibre composites are therefore an important vision for future wood thermoset composites.

When cured, the PF and MF resins form a cross-linked structure that is stiff and brittle, resulting in stiff and brittle wood composites. Phenol, melamine and formaldehyde are soluble in water and in the wood fibre cell wall, and swell the cell wall more than water does. Therefore the volume fraction of fibres in formaldehyde matrices can be quite high, which serves to improve the elastic modulus and strength of the composite. The fibre–matrix interface is also very strong because the resin polymer enters the fibre wall. When the composite is loaded, failure occurs in the brittle matrix.

Figure 12.5 shows a micrograph of a composite structure where paper has been impregnated with an UP and cured. The polymer has entered in the lumen of fibres. The volume fraction of fibres is relatively low since UP cannot swell the wood cell wall. One can see that the microstructure of a real composite is quite different from what was assumed above in the micromechanics elastic modulus model calculations.

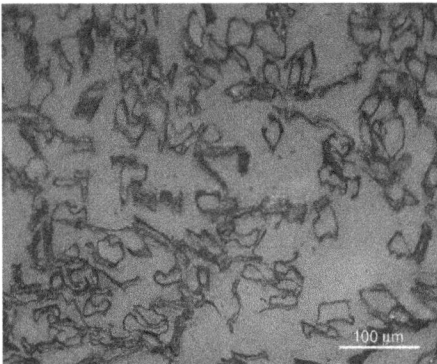

Fig. 12.5: Optical micrograph of a cross section of a composite of wood fibres embedded in an unsaturated polyester matrix; courtesy of L.O. Nordin, Luleå University of Technology.

The elastic moduli of various wood fibre composites are compared in Fig. 12.6. One can make several observations. First, paper impregnated with thermoset resins shows very high elastic modulus, 16–25 GPa, when the fibre content is high. These high-modulus materials have strongly oriented paper and phenol–formaldehyde matrix and are studied for potential use in aircraft structures. The unidirectional UF/wood fibre composite in Fig. 12.6 is the same as in Fig. 12.5. The longitudinal elastic modulus is as high as 8 GPa although the weight fraction of fibres is only 25%. The modulus values for wood flour are generally low due to the low aspect ratio. In summary, Fig. 12.6 illustrates that low aspect ratio fibres make inefficient reinforcement, and that paper with preferred orientation in the testing direction gives highest modulus. For a more detailed analysis one would need quantitative information on fibre length, orientation, and volume fraction (rather than weight fraction), as well as the elastic modulus of the matrix (see Section 12.3).

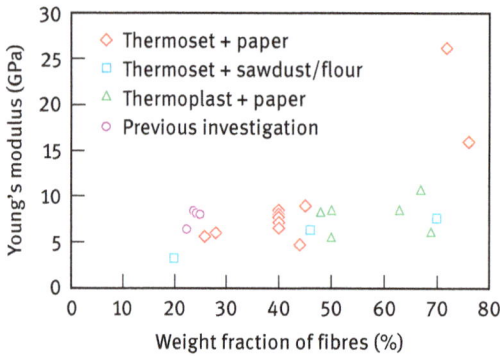

Fig. 12.6: Elastic modulus against fibre weight fraction in a variety of wood composites.

12.6 Nanocellulose materials

Plant fibres, including wood fibres, are widely used in man-made materials. The mechanical properties of paper and board materials, as discussed in the present text, are quite good. Comparisons with other polymer-based materials at similar porosities are usually favourable for plant fibre materials. This is due to the geometric shape of the fibres and the stiff and strong CNFs in the cell wall. In fact, the good strength and elastic modulus of wood fibres comes primarily from the cellulosic reinforcement of the cell wall. For this reason, the disintegration of wood into cellulosic nanofibrils provides us with a stronger and stiffer fibrous component and increases freedom in materials design because we are no longer limited by the plant fibre geometry. This is utilized in the analysis in Section 11.1.1.

New methods have been developed for disintegration of wood. The energy requirement is lower than earlier, thanks to a chemical or enzymatic pre-treatment prior to mechanical homogenization (Henriksson et al., 2008, 2007). The homogenization itself can be carried out in homogenizers commonly used for fruit juices and vegetable soups in food industry. Wood CNF is particularly interesting because of its commercial potential. The elastic modulus of the cellulose crystal is 130 GPa or higher. The elastic modulus of CNF is lower because the molecular disorder increases during mechanical disintegration. The strength of individual nanofibres is difficult to estimate, but the fine dimensions suggest that it exceeds the highest strength known for plant fibres, which is around 1 GPa. Reductions in the cellulose molar mass may reduce this value. The breaking strain of individual nanofibres in pure tension is low, probably below 1%, because of the extended molecular chain conformation that gives little opportunity for plastic deformation. The dimensions of CNF fibrils are typically 5–20 nm in diameter and 1–2 μm in length.

12.6.1 Cellulosic "nanopaper"

CNF can be used in a filtration process similar to ordinary papermaking. Although significant practical problems (difficult water removal, slow filtration) need to be solved before such a process becomes industrially feasible, the interesting properties of nanopaper structures push such a development. Figure 12.7 shows the surface of nanopaper. The extent of the pore volume fraction depends on the preparation conditions. The small size of nanofibrils is apparent; they have a diameter that is more than three orders of magnitude smaller than ordinary wood fibres (5–20 nm instead of 20–30 μm). The tensile properties are reported in Table 12.9. One may note the high tensile strength (≈220 MPa) and elastic modulus. Comparison of the breaking strain (≈7%; even higher values can be reached) with conventional paper is interesting. The most likely deformation mechanism that enables large strain is that individual CNF fibrils can slip with respect to each other without damage creation. This is possible because of favourable interaction of CNF nanofibrils at 50% relative humidity, giving rise to friction that results in distinct strain-hardening behaviour. The small size of CNF is helpful because individual CNF fibrils can fail without creating large micro-cracks. Besides, the inherent strength of CNF is much higher (1–4 GPa) than that of the wood fibres used in conventional paper and board products.

Fig. 12.7: Field-emission scanning electron micrograph of a nanopaper made of cellulose nanofibrils (CNF); courtesy of Dr H Sehaqui, Wallenberg Wood Science Center, KTH Royal Institute of Technology, Stockholm.

The strain-hardening coefficient n in Table 12.9 is the ratio of the secondary slope of the stress–strain curve (in the plastic region) to the elastic modulus (initial slope of the curve). The tensile energy absorption is the area under the stress–strain curve, measuring the total deformation energy per unit volume that is consumed when straining the material to failure. This property is remarkably high.

Table 12.9: Mechanical properties of cellulose nanopapers dried from different liquids to reach different porosities. The nanofibrillar cellulose is made from pre-treated wood fibres of high purity (Henriksson et al., 2008).

Liquid	Porosity (%)	Elastic modulus (GPa)	Strain-hardening coefficient, n	Yield stress (MPa)	Tensile strength (MPa)	Breaking strain (%)	Work of fracture (MJ/m³)
Water	19	15	1.9	91	205	6.9	9.8
Methanol	28	11	1.4	76	114	5.4	5.3
Ethanol	38	9.8	1.1	58	106	4.7	3.6
Acetone	40	7.4	0.83	48	95	6.2	4.2

12.6.2 Nanocomposites

The group of Professor Hiroyuki Yano at Kyoto University in Kyoto, Japan, pioneered the use of nanopaper structures in combination with the water-soluble PF resin. The elastic modulus of the nanocomposite is linearly proportional to the volume fraction of CNF. The same has been observed for other polymer matrices. Linear extrapolation then yields an estimate for the longitudinal elastic modulus of CNF in the limit of 100% CNF content. A typical result is in the range of 30–60 GPa for the longitudinal elastic modulus of CNF. This is a significantly lower value than the 130 GPa in the axial direction of the cellulose crystal, and the observation needs further study. The elastic modulus of wood fibre/PF composites is similar to CNF/PF composites. This is expected because elastic theories are independent of the size of the reinforcement as long as the fibre volume fraction, matrix porosity, and fibre orientation distribution remain unchanged.

The tensile strength of biocomposites with isotropic in-plane fibre orientation distribution cannot be predicted from simple micromechanical considerations. Empirical phenomenological models are available, but they do not help to understand the failure processes. Nevertheless, the experimental observations in Table 12.10 are interesting. If one compares the CNF/PF composite with a wood fibre/PF composite, the CNF composites have significantly higher strength and breaking strain. The probable reason is that defects are much smaller in the CNF composite.

The high breaking strain of the nanopaper made of CNF can be maintained in the composite if a ductile polymer is used; see the data for glycerol plasticized starch and HEC as matrices in Table 12.10. In contrast, brittle matrices (PF and MF) lead to a low breaking strain and lower tensile strength because the strain-hardening region after the yielding point is very limited. Note that all the CNF composites in Table 12.10 are

Table 12.10: Mechanical properties of cellulosic nanopaper and nanocomposites based on nanopaper. CNF, cellulose nanofibrils; PF, phenol–formaldehyde; MF, melamine–formaldehyde; HEC, hydroxyethyl cellulose.

Matrix	CNF content (% by volume)	Elastic modulus (GPa)	Tensile strength (MPa)	Breaking strain (%)	Reference
–	80 (nanopaper)	13 GPa	220	10	Henriksson et al. (2008)
PF	80 (CNF)	15 GPa	200	5	Nakagaito and Yano (2008)
PF (with pulp fibre)	80 (pulp fibre)	≈15 GPa	130	2.5	Nakagaito and Yano (2008)
MF	80	15	140	3	Henriksson and Berglund (2007)
Starch/glycerol 50/50	60	6	80	8	Svagan et al. (2007)
HEC	60	10	180	20	Sehaqui et al. (2011)

based on a nanopaper that has been impregnated with the polymer matrix. In the case of CNF/HEC, the material is also porous, which seems to facilitate ductility.

CNF has also been used in novel materials, such as CNF-reinforced polymer foams, CNF foams, and aerogels, and inorganic hybrids, such as clay nanopaper, and magnetic nanopaper (Berglund and Peijs, 2010). They represent interesting new material opportunities based on forest products, with intended uses outside the market for existing forest product markets.

12.7 Comparison of fibre network materials

There is significant potential for new wood fibre materials, due to the increased importance of eco-friendly materials, and the development of nanocellulose. Wood fibres are still much easier than CNF fibrils to industrially process into fibre network materials. A comparison between materials from cellulosic wood fibres and CNF materials is therefore interesting (Yang and Berglund, 2020). Furthermore, the effect of porosity on mechanical properties of fibre network materials is important.

As a case study, we consider holocellulose fibres and CNF prepared by mild delignification of wood. By careful processing, such fibres are characterized by negligible mechanical damage and high hemicellulose content (28%). In Fig. 12.8, stress–strain curves are presented for: paper with a porosity of 49%, compression moulded fibres (porosity 21%) and dense CNF nanopaper (2% porosity). Microscopy images of the different structures are also included and show the fine structure of CNF films. Holocellulose fibres are strong, due to high cellulose molar mass, which means that the CNF fibrils in the fibre structure are very long. The same is then also true for CNF disintegrated from holocellulose fibres.

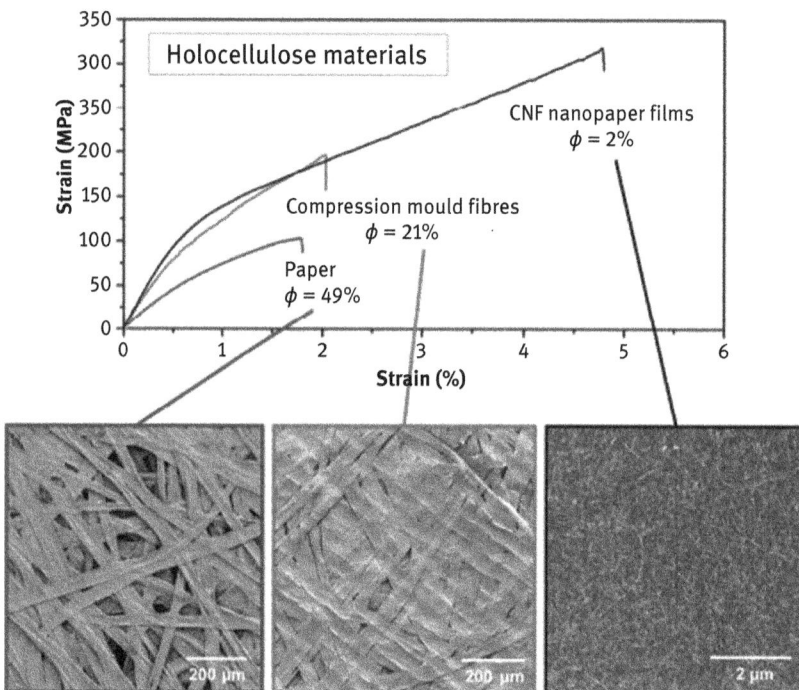

Fig. 12.8: Stress–strain curves of holocellulose materials. Electron micrographs of morphologies from left: paper, compression moulded fibres, and CNF nanopaper films. Porosity (ϕ) is reported. Reproduced (Yang and Berglund, 2020) with permission from John Wiley & Sons.

The high hemicellulose content in holocellulose is important. Young's modulus of the CNF nanopaper film is 21 GPa, which is probably the highest reported value for random-in-plane CNF films. Clearly, the xylan hemicellulose molecules provide improved fibril–fibril stress transfer, which contributes to improved modulus. The hot-pressed, compression-moulded fibres show a modulus of 18 GPa and a tensile strength of 195 GPa at 21% porosity. A simple rule of mixtures approach suggests that the modulus of a similar, non-porous material can be predicted by dividing 18 GPa by the volume fraction of fibres, 0.79, which results in $E = 22.8$ GPa. This is similar to data for the CNF nanopaper film, and confirms that the small dimension of nanofibrils have no effect on modulus.

Interestingly, also the properties for higher 49% porosity "paper" are very high, without any fibre beating, and xylan hemicelluloses instead provide fibre–fibre adhesion. The preparation of CNF films is much slower than the preferred conventional paper processing. A subsequent compression moulding step can be readily added in order to reduce paper porosity and improve properties. For this reason, it is encouraging that well-preserved compression moulded fibres are so close to CNF nanopaper in terms of mechanical performance (see Fig. 12.8).

An important mechanical property advantage of the CNF films is the high ultimate tensile strength. This is due to large-scale strain hardening, small size of defects and high strain to failure. In some applications, however, plastic deformation may not be relevant. Examples include multifunctional structures designed for stiffness, and possibly impact-loaded products. Cellulose fibre materials have lower cost and lower embodied energy than nanocellulose materials, although there are obviously other applications where CNF is the preferred constituent and nanocomposites the preferred materials.

12.8 Concluding remarks

Wood composites traditionally include established forest products such as particleboard, fibreboard, and plywood. The interest for forest products has been limited in the community of materials science and engineering. One reason may have been that the creative possibilities with fibres, particles, and veneer are somewhat limited. This is even more correct if the current cost levels and production technologies are set to constrain creativity. The use of the term *material* instead of forest products is a good start for change, since it implies that the products have a microstructure that can be altered to extend the property range.

The full potential of wood fibres is not utilized in many of the present forest products. The combination of separate long wood fibres with polymers is of interest in this context. The example of PA6 combined with high-purity wood fibres for melt-processed biocomposites is one such example. Papermaking processes have the advantage that they preserve the fibre length and make it possible to combine high wood fibre content with a polymer matrix. The existing materials include wood fibre/thermoset laminates, but in the future the flora of materials may increase to include thermoplastic matrices.

The micromechanical mechanisms of biocomposites are the same as for synthetic fibre composites. The mechanical properties of short-fibre composites depend on the fibre volume fraction, matrix porosity, inter-facial shear strength, fibre length distribution, and fibre orientation distribution. The difference is in the application potential, since wood fibres have highly non-ideal shape and variability of properties. For this reason, the application of theoretical predictions requires some caution. If the precision in the measurement of constituent properties is poor, improved modelling accuracy does not help.

The development of nanocelluloses, CNF and MFC, can potentially change the existing paradigm. It is a new material component of a very small size. It has appealing properties such as high elastic modulus, high tensile strength, low thermal expansion, and potential for optical transparency; at the same time, it preserves or even improves the network bonding characteristics of wood pulp fibres. It can be used in nanopaper films, coatings, fibres, foams, aerogels, and organic/inorganic hybrids, as well as in tough mouldable biocomposites (Berglund and Peijs, 2010). The improvements obtained in the strength and tensile energy absorption compared with conventional wood composites, paper, and board materials are impressive. A remaining challenge is to develop processing technologies.

Literature references

Agarwal, B.D. and Broutman, L.J. (1990). Analysis and performance of fiber composites, NY: Wiley-Interscience.

Ashby, M.F. (1999). Materials selection in materials design (Butterworth-Heinemann).

Berglund, L.A. and Peijs, T. (2010). Cellulose biocomposites-from bulk moldings to nanostructured systems. MRS Bull. 35, 201–207.

Henriksson, M. and Berglund, L.A. (2007). Structure and properties of cellulose nanocomposite films containing melamine formaldehyde. J. Appl. Pol. Sci. 106, 2817.

Henriksson, M., Berglund, L.A., Isaksson, P., Lindström, T. and Nishino, T. (2008). Cellulose nano-paper structures of high toughness. Biomacromolecules 9, 1579.

Henriksson, M., Henriksson, G., Berglund, L.A. and Lindström, T. (2007). An environmentally friendly method for enzyme-assisted preparation of microfibrillated cellulose (MFC) nanofibers. Euro. Pol. J. 43, 3434.

Hull, D. (1981). An introduction to composite materials, Cambridge, UK: Cambridge University Press.

Jacobson, R., Caulfield, D., Sears, K. and Underwood, J. (2002). Low temperature processing of ultra-pure cellulose fibers into nylon 6 and other thermoplastics, 6th Int. Conf. on Woodfiber-Plastic Composites, Forest Prod Soc, Madison, WI, 127–133.

Nakagaito, A.N. and Yano, H. (2008). Toughness enhancement of cellulose nanocomposites by alkali treatment of the reinforcing cellulose nanofibers. Cellulose 15, 323–331.

Sehaqui, H., Zhou, Q. and Berglund, L.A. (2011). Compos. Sci. Tech. 71(13), 1593–1599.

Stark, N.M. and Rowlands, R.E. (2003). Effects of wood fibre characteristics on mechanical properties of wood/polypropylene composites. Wood. Fib. Sci. 35, 167–174.

Svagan, A.J., Samir, M.A.S.A. and Berglund, L.A. (2007). Biomimetic polysaccharide nanocomposites of high cellulose content and high toughness. Biomacromolecules 8, 2556.

Wainwright, S.A., Biggs, W.D., Currey, J.D. and Gosline, J.M. (1982). Mechanical design in organisms, Princeton, NJ: Princeton University Press.

Wegst, U.G.K. and Ashby, M.F. (2004). The mechanical efficiency of natural materials. Phil. Mag. 84, 2167–2181.

Yang, X. and Berglund, L.A. (2020). Structural and ecofriendly holocellulose materials from wood: Microscale fibers and nanoscale fibrils. Adv. Mater., 2001118.

Index

https://doi.org/10.1515/9783110619386-013

www.ingramcontent.com/pod-product-compliance
Lightning Source LLC
Chambersburg PA
CBHW080926220326
41598CB00034B/5690